The Letters of
Sir Joseph Banks

A Selection, 1768 – 1820

To David Stam,

from Neil Chambers,

with very warmest

regards. 2014.

The Letters of
Sir Joseph Banks

A Selection, 1768 – 1820

edited by

Neil Chambers

Research Curator, The Natural History Museum

Foreword by

Professor David Mabberley

Honorary Director of The Banks Archive Project

Introduction by

Mr. Harold Carter

Distinguished Research Curator

ICP

Imperial College Press

Published by

Imperial College Press
57 Shelton Street
Covent Garden
London WC2H 9HE

Distributed by

World Scientific Publishing Co. Pte. Ltd.
P O Box 128, Farrer Road, Singapore 912805
USA office: Suite 1B, 1060 Main Street, River Edge, NJ 07661
UK office: 57 Shelton Street, Covent Garden, London WC2H 9HE

Library of Congress Cataloging-in-Publication Data
Banks, Joseph, Sir, 1743–1820.
[Correspondence, Selections]
The letters of Sir Joseph Banks : a selection, 1768–1820 / editor, Neil Chambers.
p. cm.
Includes bibliographical references and index.
ISBN 1-86094-204-0
1. Banks, Joseph, Sir, 1743–1820--Correspondence. 2. Scientists--Great Britain--Correspondence. 3. Naturalists--Great Britain--Correspondece. 4. Scientists--Great Britain--Biography. I. Chambers, Neil. II. Title.

Q143.B17 A4 2000
508'.092--dc21
[B] 99-089844

British Library Cataloguing-in-Publication Data
A catalogue record for this book is available from the British Library.

Printed in Singapore by FuIsland Offset Printing

Que ne puis-je peindre ce que je sentis en voyant pour la première fois ce vénérable compagnon de Cook! Illustre par de longs voyages; remarquable par une étendue d'esprit et par une élévation de sentiments qui le font s'intéresser également aux progrès de toutes les connaissances humaines; possesseur d'un rang élevé, d'une grande fortune, d'une considération universelle, Sir Joseph a fait de tous ces advantages le patrimoine des savants de toutes les nations. Si simple, si facile dans sa bienveillance, qu'elle semble presque pour celui qui l'éprouve, l'effet d'un droit naturellement acquis; et en même temps si bon qu'il vous laisse tout le plaisir, toute l'individualité de la reconnaissance. Noble exemple d'un protectorat dont toute l'autorité est fondée sur l'estime, l'attachement, le respect, la confiance libre et volontaire; dont les titres consistent uniquement dans une bonne volonté inépuisable et dans le souvenir des services rendus; et dont la possession longue et non contestée fait supposer de rares vertus et d'une exquise délicatesse quand on songe que tout ce pouvoir doit se former, se maintenir et s'exercer parmi des égaux.

Biot, 1818

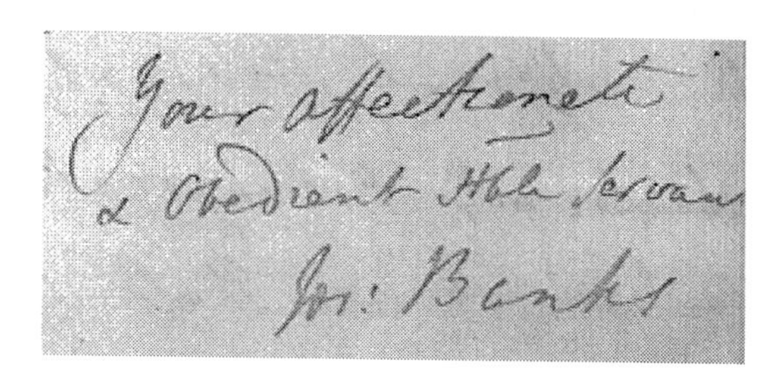
Your affectionate
& Obedient Hble Servant
Jos: Banks

[Banks's signature from his letter to Hamilton: Banks to Sir William Hamilton K.B. P.C., 4 December 1778, B.L. Egerton MSS. 2641. 130-131.]

The Banks Archive Project Executive Committee would like to thank the Pilgrim Trust for supporting the publication of this title.

The Banks Archive Project Executive Committee would like to thank the Pilgrim Trust for supporting the publication of this title.

Brief Contents

Foreword

David Mabberley
Honorary Director of the Banks Archive Project

The correspondence of Sir Joseph Banks (1743–1820) comprises a major primary source for understanding the history of the revolution in social, economic and political fields in western Europe, and its spheres of influence, through the turbulent period spanning the eighteenth to nineteenth centuries. At his death, Banks left a vast correspondence, of which some 20,000 letters survive. More than 3000 correspondents appear in these letters, including heads of state, government ministers and people prominent in science and the arts, agriculture and public life. Not only is this correspondence the tangible result of a network permeating the society of its day, but it demonstrates the historical and literary worth of Banks's writing, which has been examined by Neil Chambers in his paper, 'Letters from the President: the correspondence of Sir Joseph Banks', *Notes and Records of the Royal Society of London,* 53 (**1**), 27–57, 1999.

But the Banks correspondence was broken up and sold. Occasionally, material appears on the open market, but most is now scattered throughout the world, preserved in collections both public and private. The Banks Archive Project was set up in 1989 as the initiative of my predecessor as Honorary Director, Harold B. Carter, A.O., and Dr. Desmond King-Hele, F.R.S., who has been chairman of the Executive Committee from the outset. The aim of the project is to re-assemble the collection through copying and cataloguing the letters. Only in this way can an invaluable research tool become available once more. The Archive is housed in the Department of Library Services in The Natural History Museum in London, but is an independent entity, co-sponsored by the Museum and the Royal Society. It raises its own funding from public and private sources. This volume, for example, has been supported by a donation from the Pilgrim Trust, support for which we are most grateful.

Since the appearance of Carter's biography, *Sir Joseph Banks* (1988), substantial progress has been made in preparing the correspondence for publication. Carter himself published *The sheep and wool correspondence of Sir Joseph Banks 1781–1820* in 1979 and, now, the letters associated with Iceland, 1772–1820, are to appear shortly. The aim of the project is to produce the remainder of the correspondence in similar thematic volumes. Work is now active on other letters referring to Australia and the Pacific, to China and Japan, and to the Banks Estates in England. Neil

Chambers is currently preparing the Scientific Letters for publication, and is also researching Banks's career as President of the Royal Society.

With such a large project in progress, it seemed appropriate to the Project's Executive Committee to offer the general public a preview, by publishing a selection of letters to demonstrate the breadth and depth of Banks's involvement in British life, and to show how this involvement was not that of a dilettante, but of a hard-headed pragmatic man, whose influence was felt throughout the society of his day.

Introduction

Harold Carter

Joseph Banks (1743–1820) is now recognized as one of the most influential Englishmen of his day. As President of the Royal Society from 1778 to 1820, as a friend of King George III from 1772, and as a Privy Councillor, he had a hand in nearly all the scientific initiatives of the time, particularly the voyages of exploration.

Joseph Banks was born of Derbyshire and Yorkshire descent at 30 Argyll Street, St. James's, London, on 13 February 1743. His extensive county estates were centred on Revesby Abbey near Horncastle, Lincolnshire, which he inherited after the death of his father William Banks in 1760, and later Overton Hall, Ashover, Derbyshire, when his uncle Robert Banks died in 1792.

Successively tutored at Revesby Abbey, prepared at Harrow, disciplined at Eton, and liberated at Christ Church, Oxford, Joseph Banks was the first of his family to enter a university with full educational benefit. He matriculated as a gentleman commoner in December 1760, and came down in the summer of 1764 in full command of his inheritance. He had no formal degree, but was a serious student in natural history with a well-developed scientific curiosity focused mainly on botany and entomology. He pursued these interests at the British Museum and became a close friend of the Keeper of Natural and Artificial Productions, Dr. Daniel Solander, an apostle of Linnaeus direct from the University of Uppsala. In May 1766, Banks was elected a Fellow of the Royal Society as a man 'versed in Natural History'. By this time, however, Banks was at sea in a 32-gun naval frigate as a sort of post-graduate biologist for a summer of fisheries patrol off the coasts of Newfoundland and Labrador. He returned to London in January 1767 as an experienced collector in natural history, with the first properly documented specimens from that region of eastern America — emphatic evidence of his scientific merit at the age of 24. To accommodate his growing library of natural history specimens, he acquired a house in London, at 14 New Burlington Street.

Early in 1768 the Council of the Royal Society was preparing its "Memorial" to the King for a grant of £4,000 towards a Royal Society expedition to the South Seas to observe the transit of Venus in June 1769. Banks committed himself to the voyage in February 1768, on his own initiative and at his own expense, three months before Lieutenant James Cook was appointed commander.

On 16 August 1768 he left London to join Cook and H.M.S. *Endeavour* at Plymouth. On 13 July 1771, after their epic world voyage, he returned with the first natural history and ethnological collections from the South Seas to be seen in Great Britain. As Fellows of the Royal Society, Banks and Solander had carried the biological sciences into the South Pacific for the first time, and opened new fields of enquiry.

Banks withdrew at a late stage from Cook's second voyage to the Pacific. Instead, he embarked on his third and last maritime exploration at his own expense and, this time, under his own command. His voyage to Iceland in the summer and autumn of 1772 set the northern limit to Banks's collecting sites, while Tierra del Fuego had marked the southern — a range in latitude no naturalist before him had achieved. As an intellectual entrepreneur at 29 he had done more than most of his contemporary Fellows of the Royal Society "for the Promotion of Natural Knowledge".

For Banks his three "voyages of curiosity" in pursuit of "Natural Knowledge" were the foundation of his future as a dedicated servant of science: comprehensive in their vision; thorough in their preparation; professional in their collections; detailed in their documentation; well conserved at their end for the scientific world to study, even to the present day, notably with the botanical material. These were the far-sighted enterprises of an intellectual man of action — a man of the study as well as of the field.

In August 1777 Banks moved to a larger house, No. 32 Soho Square, and by 1778 had created there a well-organized "Academy of Natural History", pre-eminent in its botanical specimens and the conditions for their study, and backed by a library unique in its day for its coverage in natural history. Superbly catalogued in 1796–1802, and regularly updated in manuscript side-entries, it was a pioneer in scientific bibliography, and still continues as a separate unit of the British Library in London. So also the dried plant specimens at 32 Soho Square formed one of the most comprehensive reference collections available to European botanists, and became the nucleus of the General Herbarium at The Natural History Museum from 1881. The method of mounting and storage in cabinets of Banks's original design became the standard for museum collections worldwide.

The original shell collection of Banks and Solander, containing 1120 specimens of 392 species, is now lodged in the mollusca section of the Department of Zoology in The Natural History Museum, backed by Solander's descriptive manuscripts and early classification. By contrast, the original elegantly mounted entomological specimens have not survived so well. The remnants are to be found in the Department of Entomology at The Natural History Museum.

When Banks was elected President of the Royal Society of London in the old rooms at Crane Court in 1778, he had been active in its affairs for nearly seven years as a regular attendant at the meetings and as a member of Council, 1773–75 and 1777–78. In 1780, the Royal Society moved to its new rooms in Somerset House, and here Banks was annually elected President until his death in June 1820 — the longest serving of all Presidents. Here, he was at the hub of scientific and technical progress during a most active period of geographical discovery, war and revolution — a notable figure of strong personal independence in touch worldwide with all kinds and conditions of men on matters of science, manufacture, politics and trade.

Before he became President, in 1775–78, Banks had a recognized mistress in Sarah Wells, who lived at a separate address south of St. James's Park. But in March 1779, at the age of 36, Banks married Dorothea Hugessen, a 20-year old Kentish coheiress, at St. Andrew's Church, Holborn. Later that year, he leased the villa of Spring Grove near Hounslow with its 49 acres, where in time he established his experimental farm and botanic garden. His sister, Sarah Sophia, soon became a permanent member of the Banks household, to complete a childless trio well known in London society and county travels for the next 40 years. With his marriage and settlement at 32 Soho Square and Spring Grove, the pattern of Banks's life from 1779 was set, governed largely by the meetings of the Royal Society, but also by his personal service to the King.

Not since John Evelyn, and the foundation of the Royal Society under Charles II, had the Society been so close to the reigning monarch. Banks first met King George III in 1771 at St. James's Palace and the White House, Kew, to discuss the natural history harvest from the *Endeavour* voyage, and their active friendship of 40 years was stifled only by the last prolonged illness of the King from 1811. The first evidence of this friendship was the despatch of Francis Masson to the Cape of Good Hope in 1772 as the first of many Kew collectors — Anton Hove, William Kerr, James Bowie, Allan Cunningham, and Alexander Moon followed. For more than 40 years, from 1773 to 1820, Banks was in effect honorary director of what has become the Royal Botanic Gardens, Kew. Himself a pioneer breeder of Spanish Merino sheep at Spring Grove, he also established and managed a Royal stud flock of the same breed at the King's Marsh Gate Farm, Kew, and in the Little Park, Windsor, 1787–1820. Together, these small private enterprises laid foundations for the massive growth of the 19th century British colonial wool trade and home manufactures. It was Banks also, more in his role as President of the Royal Society, who dispensed the Royal funds for the telescopes of William Herschel F.R.S. from 1781 to 1820, the survey instruments of Major-General William Roy F.R.S. from 1783 to 1791, and the first stages of the Ordnance Survey of the United Kingdom, accounting to the King directly for this use of royal wealth in the national interest.

On scientific and related matters, Banks was thus the voice of the Royal Society advising the King, in friendly conversation while walking by the Thames at Kew and Richmond, or at Windsor, and also through his written advice to the King's ministers. Informally, Banks was the knowledgeable man of affairs translating the everyday world to the King, but he was silent on matters of policy or political debate. Indeed, he felt the King was sometimes too free in their conversations, and he refused to seek Royal favours either for himself or others. He accepted admittance to the Order of the Bath in 1795 (the first civilian recipient of this honour) only for its relation to his place as President of the Royal Society, free from political or personal elements. As such, it marked the rising status of science in the national scale of values, and as such Banks wore the red ribbon in the chair at Somerset House.

After many years as an unofficial adviser, Banks was from March 1797 a sworn and active member of the Privy Council, with all that this implied then in the affairs of the nation and in his relations with the King. His little study at 32 Soho Square became effectively another office of 'the great Council' with its own weight and influence, quite distinct from his place in the chair at Somerset House. In the Council meetings, Banks sat as the widely travelled and knowledgeable man of affairs of notable integrity on whom successive Governments relied, from the first of Pitt the younger, beginning in 1784, to that of Lord Liverpool, 30 years later. With his presence *ex officio* on other public bodies, such as the British Museum, the Board of Longitude and the Board of Agriculture, he developed a network of international communication sustained through 20 years of global war. For 30 years or more he served, in modern terms, as an honorary Permanent Secretary in a Ministry of Science and Technology.

For nearly half a century, he was closely involved with all the important voyages of Pacific exploration that followed the pioneering Royal Society venture of H.M.S. *Endeavour*. He managed in detail, 1780–84, the published account of Cook's last voyage, 1776–80. He proposed in 1779 and 1785 a settlement on the east coast of Australia. He advised on the mounting of the "First Fleet" under Captain Arthur Phillip to found the convict colony at Sydney Cove, 1787–88. At the same time, he proposed and supervised the two breadfruit voyages of Captain William Bligh in 1787–93. He was consulted on the early fur-trading ventures to the west coast of America, 1785–90, and especially the survey voyage of Captain George Vancouver with the naturalist Archibald Menzies, 1790–95. The Government and the East India Company drew heavily on Banks for advice in the mounting of the Macartney Embassy to China, 1792–94, as they had done in the developing affairs of the sub-continent since 1784. He stimulated and guided the coastal surveys of Australia, notably those of Captain Matthew Flinders, organizing for the Admiralty the voyage of H.M.S. *Investigator* to the final publication of its results, 1795–1814. He strongly supported the Arctic explorations of Constantine Phipps, 1773; William Scoresby Jnr., 1807–23; Captain Sir John Ross, 1819; and Captain Sir William Edward Parry, 1819–21, in the search for a north-west passage.

He was a central figure in founding: the Linnean Society of London, 1788; the African Association, 1788, later to become the Royal Geographical Society, 1832; the Board of Agriculture, 1793; the Smithfield Club, 1798; the Royal Institution, 1799; and the Horticultural Society, 1804, also known as the Royal Horticultural Society since 1861.

As a Privy Councillor from 1797 to 1820, he was a tireless member of the Committee for Trade and Plantations on matters affecting trade in general, agriculture, coinage and the affairs of the Royal Mint. As a county landowner he was involved from 1762 to 1820 in the embankment, drainage, navigation, survey and accurate mapping of Lincolnshire, and as a Fellow of the Society of Antiquaries from 1766 he was an ardent historian and student of the county antiquities. The estate of Overton Hall, near Ashover, and the family investment in the nearby Gregory Mine, ensured his continuing concern with the coal and lead mines of Derbyshire and their geological mapping, hence his strong support of William Smith and the growth of stratigraphical geology in Great Britain.

With his voyage in H.M.S. *Endeavour*, Banks provided a role model for several later biologists, such as Robert Brown, F.R.S. 1811, who sailed on the tragic but productive voyage of H.M.S. *Investigator*, 1801–03, and opened new vistas of botanical science. Brown links with Charles Darwin, F.R.S. 1839, whose voyage on H.M.S. *Beagle*, 1831–36, was to extend the intellectual boundaries of the biological sciences. Two of the voyaging biologists became Presidents of the Royal Society: Joseph Hooker, F.R.S. 1847, P.R.S. 1873–78, pre-eminent as a botanist, sailed on H.M.S. *Erebus*, 1839–43, in the Southern Hemisphere; and Thomas Huxley, F.R.S. 1851, P.R.S. 1883–85, on H.M.S. *Rattlesnake*, 1846–50. Finally, there is the voyage of H.M.S. *Challenger*, 1872–76, with Henry Moseley, F.R.S.1877, on board.

Banks's scientific reputation in his own time is derived from his long tenure as President of the Royal Society, and from the material evidence of his collections, manuscripts and library. He wrote few scientific papers and no books, but his influence was immense, and was exercised in no small part through his correspondence. Probably no man as an independent private citizen ever gathered so many far-flung correspondents into his sphere of influence on so many subjects and to such widespread practical effect — they ranged from crowned heads and ministers of state, through those eminent in public affairs, science, manufacture, the arts and literature, to those who are known to us only as names.

It was a long time before the interest and importance of his correspondence was appreciated, and many letters have been lost. For more than a century, the main source was the sample transcribed for Dawson Turner from originals loaned to him between 1832 and 1845. These copies were not generally available until lodged in the Department of Botany library at The Natural History Museum. Known as the Dawson Turner copies, they comprise the text of the Banks *Endeavour* journal and some 2032 individual letters, of which 504 are written by Banks himself. After the original manuscripts had been offered to the British Museum and largely refused, there were sales in 1880 and 1886 which dispersed much of the correspondence.

In 1958, Warren Dawson in *The Banks Letters* summarized the texts of a little over 7000 letters of the general correspondence to be found at that date in United Kingdom repositories. The global total has now grown to more than 20,000, of which about 6000 are by Banks himself, covering the years 1766–1820.

The average length of a Banks letter is about 400 words. Over three-quarters of his letters are of less than 500 words, and only about one in 40 above 1000 words. He wrote almost always in English, occasionally in French, rarely in Latin. In the surviving correspondence there are very few letters to or from women, even of his own family, though in his last 10 years he wrote quite often to Charlotte, eleventh Duchess of Somerset.

His correspondence was organized by himself as a business operation, not as a self-indulgent literary pursuit. This systematizing began in 1771, but it was not until August 1777, with his move to 32 Soho Square and the arrival of Jonas Dryander, that a firm working regime was established. This was consolidated in or about 1780 with the engagement of William Cartlich as his office clerk and copyist for the next 35 years.

The pattern of Banks's working year, and hence of his correspondence, was dominated after 1778 by his Royal Society engagements after his election as President. After the end of the Royal Society year, usually in June, he would travel to his estates. Thus his letters are mainly from 32 Soho Square during the winter and early spring each year; from then until the end of August often from his estate of Spring Grove in Middlesex; in September/October from Revesby Abbey in Lincolnshire, but also from Overton Hall in Derbyshire in September over the years 1793–1813. Wherever he was, his correspondence followed him through the diligence of Jonas Dryander as his curator and secretary at Soho Square. Banks replied from where he was at the time, often immediately, so that his reply was received in London mostly within a week of the mail coming to his hand.

Editorial Principles

Neil Chambers

The surviving Banks correspondence amounts to over 20,000 letters. Approximately 14,000 are to Banks, while about 6,000 are by him in one form or another. Some of the latter are autograph originals, signed and sent by Banks. Others are copies made by an amanuensis, the drafts of which he sometimes retained. This selection has been made from the 6,000 known letters by Banks as a sample of what the larger mass contains. It is therefore just a glimpse of a much greater whole, but every effort has been made to ensure that it is a representative glimpse.

Banks's correspondence is scattered through numerous repositories across the world. Among the repositories which possess Banks papers are some which have particularly rich collections: the Mitchell Library, Sydney; the Sutro Library, California, and the Yale University Libraries, Connecticut. All have valuable concentrations. There are more in Britain at: the Department of Western Manuscripts, the British Library; the Royal Society, London, and the Library and Archives of the Royal Botanic Gardens, Kew. I have plagued the librarians and archivists at each for help with my work, and in return they have allowed free access to original letters. A primary centre for Banks studies is at The Natural History Museum, London, and I have exploited this too. In the Botany Library there are the well-known Dawson Turner Copies. These appear again and again in works to do with Banks, and the current one is no exception. In addition, the Banks Archive at The Natural History Museum is unique for the overall access it gives to the uninterrupted correspondence. Under Harold B. Carter's initial direction, copies of Banks documents have been collected, and are being collated there.

The letters in this selection are arranged in chronological order with no breaks. Every one has been treated in the same way. Spelling, capitalization and abbreviation have been preserved. However, punctuation has been altered. Banks tended not to punctuate clearly. When he did punctuate he usually favoured a comma, but even these do not always appear where a modern reader would expect to find them. Sometimes a comma served as a full stop, and full stops are rare indeed. Furthermore, Banks tended to use capital letters erratically. Not all sentences begin with one, and many words have a capital letter whether they need it or not. In longer sentences, and particularly with more complex phrases, this presents a problem. For those familiar with his letters, it is not much of one. However, in order to introduce Banks as a writer to a wider audience, it was felt necessary to add full stops and commas. This was done silently after much consideration, and in the hope that those who favour different methods would understand why such an approach was adopted with Banks.

The notes include details of the original manuscript or any valid copies used to transcribe each letter. The particular source text used comes first in the square brackets after each letter, and any other versions of the letter follow. Readers wishing to know how the source text appeared may consult the reference given, although there will not be any significant difference from the published versions here, except in punctuation. Needless to say, re-punctuating eighteenth century prose is a difficult task, and with Banks sometimes an almost impossible one. Much care was taken to respect the sense of Banks's letters. My aim was always to make his lucid prose style immediately apparent, and his meaning available to all. When a manuscript has not been located, I have relied on published material. The volume or paper used is stated in such cases, but I have not listed all the previous occasions when a selected letter has appeared in print, since most transcripts were made from other sources.

Banks's spelling may seem a little peculiar at first, but words are rarely unrecognizable. For instance, Banks did not always bother with double letters. To him the Prime Minister was not *Mr. Pitt*, but *Mr. Pit.* Meanwhile, other words sometimes gained an extra letter. Thus, on occasion, Banks might have written *figgs* or *cabbin*. The overall effect is not at all confusing though. Most irregularities were minor and characteristic quirks. However, in some letters he produced words which are hard to understand, or which make no sense at all. Rather than use footnotes for these, I have clarified the text with an editorial insertion in square brackets. All editorial insertions are in square brackets. Place and personal names are exceptions to this rule. Banks frequently misspelled both, and this causes obvious difficulties. In keeping with common editorial practice, places and people are therefore always explained in a footnote.

Banks tended to abbreviate certain words. He would ignore the *e* in the *ed* ending of verbs in the past tense. Thus, *offered* was *offerd*, *learned* was *learnd* and so on. As before, these are unlikely to trouble a reader, and in any case were acceptable forms in the eighteenth century. So were superscripts, but these have been eliminated. Where Banks wrote wch for *which*, the abbreviation will be *wch.* Some abbreviations, such as *Commee*, *R.S* and *B. of A.*, have occasionally been expanded with an editorial insertion to acquaint readers with them. Banks meant Comm[itt]ee, R[oyal] S[ociety] and B[oard] of A[griculture]. He also regularly abbreviated naval and army titles. These have been left alone as it is still conventional to do so.

Crossings out and insertions show Banks at work, planning each remark in the light of his abundant experience as a writer. Crossing out is presented in italics enclosed in obliques: /*deletion*/. Insertions are in plain text enclosed in obliques: /insertion/. Marginal comments by Banks have been included in a letter only when a clear indication was given that they were to be read as part of the text. These are treated as insertions. Marginal comments which cannot be placed in this way are given as footnotes. Words or phrases which Banks underlined have been set in italics.

Footnotes have been provided to assist readers, but the letters can be appreciated without them. The first mention of an individual by Banks receives a footnote, with full name, dates and a brief biographical comment. Thereafter, each individual appears in the footnotes, but further details have been provided only when a better understanding of a letter demands them. Places have been identified wherever possible, and the Appendix gives simple maps illustrating the course taken by ships Banks sailed in. These maps are confined to those parts of a voyage which may seem unclear in the letters. Otherwise, the routes of important missions Banks participated in or organized are described in a footnote in the usual way.

The Biographical Register of Recipients follows the letters, and refers to Banks's correspondents in this volume. A more complete account is given there than is usual in the footnotes. Where helpful, remarks on the official positions held, and particulars of the individual's relationship with Banks are included. However, the entries are not intended to be fully comprehensive. The principal titles and posts of each correspondent are indicated at the head of each letter with the name.

Wider historical events have been explained, as have the particular circumstances in which the letters were written. The act of selection made both a necessity. Letters taken out of a natural sequence sometimes make little sense to readers who, unlike Banks, have not seen the correspondence he answered or subsequently received. For these reasons, and because of the tremendous scope of Banks's daily business, the footnotes contain manuscript references for letters to and from Banks. These are a starting point for readers wishing to know more about Banks and his correspondence. So is the Selected Reading section at the end. It lists some of the excellent works I have used in preparing this volume.

I hope that the letters speak for themselves. Banks was an assured writer, with a firm prose style. His correspondence is remarkable for the breadth of subject matter dealt with. Students of literature and history alike will therefore find it a rewarding and frequently surprising source of ideas and facts about life in the eighteenth and early nineteenth centuries. I also hope general readers with no special academic interest will enjoy the letters. They are a fascinating insight into an exceptional life, and prove that English letters flowered in Banks's hands.

Acknowledgements

During the production of this book John Thackray died. John is acknowledged below in his role as archivist at The Natural History Museum, London, but he also served on the Banks Archive Project Executive Committee. His death caused great sadness to all his friends and colleagues, and I would like to commemorate his kindness, patience and determination.

Institutions and Staff

I have already mentioned some of the libraries and repositories I received help from in the eight months allowed to select and prepare these letters. This is an opportunity to repeat my thanks to them, and indeed to all of the institutions and staff whose patience and knowledge contributed to this work. Most are cited in the References and Abbreviations for the Letters below. I have space, however, to record special thanks. These are entirely due to: the staff of the Department of Library and Information Services, The Natural History Museum, London — especially Malcolm Beasley and his team in the Botany Library, Carol Gokce and hers in General Library, and John Thackray, formerly Museum Archivist; Mary Sampson, formerly at the Royal Society, London; Gina Douglas, The Linnean Society, London; Louise Anemaat and Alan Ventress, Mitchell Library, Sydney; Martha Whittaker, Sutro Library, California; Graeme Powell, National Library of Australia, Canberra; Leslie Price, the Royal Botanic Gardens, Kew; Stephen Parks and Sarah Knight, Beinecke Rare Book and Manuscript Library, Yale University, Connecticut; all the kind staff with whom I corresponded at the Lincoln Central Library, Lincolnshire; the Banks Archive Project, The Natural History Museum, and its Executive Committee, especially Desmond King-Hele and David J. Mabberley, both of whom offered many useful suggestions which improved the text. Desmond carefully proof read the manuscript before printing, and I would like to thank him for this as well.

Individuals

Many other people have been extremely kind with time and expertise: Anna Agnarsdóttir; Per Erik Ahlberg; Charles Aylmer; Stuart R. Band; Keith Clancy; Scott De Haven; Adrian Doyle; Walter and Karin Eyles; Teresa Farnham; Gail Fordham; Timothy Fulford; Eric Groves; Alan Hart; Daphne Hills; Lindsey Hughes; Mary Hyde; Charlie Jarvis; Elias Johnson; Anita L. Karg; Goulven Keineg; Johann Liebenberg; Elisabeth Mansen; Anne Marshall; David T. Moore; Patricia Rees; Tony Rice; Markus Seinsche; Nancy M. Shawcross; Chris Smout; Charlotte Stockley;

Orsolya Szakaly; John Taylor; Roy Vickery; Julian Ward; Gary West; Ben Williamson; Frances Wood. I very much hope I have not forgotten anyone here, and apologize deeply if I have.

Without help from these people I would not have completed the text on time: Andrew Swindells, and *especially* Samantha Quash.

To Mum

Neil Chambers
2000

References and Abbreviations for Notes to the Letters

The abbreviations in the notes for the letters are explained below. They relate to the repositories where manuscripts and collections are currently located. Many are, of course, particular to the recipient of Banks's letter. Original autograph manuscripts or valid copies have been used for transcription in most cases. Where the published text of a Banks letter has been referred to instead, the book or paper is also included below. Many letters were drafted at least once, and then copied to be sent. Consequently, versions of the same letter have sometimes survived in more than one place. Certain letters will therefore be cited in two or three locations.

A.P.S.	American Philosophical Society, Philadelphia, Pennsylvania, USA.
A.T.L.	Alexander Turnbull Library, National Library of New Zealand, Wellington.
B.C.A. (Boulton Papers.)	Birmingham City Archives, Central Library, Chamberlain Square, Birmingham.[1]
	Beaglehole, J.C., (Ed.), *Endeavour Journal of Joseph Banks 1768–1771*, London (1962).
	Berlinische Monatschrift, September 1785.
Biblio. du Mus. Nat. d'Hist.	Bibliothèque du Muséum National d'Histoire, Paris, France.
B.L.	British Library, London.
B.P.L.	Boston Public Library, County Hall, Boston, Lincolnshire.
B.P.L.M.A.	Boston Public Library, Massachusetts, U.S.A.
B.R.O.	Buckinghamshire Record Office, Aylesbury, Buckinghamshire.
	Brougham, Lord, (Ed.), *Lives of Men of Letters and Science who flourished in the Time of George III*, London (1845–1846).
C.C.M.M.	Captain Cook Memorial Museum, Whitby, Yorkshire.

C.K.S.	Centre for Kentish Studies, Maidstone, Kent.
C.U.L.	Cambridge University Library, Department of Manuscripts and University Archives, West Road, Cambridge.
	Dawson, Warren R., *Banks Letters: A Calendar of the manuscript correspondence*, The Natural History Museum, London (1958).
de Beer Coll.	'Collection of Banks Correspondence belonging to Sir Gavin de Beer, F.R.S.'[2]
D.L.	Dixson Library, Sydney, Australia.
D.R.O.	Derbyshire Record Office, Matlock, Derbyshire.
Dryander Corr.	The Natural History Museum, London. Correspondence of Jonas Dryander, F.R.S., Botany Library, Special Collections.
	Duyker, E., and Tingbrand, P., (Eds.), *Daniel Solander Collected Correspondence 1753–1782*, Stockholm (1995).
F.M.C.	Fitzwilliam Museum, Cambridge.
	Gentleman's Magazine, XC, vol. II, (August, 1820).
Gunther	The Museum of the History of Science, Oxford.
Hawley	'Banks papers in the possession of Major Sir David Henry Hawley, 7th Bart., of Mareham-Le-Fen, Lincolnshire.'[2]
H.B.D.	Hunt Institute for Botanical Documentation, Pittsburgh, Pennsylvania, U.S.A.
H.C.	Hyde Collection, Four Oaks Farm, New Jersey, U.S.A.
	Hollandsche Maatschappij der Wetenschappen, at the Rijksarchief in North-Holland.
Hooker Corr.	Correspondence of Sir William Jackson Hooker, F.R.S., in the Library of the Royal Botanic Gardens, Kew.
H.R.N.S.W.	Bladen, F.M., (Ed.), *Historical Records of New South Wales,* 7 vols., Sydney (1893–1901).
H.R.O.	Hereford Record Office, Hereford.
H.S. Penn.	Historical Society of Pennsylvania, Philadelphia, U.S.A.
I.A.C.R. Rothamsted	Rothamsted Experimental Station Archive, Harpenden.
I.C.C.L.	Imperial College Central Library, South Kensington, London.
	Journal of the Royal Horticultural Society, London, vol. 79, (1854).
L.A.O.	Lincolnshire Archives Office, Lincoln.
L.C.L.	Lincolnshire County Library, Lincoln.

L'Héritier Coll.	The Natural History Museum, London. Correspondence of Charles Louis L'Héritier de Brutelle, F.R.S., Botany Library, Special Collections.
L.S.	Linnean Society of London, Burlington House, London.
	Mackaness, G., *Life of Vice-Admiral William Bligh*, Sydney (1931).
M.L.	Mitchell Library, Sydney, Australia.[3]
N.H.M. B.C.	The Natural History Museum, London. Banks Correspondence, Botany Library, Special Collections.
N.H.M. B.L. D.T.C.	The Natural History Museum, London. Dawson Turner Correspondence, Botany Library, Special Collections.
	Nichols, J., and Nichols, J.B., *Illustrations of the Literary History of the Eighteenth Century*, London (1817–1858).
N.L.A.	National Library of Australia, Canberra, Australia.
N.L.W.	National Library of Wales, Aberystwyth.
N.M.M.	National Maritime Museum, Greenwich, London.
P.M.L.	Pierpont Morgan Library, New York, U.S.A.
P.R.O.	Public Record Office, Kew, Surrey.
R.A.S.	Royal Asiatic Society, Calcutta, West Bengal, India.
	Rauschenberg, R.A., 'A Letter of Sir Joseph Banks describing the life of Daniel Solander', *Isis*, vol. 55 (1964).
	Rauschenberg, R.A., 'The Journals of Joseph Banks's voyage up Great Britain's West Coast to Iceland and to the Orkney Isles July to October 1772', *Proceedings of the American Philosophical Society*, vol. 117 (1973).
R.B.G. Kew Archive B.C.	Banks Collection, Royal Botanic Gardens Library, Kew, Surrey.
R.H.S.	Royal Horticultural Society, Lindley Library, London.
R.I.	Royal Institution of Great Britain, Albermarle Street, London.
R.S.	The Royal Society, Carlton House Terrace, London.
Rylands	The John Rylands University Library, University of Manchester, Manchester.
(S.C.)	Copies printed from microfilms of Sutro Library manuscripts: *Wo*, Banks Collection, in the same sequence as the originals. Also, the microfilm copy of the Sutro Library collection of Banks papers: 30 reels. These are held at the The Natural History Museum, London, in the Banks Archive.

S.I.L.	Smithsonian Institution Libraries, Washington, USA.
S.L.	The Sutro Library, University of San Francisco, California, USA.
	Smith, E., (Ed.), *Life of Sir Joseph Banks*, London (1911).
	Smith, J.E., (Ed.), *Selection of the correspondence of Linnaeus and other naturalists, from the original manuscripts*, London (1821).
	Smith, Lady, (Ed.), *Memoir and Correspondence of the late Sir James Edward Smith M.D.*, London (1832).
	Sotheby's catalogue: sale, 24 June 1975, lot 261 for £240. Robert E. Levitt, Durban, South Africa was the owner.
	Stamp, T. and C., *William Scoresby: Arctic Scientist*, Whitby (1975).
Ub. Uppsala Ur.	Universitetsbiblioteket Uppsala Ur: Wallers autografsamling Amerika och England.
U.G.	Niedersächsische Staats-und Univeritätsbibliothek, Göttingen, Germany.
U.L.K. Wedgwood. Papers.	University Library Keele, Keele Staffordshire.
	Upfostrings-Salskapets Tidningar, No. 14, Stockholm d. 21 February 1785.
U.P.L.	University of Pennsylvania Library, Philadelphia, Pennsylvania, USA.
U.W.	University of Wisconsin, Memorial Library, Special Collections Department, Madison, U.S.A.
U.Y.	University of Yale, New Haven, Connecticut, USA. Manuscripts in the Banks collections of the Sterling Memorial Library, and the Beinecke Rare Book and Manuscript Library. These will be differentiated by 'U.Y.S.L.' for the Sterling Memorial Library, and 'Beinecke' for the Beinecke Rare Book and Manuscript Library.
Windsor R.A.	Windsor Royal Archives, Windsor, Berkshire.
W.M	Whitby Museum, Whitby, Yorkshire.

[1]The Matthew Boulton papers at Birmingham City Archives will shortly be re-catalogued as part of a comprehensive project to organize and preserve them for future use.
[2]References as cited in *Banks Letters: A Calendar of the manuscript correspondence*, Warren Dawson, The Natural History Museum (1958).
[3]All previous shelf and volume numbers were superseded in a recent project to catalogue and digitize the Banks collections at the Mitchell and Dixson libraries, Sydney. A series number is given for the letters now, and this is usually cited.

However, some letters retain their old citation, which will be revised in the near future. Others were never part of a Banks collection. Such exceptions are all designated by an asterisk. Ongoing work of this kind, along with the volume and distribution of the correspondence, did not permit a complete listing of the various locations for every letter published in this selection. There will be drafts or copies of some which have not therefore been given.

List of Illustrations

Appendix Contents: Maps

Sir Joseph Banks

Soho Square
Decr 4 — 78

130

Dear Sir

I have from time to time postpond an answer to your favor in hopes that I should be able to pick up some literary news for your amusement but I fear our very Scientifick ones are too much involvd in contemplating the melancholy situation of poor Mrs Britain like the upholsterer to mind their own concerns

They did me the honor however last Monday to Elect me unanimously President of the Royal Society in the

Banks to Sir William Hamilton K.B. P.C., 4 December 1778, B.L. Egerton MSS. 2641, ff. 130–131. In this Banks announces his election as President of the Royal Society, envies Hamilton being so close to an erupting volcano, and describes the political scene in England as 'our Billingsgate Senate'.

room of Sr Jno Pringle who resigned that
office finding I suppose Newtons Chair
not so easy a one as his own Fire side
Elbow

That I Envy you your situation
within two Miles of an Erupting Volcano
you will easily guess I read your Letters
with that kind of Fidgetty anxiety which
continualy upbraids me for not being in
a similar situation I envy you I pity
myself I blame myself & then begin
to tumble over my Dried Plants in hopes
to put such wishes out of my head which
now I am tied by the leg to an Arm Chair
I must with diligence suppress

So great is our [illegible]

131

Philosophick news that I am oblig'd to
have recourse to politicks to amuse you
but as I do not value myself upon
my political performances I have chosen
an abler Pen Mr. Tickel the Author
of the Pamphlet you receive with this
has lived a good deal with Mr. Brummel
Ld Norths Secretary from that School he
has got his Knowledge & some do not
scruple to say that the head master
gave him Lessons which have enabled
with so much real humor to Caricature
the different styles of Eloquence ~~with~~
which our Billingsgate Senate affords
Certain it is that the Laugh is so
much with him that scarce a man

for whom he has spoke can keep his
speech clear of some expressions to
be found in anticipation & the house
no sooner hears them than a titter
confounds the poor orator

Luttrel mentioned Egypt the other
Day Mulgrave burst into a Horse-
laugh & Luttrel tore it so ill that
he called him a Bear

Do me the favor to make my respects
acceptable to Lady Hamilton & believe
me at all times

Your affectionate
& Obedient Hble Servant
Jos: Banks

List of Letters

The Letters:

The Letters

1768

To William Philip Perrin F.R.S.

[New Burlington Street]
16 August 1768

My Dear Perrin,

I am now on the Brink of Sailing on the Expedition you hinted at in your Last.[1] I shall therefore send as ample an account of the Expedition as I dare trust to paper, hoping it will be some satisfaction to you to guess at the Station of a freind from whoom, or of whoom, you will not hear any more for three years at least.

In march Last the Goverment, at the Instance of the Royal society, resolved to send out a ship to any part of the world which should be found most Convenient for an observation on the Transit of Venus,[2] which the Latter was to Supply with Proper instruments & observers. The Place was soon fix'd upon somewhere in the South Sea, over a Large part of which the Limits Convenient for Such an observation are Extended.[3]

Upon Considering the plan of this Scheme, it immediately occurd to me that it would be a most desirable one for me to Engage in. The Whole tract of the South Seas, &, I may say, all South America is Intirely unknown to a Naturalist. The South Sea at least has never been visited by any man of Science in any Branch of Literature.[4]

Upon looking at the Plan of the Voyage, it might Easily be seen that this would not be the Extent of it. A Ship in the midst of the South Seas would never attempt to return against the S[outh] E[ast] Trade. She must therefore necessarily go forwards & visit the Ladrones,[5] some parts of the East Indies, & the Cape of Good Hope, all places much worth the attention of a Naturalist. This [is] the Least of the plan. She may do much more, as if you look upon a Chart you may see.[6]

I was much Encouraged in this Scheme by our Freind Solander,[7] who so heartily agreed in the Excellence of it that he promis'd to make application to the trustees of the Museum,[8] & if Possible get Leave to accompany me, which [he] has done, & got the nesscessary leave of absence signd, & is now going with me. I take also, besides ourselves, two men to draw, & four more to Collect in the different branches of Nat[ural] Hist[ory], & such a Collection of Bottles, Boxes, Baskets, bags, nets &c. &c. &c. as almost frighten me who have prepard them.

England you say is behind hand with the rest of Europe in Undertakings of this Kind. You will not wonder at it when I tell you that on application to the first Lord of the Admiralty, when I had stated the Case, & told him what I meant to do, his answer was, 'You sir are very welcome to go, & it shall be my care that you have Every Convenience which I Can Procure for you, but we Cannot find room for people skilld in Botany & drawers of Plants.' This at first hurt me much, & I had almost given over my Plan, but upon Application to the Secretary of the Admiralty, he undertook to do it all without any more trouble to me. So I have not been near Sr E: H:[9] since, but made Every application to the Secretary,[10] who has done Everything we wanted with as much alacrity & spirit as could be wishd.

Adieu. I thought to have made this much Longer, but am sent for by Express to Join the Ship[. I] will write again From Madera. Till then, beleive me,

Sincerely Yours,

J Banks.

[D.R.O. FitzHerbert MS. D239M/F15883.]

[1]The last letter Perrin sent: Perrin to Banks, 15/4/1768, N.L.A. MS. 9/129.
[2]The Transit of Venus occurs when Venus passes in a direct line between the Sun and the Earth. Scientists hoped that a successful observation would allow an exact calculation of the distance between the Sun and Earth. They argued that navigation would be improved because longitude could then be accurately determined using the Moon's position. The transit could be observed from certain sites in June 1769.
[3]Following discussions in 1766, a committeee of Royal Society Fellows was established in 1767 to consider sending astronomers to observe the Transit of Venus. They proposed an expedition into the Pacific for late 1768 or early 1769, and in the New Year of 1768 a memorial to George III (1738–1820) was drawn up, which received royal approval. The King gave £4000 as personal grant, and ordered that the Navy Board should provide a suitable vessel.
[4]This is not strictly true. For instance, voyages to the South Seas had been, or were being made. Two were made in HMS *Dolphin*, 1764–1766 and 1766–1768, under John Byron (1723–1786) and then Samuel Wallis (1728–1795) respectively. Wallis was closely followed by Louis-Antoine de Bougainville FRS (1729–1811) on the *Boudeuse*, accompanied by the store-ship *Étoile* for part of the journey. Philibert Commerson (1727–1773), a naturalist, and Pierre Antoine Véron, an astronomer, went with Bougainville.
[5]Ladrones, a Pacific island chain, also called the Marianne or Mariana Islands. They were discovered in 1521 by Ferdinand Magellan (c.1480–1521).
[6]At 3 o'clock in the afternoon on 25 August 1768 HMS *Endeavour* left Plymouth Sound, and a summary of its route follows. HMS *Endeavour* visited Madeira before sailing for the coast of South America, where it stopped at Rio de Janeiro. It subesequently moved on to stormy Tierra del Fuego, and passed Cape Horn late in January 1769. In the Pacific the *Endeavour* sailed to the Society Islands, and the Transit of Venus was observed as part of a three month stay at Tahiti (King George III's Island, Otaheiti). Having

circumnavigated the islands of New Zealand, a course was set towards Australia, and along its East Coast. This epic portion of the voyage was completed by August 1770. Afterwards, HMS *Endeavour* passed through the Endeavour Strait, Torres Strait, and, following a brief stay at Irian Jaya, steered for Java Head, the Sunda Strait, and then tragedy in Batavia. The way home led to Cape Town on 14 March 1771, and from St. Helena to Deal on 12 July.

[7]Banks's party consisted of: Dr. Daniel Carl Solander FRS (1733–1782) as Banks's companion botanist; Sydney Parkinson (c.1745–1771) and Alexander Buchan (d.1769) as artists; Herman Dietrich Spöring (1733–1771) as secretary and assistant draftsman; Peter Briscoe (1747–1810), James Roberts (1752–1826), Thomas Richmond (d.1769) and George Dorlton (d.1769) as servants and field workers.

[8]The British Museum.

[9]Sir Edward Hawke (1705–1781): First Lord at the Admiralty, 1766–1771.

[10]Sir Philip Stephens FRS (1725–1809), 1st Baronet: First Secretary at the Admiralty, 1763–1795.

Letter 2

1768

To The Portuguese Viceroy of Brazil, Don Antonio Rolim de Moura, Conde d'Azambuja

[HMS *Endeavour*], Rio de Janeiro
17 November 1768

The Memorial of Joseph Banks Esqr.
to His Excellency Count Rolim,
Vice-Roy and Captain General of
the Estates of Brazil.

The very disagreable situation to which Your Excellency's most unprecedented behaviour[1] has reduced me makes it necessary for me to state in writing the facts relating to it, that I may be convinced by Your answer that those unexampled Orders, which are issued against me in particular and the whole Ship in general, are not the effect of [a] mistake or misrepresentation, which even at this time I cannot help suspecting.

Your Excellency has before now been acquainted with the nature of the favours I ask,[2] which, tho' I call them favours, appear to me to be of such a nature as never were before denied, even to the meanest subjects of a Crown in Peace and Amity with His Most Faithfull Majesty;[3] notwithstanding which You have thought proper to deny me every one, not even permitting me to go on shore, but ordering our Ship to be guarded in the same manner as would have been done to His Most Faithfull Majesty's declared and inveterate enemies.

Disagreable as it is for any man to declare his own rank and consequence, my situation makes it necessary. I am a Gentleman, and one of fortune sufficient to have at my own expence fitted out that part of this Expedition under my direction which is intended to examine the Natural History of the Countries where we shall touch. For the execution of this undertaking I have with me proper people, who, as well as myself, have made that Science their particular study. To all these His Britannic Majesty was graciously pleas'd to allow conveniencies and accommodations on board His Ship in consideration of the use which from such researches might accrue to Mankind in general.

I ask, therefore, leave to go on shore, taking with me proper People who may assist me in collecting and examining such Trees, Shrubs, Plants, Birds, Beasts, Fishes and Insects as I may meet with. The Collection and examination of such things being the sole business I have undertaken in this Voyage, this is the only indulgence which I ask.

To prevent any suspicion of my acting otherwise, it is also my desire that I may in the execution of this be attended by any Person or persons whome Your Excellency shall chuse, who may be eye-witnesses of every thing which I do, and may serve to convince You that nothing was meant in the fitting out of this Ship but the promotion of Learning in general.

If Your Excellency should have any objection to my coming in to the Town or Forts, I here publickly declare that I have not the least business, nor have I a wish to enter within the walls of any of them. My business is best carried on in places far remov'd from Men and Houses. Wild and desart places a league or two from the Town would suit my purposes much better. There Nature is to be seen in her primitive beauty, which alone I endeavour to study and enjoy.

It may be unnecessary to remind Your Excellency that His Most Faithfull Majesty's Subjects have always been treated in different manner in every part of His Britannic Majesty's Dominions in Europe, Asia, Africa and America, where I am certain that His Most Faithfull Majesty's Subjects have always received from His Britannic Majesty's Officers every mark of politeness and friendship. Such behaviour is esteem'd by Englishmen a debt due to every Subject of a King at Peace and in Amity with their Master, returns of which they think they have an undoubted right to expect.

Should Your Excellency still persist in Your refusal, I must insist upon having Your reasons return'd to me in writing that I may be able to lay them properly before my own Court; a duty which every Englishman thinks he owes to his King and Country.

Dated on board His
Britannic Majesty's
Ship of War, Endeavour,
in the Port of Rio Janeiro.

[N.L.A. MS. 9/2. (drafts and fragments: MS. 9/2c, MS. 9/2d, MS. 9/2e, MS. 9/2f); B.L. Add. MS. 34744, ff. 41–42; Also: J.C. Beaglehole (Ed.), *The Endeavour Journal of Sir Joseph Banks 1768–1771*, Sydney (1962), vol. II, pp. 315–316, and the draft on pp. 317–318, (pp. 318–320 include the remainder of the letters between the two men). See Warren R. Dawson, *The Banks Letters: A Calendar of the Manuscript Correspondence*, London (1958), p. 710. Dawson noted: 'The originals of this and the following letters are now in Yale University Library.']

[1]HMS *Endeavour* anchored in the harbour at Rio de Janeiro, where Banks and Dr. Daniel Solander were frustrated in their attempts to go ashore by the Portuguese Viceroy, Don Antonio Rolim de Moura, Conde d'Azambuja. Only the commander and sufficient men for organizing supplies were allowed to leave the ship. From 13 of November to 7 December Banks and Solander could do little more than look at a South American continent rich with largely unexplored flora and fauna.

[2]Banks exchanged more than one letter with the Viceroy, but to no avail: Rolim de Moura to Banks, 18/11/1768, N.L.A. MS. 9/2h, B.L. Add. MS. 34744, f. 43; Banks to Rolim de Moura, 19/11/1768, N.L.A. MS. 9/2l, B.L. Add. MS. 34744, f. 43v; Rolim de Moura to Banks, 20/11/1768, N.L.A. MS. 9/2p, B.L. Add. MS. 34744, f. 44.

[3]The Portuguese monarch is described as 'Faithfull' throughout. He was José I (1714–1777).

Letter 3

1768

To William Philip Perrin F.R.S.

[HMS *Endeavour*], Rio de Janeiro
1 December 1768

My Dear Perrin,

Before you receive this you will have traveld[1] over Alps & Appenines,[2] & seen the customs of many nations & people[s], but never, I will venture to say, met with so illiterate, unhumanizd [and,] I may say, Barbarous a set of people as I am now in the Possession [of]. Three weeks have I been laying at an anchor in this river, the banks of which are crowded with plants, animals &c. such as I have never seen before. All this time have I not been permitted to set my foot upon the land because forsooth the Gentry here think it impossible that the King of England could be such a fool as to fitt out a ship merely to observe the transit of Venus, from hence they Conclude that we are Come upon some other Errand, which they think to disapoint.

O, Perrin, you have heard of Tantalus[3] in hell, you have heard of the French man laying swaddled in linnen between two of his Mistresses, both naked [and] using every possible means to excite desire, but you never heard of a tantalizd wretch who has born his situation with less patience than I have done mine. I have cursd, swore, ravd, stampd & wrote memorials to no purpose in the world. They only Laugh at me, & exult in their own penetrations to have defeated so deep laid a scheme as they suppose ours to have been.

Except this accident, than which worse could not have happnd, every thing has been favourable. The winds & Seas have Combind to make our passages pleasant. 18 days brought us to Madeira, where, tho we staid only 5, we collected above 300 species of plants, 200 of insects &c. &c; & this late in their autumn, the worst time of the year for vegetation.

Two months more brough[t] us here over seas mild & calm as they, you Know, always are between the tropicks so that [a] great part of our time has been spent in fishing for fish, mollusca &c., & not without great success as new genera in the last of these Classes are as common as new species in any other part of Nat[ural] Hist[ory]. Drawing has gone forward Every day so that I have seldom spent time more to my satisfaction than these two months, buoyd up with the hopes of what we should find here, hopes not without foundation, which nothing but the unexampled Barbarity of these Rascaly Portugese could have disappointed.

You know that I am a man of adventure, & as I scapd hanging[4] in England have now taken it for granted that I am born for a different [fate.] In pursuance of this opinion I have venturd ashore once, evading a boat load of soldiers who look after us, & found such things as well repaid my risk. Tho the next morn the Viceroy had intelligence that such a thing had been done so I dare not venture any more.[5]

Since we came here a Spanish packet boat arrivd here from Buenos Ayres. You as well as myself Know that the Spaniards are the natural enemies of the Portugese, yet they were receivd with uncommon marks of politeness. This, I must confess, almost drove me mad, but I was without remedy.

The people here live in a state of Slavery hardly to be equald, I beleive, in the world. I hope sometime or other to be able in person to tell you what little I have learnt of them, which appears realy more like fable than reality. Their town is as large [as] Bristol or Liverpool, [and] as ill defended as is possible. [There is] scarce Even an appearance of Fortification, & What there is mounted with rusty gunns falling off from their Carriages. So much I have learnt from my Glass, for we lay Just under Ilhoa de Cobras,[6] on which is the chief fortification that defends the town. As for Sta Cruz,[7] which Guards the Entrance of the river, even if it was good, which it is not, the sea breeze that blows here every day from 12 till 6 or 7 would Carry a fleet of ships by it without danger. If they chuse their time well, they may go in at the rate of 6 or 7 knots an hour. The harbour we are in is certainly a very fine one, /& capable of containing any number of ships which may/ heave down close under the Fort in four or five fathom[s] of water. Surely, if the Portuguese continue to treat us as they have for some time done, a fleet of ships sent here would be a medecine very easily administerd, & very likely to make a compleat Cure. But this is a measure so contrary to the present politicks of our countrey that, much as I should from revenge wish it, I cannot hope to See it.

My time here is now unexpectedly shortend, for we have so broad a hint given us as to think it nesscessary to Leave the port tomorrow morn. So you must excuse my Shortning my letter more than I should have wishd as it is probably the last you will receive, or any of my Freinds, till we have met. Adieu. Success attend your Endeavours & undertakings shall be my warmest wish even at the antipodes,

Your Sincere & affectionate
Freind,

Jos: Banks.

P.S. Dr. Solander, who is included in the 'we' so often repeated in this letter, desires his best Compts, & Joins in my wishes.

[D.R.O. FitzHerbert MS. 239M/F15882.]

[1]Perrin had been touring parts of Europe, starting in 1767: Perrin to Banks, [January 1767], N.L.A. MS. 9/129.
[2]The Apennines, Italy.
[3]Of Greek legend, Tantalus was a son of Zeus and the Titaness Pluto. When he became the king of Phrygia Tantalus revealed the secrets of the gods, and was condemned to stand in Tartarus up to his chin in water. The water receded each time he tried to drink, and there were branches of fruit above him which he could not grasp.
[4]An oblique reference, as a joke, to a youthful misdemeanour at Eton when Banks and a friend, perhaps Perrin, shot a swan and ate it in a pie. Killing swans was illegal.
[5]On 22 November Peter Briscoe and James Roberts spent the day ashore secretly collecting plants and insects for Banks. They returned for more on 24 November. The next day Dr. Daniel Solander went to the town of Rio de Janeiro disguised as the ship's surgeon visiting a friar. Finally, Banks landed to botanize. He also met local people, and went to their houses to purchase stock for the ship.
[6]îlha das Cobras, Rio de Janeiro.
[7]îlha de Sta Cruz, Rio de Janeiro.

1768

To James Douglas P.R.S., 14th Earl of Morton

[HMS *Endeavour*], Rio de Janeiro
1 December 1768

My Lord,

I should Certainly have wrote to your Lordship[1] from Madeira to have given you an account of our proceedings there had not the Shortness of our stay renderd it impossible. I shall not, however, let the present opportunity pass as I imagine it will be the only one I shall have of a[c]quainting your Lordship with the success of a scheme which I am convincd you have honourd with no small share of good wishes.

We set sail from Plymouth (after having been confind there by contrary winds for almost ten days) all in high spirits; myself more particularly as I was after many delays at last fairly Embarkd in an undertaking from which I promisd myself three years uninterrupted enjoyment of my Favourite pursuit.

My Sea sickness was more than usualy favourable so that in about a week I was able to begin as we all were. Light winds giving us an opportunity, we made our first Essay on the inhabitants of the Sea calld by the seamen blubbers.[2] In these we hop'd to make great additions to natural history, as probably nobody but ourselves ever had so good an opportunity of taking & preserving them. Nor were we mistaken, for our first Essay produced an animal whoom we could not refer to any known genus, whoose singularity of structure I hope Some time or other to shew to your Lordship. Of these we have already taken twelve species, & made Drawings & descriptions of all, as well as preservd in spirit as many as we could.[3]

18 days brought us to Madeira, where we were to stay no more than five days, one of which was spent in getting leave of the Governor to range about in search of what we could find. Notwithstanding this delay we Collected above 300 species of Plants, 200 of Insects & about twenty of fish, many of all these three Kinds such as had not before been describd. For our extraordinary success we were not a [little indebted] to the assistance of a very ingenious gentleman, [the] brother to Dr. Heberdene,[4] who has long been settled in the Island. He was indeed indefatigable in procuring us all the assistance in his power from the people of the Island, as well as in Communicating some very ingenio[u]s observations of his own on the trees that are found there.

From Madeira we saild happy in having Collected sufficient to keep us employd till we arrivd at this place, where, from the Situation &c., we did not doubt of making very great acquisitions. The Voyage was pleasant, & took us up two Months, the whole in fine weather so that never a day passd but the business of drawing went on without interruption. The Calms of the line lasted about a fortnight, in which time I was almost Constantly in a boat rowing about the Ship, & seldom returnd empty so that upon the whole the voyage was profitable to the undertaking, tho we had uncommonly bad success in fishing.

What will most likely surprise your lordship I have yet to tell, which is the reception that we met with here. As it is quite extraordinary in its nature, I shall give a minute detail of the Particulars.

On the 13th. of this Month we arrivd here, having saild up the river with a very light breeze, & amusd ourselves with observing the shore on each side coverd with Palm trees, a production which neither Dr. Solander[5] or myself had before seen, & from which, as well as every thing else which we saw, we promisd ourselves the highest satisfaction. As soon as we came to an anchor a boat full of armd soldiers came from the town, &, without saying a word, stationd themselves near our ship. Soon after another came off bringing a Colonel & officers of the Portuguese, who askd many questions, but seemd satisfied with our answers, & told us that the next morn we might come ashore. In the morn the Captn.[6] went on shore Early, & we prepard to have followd him, but he returnd & brought with him an officer, who after some time told us that we should not be allowd to Come ashore at night. However, we at[t]empted it, but were stopd & Sent aboard again.

Your lordship Can more easily imagine our situation than I can describe it. All that we so ardently wishd to examine was in our sight. We could almost but not quite touch them. Never before had I an adequate Idea of Tantalus's punishment, but I have sufferd it with all possible aggravations. Three weeks have I staid aboard the ship regardless of every inconvenience of her being heeld down &c. &c., which on any other occasion would have been no small hardships, but small evils are totaly swallowd up in the Larger. Bodily pain bears no comparison to mine. In short, the torments of the damnd must be very severe indeed as doubtless my present ones Cannot nearly Equal them. I twice remonstrated to his excellency, letting him Know my business & who I was, offering to submit myself to any precautions he should think necessary, & to be attended wheresoever I should go by whoever he should a[p]point, but all to no purpose. His answers were so little to the purpose that I am forced to submit to the necessity of my situation.

I have taken the liberty to Enclose to your lordships the Memorials as well as the answers which I have receivd, which I shall forward by a Spanish pacquet now laying in the harbour. They may serve to convince your Lordships that I have left no stone unturnd to get the liberty I askd, tho that is a point which I fancy your lordships will not much doubt.

The people here are ignorant to a degree of wonder, of which I shall give only one instance. When the Captn. first went ashore, the Viceroy,[7] upon being told that the Ship was fitted out to observe the Transit of Venus, gravely askd whether that was the Passing of the North Star to the South Pole. This alone will, I think, Sufficiently shew your Lordship the State of Learning in this place.

The Captn. means to write to the [Royal] Society an account of this transaction, which, I fancy, will be very full so I shall not trouble your Lordship with any more particulars, but Content myself with assuring your Lordship that I am your affectionate & much Obligd,

Hble Servant,

Jos: Banks.

[N.L.A. MS. 9/3; B.L. Add. MS. 34744, ff. 39–41 (contemporary copies of this letter and the memorials are enclosed). See Warren R. Dawson, *The Banks Letters: A Calendar of the Manuscript Correspondence*, London (1958), p. 272. Dawson noted: 'The original was sold at Sotheby's, 13 May, 1929, Lot 1, where it was wrongly described as written to Lord Sandwich; it is now in the possession of R. de Nankivell Esq.; contemporary copies of this letter and the enclosed memorials are in the West Papers, Vol. 18, in the British Museum [Library]...' Also: J.C. Beaglehole (Ed.), *The Endeavour Journal of Sir Joseph Banks 1768–1771*, Sydney (1962), vol. II, pp. 313–315. Beaglehole notes that another version of this letter, probably the original, is 'in the Yale University Library (except No. 7).']

[1]Banks could not have known that that Lord Morton died on 12 October 1768.
[2]Sailors applied this term to jellyfish and some other transparent pelagic animals.
[3]A number of animals were collected at this stage, both from the air and the sea. Banks and his party were particularly excited by different species of salp, which they collected in clusters and individually from 28 August through to 6 September 1768. On 12 September Port Santo and Madeira were in full view: J.C. Beaglehole (Ed.), *The Endeavour Journal of Sir Joseph Banks 1768–1771*, Sydney (1962), vol. I, pp. 154–157 (Hereafter called the *Endeavour Journal*).
[4]Thomas Heberden FRS (1703–1769): physician; brother of William FRS (1710–1801), who was a London physician. Banks and Solander commemorated Thomas Heberden in the name *Heberdenia* (Myrsinaceae).
[5]Dr. Daniel Solander. See Letter 23.
[6]Captain James Cook FRS (1728–1779): captain, RN; cirumnavigator; explorer. The *Endeavour* expedition was the first of Cook's three great voyages to the Pacific. The second commenced in HM Ships *Resolution* and *Adventure*, 1772–1775. The third was made in HM Ships *Resolution* and *Discovery*, 1776–1780. He was killed in Kealakekua Bay, Hawaii, during a confrontation with the islanders there.
[7]The Portuguese Viceroy, Don Antonio Rolim de Moura, Conde d'Azambuja.

i. **Joseph Banks F.R.S., F.S.A.**

Painted shortly after the return of H.M.S. *Endeavour.* Portrait, 1771, by Benjamin West P.R.A. Lincolnshire County Council, Usher Gallery, Lincoln.

Note: This portrait was rediscovered in 1986, and was offered for exhibition at Sothebys in 1987, where it was sold. In it Banks is shown aged 29, wearing a Maori cloak made of flax, surrounded by artefacts and specimens to signify his activities on the *Endeavour* voyage, 1768–1771. The style is conventional, with a classical column and drape in the background. Banks is fixed in heroic mid-stride, pointing at his cloak with its dog's hair fringe. Standing to his right is a carved Maori fighting-staff, or *taiaha*, and a canoe paddle. The basketry head-dress adorned by feathers and worn by Tahitian warriors may also be seen on this side. It is known to natives as a *fau*. On the floor to Banks's left is a Polynesian adze, a volume of botanical illustrations, and a bark cloth beater, used by Society Islanders in the manufacture of cloth from the paper mulberry tree. In 1773 the portrait was exhibited at the Royal Society under the title 'A whole length of a gentleman with a New Zealand mantle round him.' Prints were then made, and these were sold to the public in shops.

Letter 5

1771

To Thomas Pennant F.R.S.

London
13 July 1771

Dear Sir,

A few short lines must suffice to acquaint you with the arrival of Dr. Solander & myself in good health this day.[1] Mr. Buchan,[2] Mr. Parkinson & Mr. Sporing are all dead, as is our astronomer, seven officers, & about a third part of the ships crew of diseases contracted in the East Indies, not in the South Seas where health seems to have her cheif residence.[3]

Our Collections[4] will, I hope, satisfy you: very few quadrupeds; one mouse, however, (Gerbua) weighing 80 Ib weight.[5] I long for nothing so much as to see you, but must delay that pleasure some time. My relations are dispersd almost to the extremities of the Kingdom, & I must see them before I begin to arrange or meddle with anything. Winter, however, will soon come, & doubtless bring us together. Grass I must have in the mean time. Salt provisions & Sea air have been to me like too much hardmeat to a horse. In a few days I shall be able to write more understandably. Now I am Mad, Mad, Mad. My poor brain whirls round with the innumerable [sensations] which the return to my native countrey have excited. As soon as I enjoy a lucid interval I will write again. Till then, adieu.

Your Sincerely affectionate,

Jos: Banks.

Compts. to all freinds at & about Downing.[6] Dr. Solanders best respects to yourself.

[D.L. Q77/32.*]

[1]HMS *Endeavour* 'landed at Deal' on 12 July 1771.
[2]Alexander Buchan, who died of a fit early in the morning of 17 April 1769.
[3]Tertian malaria seems to have been the chief cause of illness at Batavia. Dr. Daniel Solander was very unwell, but survived. Seven other people died though, including the Tahitians who had accompanied Banks, Tupaia (d.1770) and Tayeto (d.1770). On the journey to the Cape of Good Hope twenty two people died of dysentery and malaria, including most of Banks's party. Herman Spöring succumbed on 24 January 1771, and was followed two

days later by Sydney Parkinson. The astronomer, Charles Green FRS (1735–1771), expired on 28 January. Banks was also unwell at this stage, suffering excrutiating bowel pains in particular.

[4]A general review of the historic collections is made by H.B. Carter *et al* in: *History in the Service of Systematics*, 'The Banksian Natural History Collections of the Endeavour Voyage and their Relevance to Modern Taxonomy', Number 1, London (1981).

[5]Kangaroos were shot by Second Lieutenant John Gore RN (c.1730–1790) in July 1770 on the east coast of North Queensland. On 14 of this month it seems he killed a Great Grey Kangaroo (*Macropus canguru*). It weighed 38Ib, and was eaten the next day. On 27 July Gore went shooting again. An 84lb Wallaroo (*Macropus robustus*) fell victim to his gun on this occasion. Banks's greyhound caught a small kangaroo or wallaby on 29 July. This weighed about 8Ib, and so may have been an immature specimen of *Macropus*, or possibly a female. Banks returned to England with the skin of this animal, some skulls and notes, from which George Stubbs (1724–1806) made a painting: See Illustration II. This interesting event has been discussed elsewhere: T.C.S. Morrison-Scott and F.C. Sawyer, *The Bulletin of the British Museum (Natural History), Zoology*, 'The Identity of Captain Cook's Kangaroo', London (1950), vol. I, no. 3, pp. 45–50; Sir R. Cilento, *Notes and Records of the Royal Society*, 'Sir Joseph Banks, F.R.S., and the naming of the Kangaroo', London (1971), vol. XXVI, no. 2, pp. 157–161.

[6]Downing, Holywell, Flintshire.

ii. **A Kangaroo.**

Painting, 1771–1772, by George Stubbs A.R.A. for Banks. From a collection at Parham Park, West Sussex, England. Captain Cook's kangaroo?

Letter 6

1771

To the Comte de Lauraguais

London
6 December 1771

My Dear Count,

The abstract of my Voyage, which I have so long Promis'd you, I at last begin to write.[1] The multiplicity of employments in wch I am engaged will, I know, with you plead my excuse for having so long delay'd it.

On the 25th. of August 1768[2] we set sail from Plymouth,[3] & on ye 12th. of Sepr. arriv'd at Madeira after a moderate Passage. Here we were receiv'd with great Civility by our Consul, & not uncivily by the Portuguese Governor, & during our stay we collected some specimens of Natural Curiosities not unworthy [of] our Notice.[4]

On ye 18th. of ye same Month we set Sail from yt Place, & on the 13th. of November arriv'd at *Rio de janerio*, where, instead of being receivd as friends & allies of his most faithful Majesty, orders were immediately issued out yt every insult possible should be offered to the officers of our ship, whose duty obligd them to Land; & as for us (*Foutres Philosophers*), we were refus'd to land on any pretence whatsoever on the peril of being sent to Portugal in Irons. A thing I verily believe their absurd Viceroy would have done had he caught either Dr. Solander or myself upon any of our little Excursions.

Notwithstanding the Vigilence of his Excellence le *Comte D'Azambusio*, however, we ventur'd a shore each of us once, & had several parcels of Plants brought off to us under the title of grass for our Cattle, as we were absolutely forbid to have them under any other Denomination.[5]

The abject slavery of the Portuguese in this Colony is beyond imagination. Suffice it to say that to prevent any attempt against Government, every officer & other person of any Distinction is oblig'd to attend ye Levée of ye Viceroy twice every day under penalty of his displeasure, which is follow'd by an Instant excommunication from all Society, for whoever speaks to a man under these circumstances is instantly himself under ye same.

From these unfriendly & illiberal people we departed on ye 7th. of Decr., not forgetting in our way out of ye Harbour to land upon a small Island Call'd *Raza* off ye mouth of it, where in a few hours we much increas'd our natural Collections.

On ye 15th. of January [1769] we arriv'd at *Terra del Fuego*, & soon anchor'd in a small bay near ye middle of ye *Streight le Maire,* which had been formerly call'd by the Nassau fleet[6] the Bay of Good Success. Here we lay some Days in a tolerable Harbour, which offered plenty of Woods & water, & an innumberable quantity of Plants incomparably different from any which had before been describ'd by any writers on Botany. The inhabitants were of a moderate size, were friendly to us, but seem'd to have no provisions to spare; nor if they had would it have suited our Palates, being generally the Flesh of Seals. We found, however, a kind of watercress (cerdamine), & a kind of Parcely (apium),[7] which we made into Soupe, & no doubt re'p'd benefit from their antiscorbutik virtues; tho' in reality none of our people were absolutly ill of the scurvy.[8]

From hence we Sail'd on ye 21st. January, & having Passed *Cap Horn*, & Pass'd sufficiently to the westward of ye Coast of America, we sail'd in almost a N.W. direction for ye Island of Otaheite, ye Taiti of Mr. Bougainville,[9] which was ye Place of our Destination. On ye 4th. of April we saw land, may be the 4 Facardin[10] of that gentleman, & from thence, passing by several low Islands, arriv'd at ye place of our Destination. On ye 13th. of the same month the inhabitants receiv'd us wth great politeness, but it was visibly the Effect of Fear. We immediately erected a small stockade [for] defence, & in yt observ'd the Transit of the Planet *Venus* over ye Sun on the [3rd.] of June 1769, for which observation we had ye most favourable weather imaginable.

The inhabitants of this Island during our whole Stay of 3 months behav'd to us wth great affability. Mr. Bougainville's account of them is as good as could be expected from a man who staid among them only 9 Days, & never, tho' a native went away with him, made himself master of their Language. This, not only myself, but several of our Company did, & of it I shall only say that Mr. Bougainville has omitted in his Vocabulary every aspirate in it, tho' the use of them is very Frequent, I suppose in Confirmity to his Mother Tongue.[11] After a stay of 3 Months we left our belov'd Islanders wth Much regret on ye 13th. of July, & sail'd to ye Westward in search of other Islands, wch. a native of Otaheite, who chose to embark wth us, offerd to direct us to. We found them with great facility. They were in number 6: Huaheine; Vehieta; Otaha; Bolabola; Maurua, & Tupi. The Natives we found to be of exactly the same manners, Customs & language as those of Otaheite. After a month's Stay among them, we left them on ye 9th. of August in order to steer to ye southward in hopes of finding a land more worthy of our Notice; tho' we were absurdly forbid to proceed [to a] higher Latitude than $40^{d.}$, into which Lati[tud]e we arriv'd by a due South course, & turning then to the westward on ye 3d. of October, fell in wth the Easter[n] side of New Zeland.[12] The extent of this Country, which extends from ye Lat[itud]e of $34^{d.}$ to that of $47^{d.}$ S., took us up six Months before we could compleat our Circumnavigation of it. In yt Time, however, we discover'd that Instead of being, as generally suppos'd, part of a Southern Continent, it was in reality only 2 Islands without any firm land in their Neighbourhood.[13]

The coast of these Islands abound in Harbours, the country is fertile, & ye Climate Temperate. The inhabitants are a Robust, lively & very ingenious People.

They always strenuously oppos'd us so that we sometimes were laid under the disagreeable necessity of effecting our Landing by Force.[14] They were, however, when subdued, unalterably our friends, & carried yt sentiment to lengths which in Europe we are unacquainted with; notwthstanding that their barbarous customs taught them to eat ye Bodies of such of their Enemies as were kill'd in Battle. But what surpriz'd us ye most, was that, notwithstanding ye distance, these people all along that large extent of Coasts spoke different dialects of ye language of Otaheite, every one of which were tolerably well understood by ye Indian who accompanied us.[15]

From these brave people we departed on the 1st. of April 1770, &, steering a Course nearly West, in the 19th. of ye same Month fell in wth the coast of New Holland in Lati[tud]e 38 S,[16] a coast which had never before been investigated by any navigator. Along this Coast we sail'd, often carrying [coming] to an anchor generally in very fine Habours, 'till on ye 10th. of June we struck upon a Rock in Lati[tud]e 15 S.,[17] nearly about ye same place were Mr. Bougainville heard ye voice of God.[18] On this Rock we lay 23 hours in ye utmost danger, & when ye ship got off, which was effected by throwing over board almost every thing heavy, we found her so leaky that she would hardly swim. We got her, however, into a small harbour, where with great difficulty in two Months we refitted her.

During our stay in this harbour we made friendship wth several of ye inhabitants, whom, from their shy dispositions, we had not before well seen. They were of a moderate size, but slender Limbed, dark brown & stark naked both sexes. Their Language is not unmusical, but different from any we have either before or since met wth. Their arms are *Assagayes*, headed wth ye Boarded [bearded] Bones of Rays. They were, however, not uncivil, tho' very timorous & Jealous of their Sooty Wives.[19]

After having repaird our ship as well as we could, we, on ye 4th. of August, sallied first into a Sea of Dangers[20] more difficult to imagine than to describe. Without, a wall of rocks ran paralel to ye shore at ye distance of 8 or 10 Leagues. Within, were Shoals innumerable, which ye smoothness of ye water, caus'd by the barrier that prevented our retreat, prevented us from discovering. In this Sea of Dangers we remaind, after having once escaped & having been driven back again into it with ye utmost hazard of our Lives, 'till we carried in Latie. 10 S., where to our great Joy we discoverd an opening to ye west of us, which seem'd to promise a passage into ye Indian Sea. We accordingly followed it, & found it indeed a Streigth between New Guinea & New Holland,[21] through which we passed & became at once easy & happy.

We now resolved to see the Coast of New Guinea in order to ascertain whether or no ye Chart has laid down that Country in a right Position, & accordingly on ye 3d. of Septr. fell in wth it about the Island of *Vleer Moyen* as it is laid in the Charts of the ingenious President de Brosses.[22] From hence we coasted along round the Cape St. Augustin, finding the land very low every where, & shoaling off so far that in 6 fathom of water we sometimes could not see it from ye deck, nor could we ever get nearer it than a league, tho' our ship did not draw above 13 feet of Water.

Nearly about the place call'd *Keer Veer* in ye Dutch Charts, we landed with our Boat, & Saw Cocoa nut trees, & a fertile or at least a Rich Soil. The natives soon attack'd us wth their arrows, & we being but 8 in number, [and] not able to bring our ship nearer than a League, or our Boat than a quart[er] of a Mile of the Shore, were oblig'd to retire, wch we did in Safety, tho' followed by near 300 of ye inhabitants, who to our great surprise threatened us with fire thronk [thrown] out of reeds I know not how, but exactly ressembling the flash of a Musquet, so much so that those who remain'd in the ship were much alarm'd.[23]

From this place then we saild immediately, & passing by Islands, which, by their situation, we judged to be *Arrow* & Timorlaut, we arriv'd in sight of Timor. From whence, passing between *Rote* & *Simau*, we fell in wth a small Island call'd *Savu*. Here we Came to an anchor, & bought from ye Natives Sheep, Goats, Buffaloes &c.; the first we had met since we left Rio de janeiro. Then Passing along ye south side of Java, & into the Streights of Sunda, we arriv'd at Batavia on ye 9th. of October, where we resolv'd to repair well our ship, which had suffer'd when lying on ye Rocks on the Coast of *new South Wales*, as we Call'd it very materially.

Tho' we had been remarkably Healthy through the great Variety of climates which we had pass'd before, yet the uncommon Malignity of ye air of Batavia, so fatal to Europeans, was not the less terrible during our stay here about two Months. Afterwards, at sea, of Distempers contracted here we lost above $\frac{1}{3}$ of our people, among whom were all my Artists, & the two poor Indians, whose loss especially I much regretted.[24] I hope[d] to have pleas'd my country men much with the answers they would have made to their questions, wch I was capable of doing having learnt tolerably well their Language.

From hence, touching at the *Cape of Good Hope* & *St. Helena*, as is ye Custom of India ships, we arriv'd in the Downs[25] on ye 13th. of July *1771* so well satisfy'd with the discoveries which we had made in ye three Kingdoms of nature, that we resolv'd to solicit the Government to furnish Ships for another undertaking of ye same Nature, which they have accordingly done; & in the Month of March 1772 we hope to enter upon our new undertaking.[26]

The Number of Natural productions discover'd in this Voyage is incredible: about 1000 Species of Plants that have not been at all describ'd by any Botanical author; 500 fish, as many Birds, & insects [of the] Sea & Land innumerable. Out of these, some considerable oeconomical purposes may be answerd, particularly with the fine Dyes of the Otaheitians, & the Plant of which the new Zelanders make their Cloth,[27] of which we have brought over ye seeds. The fine red Colour us'd by the inhabitants of the Islands situated between the tropicks in the South Sea, the tinge of which seems to be between that of Scarlet & a pink, is made by mixing the juice of the Fruit of a Fig Tree, suppos'd to be peculiar to those Islands, with the juice of the Leaves of the *Cordia Sebestena orientalis* Lennius.[28] [Linnaeus]

NB: The fig Tree is now describ'd under the name of *Ficus Tinctoria*,[29] & probably did not escape the researches of so accurate a Botaniste as Mr. de Commerson,[30] who sail'd with Mr. Bougainville,[31] is disputed [reputed] to be. — Quadrupedes we found few, & none remarkable but one Species totally different from any known kind. The full grown of it was as large as Sheep, yet went totally on its hind legs as the *Jerbua* & the *Tarsier* of De Buffon,[32] yet in every other part of its external Structure was totally different from either of these Animals.[33]

Thus, my dear Count, I have Given you an abstract account of my last Voyage, the narrative of which will appear, I hope, some time next Winter as I have put all the Papers relative to ye adventure of it into ye hands of Dr. Hawkersworth,[34] who I doubt not will do justice to ye work, which ye shortness of my Stay in England would not permit myself to attempt. In march Next we shall sail upon a new undertaking of ye same kind, in which we shall attempt the Souther[n] Polar Regions. O, how Glorious would it be to set my heel upon ye Pole, & turn myself round 360 degrees in a second ! But that as the unexplaind Secret of the creations shall Please. Whatever may Happen to me, I hope for the Pleasure of relating to you at my return, & truly sign myself,

Your Oblig'd &
Affectionate,

sign'd Jos. Banks

[M.L./D.L. Banks Papers Series 05.01; Also in J.C. Beaglehole (Ed.), *The Endeavour Journal of Joseph Banks 1768–1771*, London (1962), vol. II, pp. 324–329: 'The MS., a copy, is accompanied by a letter of 8pp. folio, closely written, evidently from Lauraguais to D'Alembert — though this is nowhere explicity stated: it is a sort of dissertation on Bougainville and Banks, sings the praises of the latter, and announces a second voyage by him', pp. 324–325.]

[1]This letter was written for Lauraguais by Banks as a private account of the *Endeavour* voyage. However, Lauraguais had it printed, much to Banks's disapproval. Banks immediately acted to stop publication. A printed version at the Mitchell Library is therefore marked 'Abstract of Endeavour's voyage written for Count Lauraguais who printed it. I seized the impression and burn'd it.': M.L./D.L. Banks Papers Series 05.01.
[2]The dates Banks provided for the events described in this letter almost always agree with those in his journal. There is one slight exception, which is 3 October 1769 when HMS *Endeavour* approached New Zealand. The coast was sighted on 6 October according to the journal: *Endeavour Journal*, vol. I, pp. 397–398.
[3]In sequence, the places Banks mentions after Plymouth, Madeira and Rio de Janeiro are: 'Raza', island off Rio de Janeiro; 'Terra del Fuego', Tierra del Fuego, islands off South America; 'Cap Horn', Cape Horn, South America; 'Taitit', Otaheiti or Tahiti, Society Islands; '4 Facardin', Tuamotu island group, otherwise known as the Low Islands or Vahitahi; 'Huaheine', Huahine, Society Islands; 'Vehieta', Raiatea, which is enclosed by a reef along with Tahaa; 'otaha', Tahaa, Society Islands; 'bolabola' or Borabora, Society Islands; 'Maurua', Maupiti, Society Islands; 'Tupi', Motu-iti, Society Islands; New Zealand, the islands were circumnavigated by Europeans for the first time on this voyage; 'New Holland' here means the east coast of Australia as shown in Dutch charts; 'Streigth between New Guinea and

New Holland', the expedition made its way from Endeavour Strait, next to Torres Strait, on 22 August, and New Guinea was reached on 29 of the month; 'Vleer Moyen', known variously as St. Bartholomeo Island, Wleermoysen or Habeeke Island, off Irian Jaya; 'Cape St. Augustin', now False Cape, after which the ship entered a large bight on the West Coast of New Guinea; 'Keer Veer' [Weer], possibly Flamingo Bay, an indentation on the western side of Irian Jaya; 'Arrow', Aru or Aroe Islands, part of the South Moluccas in Indonesia; 'Timorlaut', Tanimbar, Tenimbar or Timor Laoet Islands, east of Timor; 'Timor', an island of the Malay Archipelago; 'Rote', Roti, island off the south-west end of Timor; 'Simau', Semau or Samau, north of Roti, off Kupang; 'Savu' or Sawu, island west of Timor; Java, in the Greater Sunda group, Malay Archipelago; 'Streights of Sunda', between Java and Sumatra, leading out to the Indian Ocean; Batavia, now Jakarta, city and seaport of Java; 'Cape of good hope', South Africa; 'St Helena', southern Atlantic Ocean. See also 'Appendix: Maps'.

[4]See Letter 4, Notes 3 and 4.

[5]See Letter 2.

[6]The Nassau Fleet was named after Prince Maurice of Nassau (1567–1625). It sailed in 1636 to raid Peru, but the raid was a failure.

[7]Watercress (*Cardamine glacialis*), which Banks refers to as '*cardamine antescorbutica*' in his journal entry 14/1/1769: *Endeavour Journal*, vol. I, p. 217. He meant 'Scurvy grass', which was a loose term applied to unrelated plants with antiscorbutic properties. Parcely (*Apium australe*), a common coastal species, is referred to in the journal as well.

[8]Banks does not mention that two men died on a group expedition inland on 16 and 17 January 1769. Banks and Dr. Daniel Solander were in the party, and Solander almost gave up in the cold. Thomas Richmond and George Dorlton, two of Banks's servants, did not survive. They both froze to death during the night. The remainder of the party huddled around a small fire until morning.

[9]Louis-Antoine de Bougainville, the mathematician and explorer who circumnavigated the globe, 1766–1769. He passed through the Pacific naming many islands, including Tahiti. In 1771 he published an account of his voyage, *Voyage autour du monde*..., Paris.

[10]In the printed version of this letter a footnote is supplied here: 'During March, 1768, says Bougainville, we ran on the first sands and isles marked on the chart of M. Bellis, by the name of Quiro's Isles. On the 22d of that month, at six in the morning, we saw at once four little isles, bearing S.S.E. half E. and a little isle about four leagues West; the four isles called *Les Quatres Facardins*. Page 204.'

[11]The culture and language of the South Sea Islands fascinated Banks. He recorded many new words, their meanings and pronunciation in his journal. For instance, some are listed: *Endeavour Journal*, vol. I, pp. 370–373. He remarked that: 'Their Language appeard to me to be very soft and tuneable, it abounds much with vowels and was very easily promounc'd by us when ours was to them absolutely impracticable.' Even Banks's name proved too difficult to learn, and so he was called 'Täpánė' or 'Topane' by the islanders instead.

[12]Lieutenant James Cook had secret additional orders, which were to be followed once the Transit of Venus had been observed. They required him to explore the South Pacific by proceeding to latitude 40°, and from there westward between 40° and 35° until New Zealand was reached. Some imagined that 'a Continent or Land of great extent may be found to the Southward...': instructions, dated 30 July 1768, P.R.O. Adm. 2/1332, 160. Cook demolished the idea of Terra Australis Incognita.

[13]New Zealand was sighted on 6 October 1769, and on 1 April HMS *Endeavour* departed for New South Wales.

[14]Banks wrote this guilty comment in his journal on 9 October after one violent encounter which led to the death of at least four Maori men: 'Thus ended the most disagreable day My life has yet seen, black be the mark for it and heaven send that such may never return to embitter future reflection.' *Endeavour Journal*, vol. I, p. 400–403.
[15]Banks wrote thoughful essays following each major stage in the voyage. His theory of Pacific languages is explained in one: *Endeavour Journal*, vol. II, p. 35–36, 'Account of New Zealand'. The 'Indian who accompanied us' was Tupaia, originally a native of Raiatea, who travelled with Tayeto, his young follower. Tupaia was a priest.
[16]The coast of New South Wales was sighted at first light, 6 a.m., by Lieutenant Zachariah Hicks RN, after whom Point Hicks was named.
[17]The ship struck Endeavour Reef in the night while under sail. Good luck and frantic work freed the boat though, and by 10 a.m. it was limping towards the coast in search of a harbour. On the way the crew fought to stem the water entering through its damaged hull, and struggled to pump out what still seeped in. On 14 June 1770 a harbour was found in which HMS *Endeavour* could be repaired. This was Cook Harbour, where Cooktown now stands, and the Endeavour River drains into it. Banks and his party explored and collected extensively during the weeks ashore. They found much that was new and surprising in the flora and fauna of the area. They also encountered local inhabitants. By the beginning of August HMS *Endeavour* was seaworthy again, and the expedition continued on its way: *Endeavour Journal*, vol. II, pp. 77–100. See 'Appendix: Maps'.
[18]'Ye voice of God' was the roar of the sea on the Great Barrier Reef. A reference to Bougainville's description in *Voyage autour du monde...*, Paris (1771), p. 257.
[19]On the printed version of this letter a footnote is supplied here: 'Note of Dr. Solander. The inhabitants of New Holland, who are far from numerous, considering the extent of the country they live in, never approached in any greater numbers than from ten to fifty at one time. They were so shy, or so much afraid of us, that we could never prevail with them, except only once, to stay with us, and even then for no more than three or four days. Nor would they allow their women to hold any correspondence with us.

We left presents, consisting of nails, hooks, knives, scissars, hatchets, wedges, glass baubles, &c. in every house we came to, and wherever we discovered that the natives usually resorted; but always remarked that they never took any thing, although we were certain they had been at their dwellings, and must have seen our presence. When we presented them with any thing, they did not refuse to accept it, but soon afterwards they would lay it on the ground in the woods, not excepting even our stuffs, which at first seemed not unacceptable to them.'
[20]As HMS *Endeavour* travelled northwards along the coast, it passed through a maze of shoals, reefs and islands. The tide hid many sharp rocks and corals, making this a particularly hazardous route. The safest place was open sea. A gap in the Great Barrier Reef was discovered on 11 August, and the ship sailed through it two days later. It was named Cook's Passage. However, on 16 August a heavy swell forced HMS *Endeavour* towards the reef, and disaster was only narrowly avoided when Providential Channel was seen. Cook steered back within the Reef this way, and Banks reflected: 'By 4 we came to an anchor happy once more to encounter those shoals which but two days before we thought ourselves supreamly happy to have escap'd from.' *Endeavour Journal*, vol. II, pp. 107–108.
[21]'A Streigth between New Guinea and New Holland' was Endeavour Strait. This led to Torres Strait on 23 August. Cook claimed the coast of Australia he had just charted for George III. The name he chose was New South Wales. He did this on Possession Island, which is at the entrance to Endeavour Strait. See 'Appendix: Maps'.

[22]In the MS. and the printed version of this letter a footnote is supplied here: 'Author of L'Histoire des Navigations aux Terres Australes.' It referred to Count Charles de Brosses (1709–1777), a lawyer, magistrate, classical scholar and *philosophe* who published *Histoire des Navigations aux terres australes*..., Paris (1756).
[23]The attack took place on 3 September 1770. The local inhabitants seem to have been using tinder in hollow canes, which, when lit, produced a small explosion. Musket shots were fired in return, but Banks and his group decided to retreat back to the ship. Banks perhaps recalled earlier casualties among Maori warriors when he wrote: 'From this specimen of the people we immediately concluded that nothing was to be got here but by force, which would of course be attended with the destruction of many of these poor people, whose territories we had certainly no right to invade either as discoverers or people in want of provisions.' *Endeavour Journal*, vol. II, pp. 143.
[24]For an account of the losses see Letter 5, Note 3.
[25]The Downs, an anchorage off the east coast of Kent.
[26]Banks did not return to the Pacific with Cook. See Letter 7.
[27]New Zealand flax, also New Zealand hemp and New Zealand cloth plant, (*Phormium tenax*).
[28]'*Cordia Sebestna Orientalis Linnaei*' is the cordia (*Cordia subcordata*). This was used by Polynesians for dyeing. Banks described it in his journal: *Endeavour Journal*, vol. I, pp. 357–358.
[29]This is called 'Mati' in the Polynesian language, which Banks writes as '*Matte*': *Endeavour Journal*, vol. I, p. 357.
[30]Philibert Commerson.
[31]This sentence appears as a footnote in the printed version of the letter above.
[32]Georges Louis Le Clerc, Comte de Buffon FRS (1707–1788): founder of the Jardin des Plantes, Paris. Buffon was one of the great systematizers of the eighteenth century. He published the monumental *Histoire naturelle*..., Paris (1749–1804).
[33]For details of the kangaroo see Letter 5, Note 4.
[34]Dr. John Hawkesworth (c.1715–1773). Hawkesworth edited the official account of the *Endeavour* voyage. He used Banks's impressive journal and Cook's records for: *Account of the Voyages undertaken by order of His present Majesty for making Discoveries in the Southern Hemisphere*, London (1773).

1772

To John Montagu F.R.S., 4th Earl of Sandwich, and 1st Lord of the Admiralty

New Burlington Street
30 May 1772

My Lord,

The present situation of Things regarding the proposed Expedition[1] to the South Seas, which it was my intention and inclination to have taken an active Share in, will, I trust, render any other Apology to your Lordship for this intrusion unnecessary.

To avoid the appearance of inconsistency, and to justify my Conduct in the Eyes of the Public and your Lordship, I feel it incumbent to state the reasons by which I am influenced to decline the Expedition.

When it was first proposed to me by your Lordship to go to the South Seas again, if His Majesty should think proper to send Ships to perfect the Discoveries that had been begun in the last Voyage, I joyfully embrac'd a proposal of all others the best suited to my Disposition and Pursuits. I pledg'd myself then to your Lordship, and have since, by the whole tenor of my Conversation and Correspondence, pledg'd myself to all Europe, not only to go [on] the Voyage, but to take with me as many able Artists as the Income of my Fortune would allow me to pay, by whose means the learned world in general might reap as much benefit as possible from those Discoveries which my good Fortune or Industry might enable me to make.[2]

The Navy Board was, in consequence, ordered to purchase two Ships, [and] to fit them up in a proper manner for our reception [so] that we might be enabled to exert our utmost Endeavours to serve the Public wheresoever the course of our discoveries might induce us to proceed.

Two Ships were accordingly purchased, but when I went down to see the principal Ship I immediately gave it as my opinion that she was very improper for the Voyage, and went So far as to declare that if the alterations which I proposed would not be made I would not go in her.

In consequence of this the Surveyor of the Navy was sent to me with a Plan of the Ship. To him I stated my Proposals, and laid down upon that Plan the quantity of room that I thought absolutely necessary to be allotted to me and my people for the carrying on [of] our respective Employments.

When these Alterations, and those which were judg'd necessary also for the Accomodation of the Captain and the People were made, the Ship in falling down the River was found absolutely incapable of pursuing her intended Voyage.[3]

The Navy Board have attributed this incapacity to the alterations which had been made, and are of the opinion that when the Ship is reduced to her original Situation, that in which I before refused her, she will be the fittest that can be had for answering the Nautical purposes of the Expedition. Without suffering myself to controvert this Opinion of the Navy Board that the Ship will be very fit for Sea, although many able seaman concur with me in doubting it, I must be allowed to say that the Ship will thus be, if not absolutely incapable, at least exceedingly unfit for the intended Voyage.

We have pledged ourselves, my Lord, to your Lordship and this Nation to undertake what no Navigator before us has ever suggested to be practicable. We are to attempt at least to pass round the Globe through Seas of which we know no Circumstance but that of their being tempestuous in those very Latitudes, in passing through which, in Order to get round one Cape, the whole Squadron comanded by Lord Anson narrowly escaped being destroyed.[4] We have done more. We have undertaken to approach as near the Southern Pole as we possibly can, and how near that may be no Man living can give the least guess.

In Expeditions of this nature the Health and Accomodation of the People are essential to Success. When Sickness and discontent are once introduced it will be absolutely impossible to continue the discovery. By the Alterations made the Accomodations of the People are very much reduced. The Spar Deck being cut away, 30 of the Crew are to be removed under the Gun Deck, before sufficiently crowded, which, being very low & confined without a free Air, must infallibly in so long a Voyage produce putrid Distempers and Scurvy; and what, my Lord, ought more to be dreaded by a Discoverer than such a Calamity, which must soon Oblige him to quit his Discovery, and very probably even put it out of his Power to bring home any account of what he has done previous to its fatal influence?

The Accomodations in the Ship are much lessened by the changes which have been made in the Equipment since the first Plan. The House of Commons have thought the Undertaking of so much importance as to vote the Sum of £4,000 to enable Dr. Lind[5] to accompany us, and assist us with his extensive Knowledge of Natural Philosophy and Mechanics. The Board of Longitude have also engaged an Astronomer to proceed in each Ship, and an extraordinary Establishment of Officers was thought necessary on account of the difficulties and Dangers which we were likely to experience in the course of our Voyage.

Shall I then, my Lord, who have engaged to leave all that can make Life agreable in my own Country, and throw[n] on one side all the Pleasures to be reap'd from three Years of the best of my Life, merely to compass this undertaking, pregnant enough with Dangers and difficulties in its own Nature, after having been promised every security and convenience that the Art of Man could contrive, without which Promise no Man in my situation would ever have undertaken the Voyage, be sent off at the last in a doubtful Ship, with Accomodations rather worse than those which I at first absolutely refused, and after spending above £5,000 of my own Fortune in the Equipment upon the credit of those Accomodations which I saw actually built for me? Will the Public be so ungenerous as to expect me to go out in a Ship in which my People have not the room necessary for performing the different Duties of their Proffessions, a Ship apparently unhealthly, and probably unsafe, merely in conformity to the official Opinion of the Navy Board, who purchas'd her without ever consulting me, and now in no degree consider the part which I have taken in the Voyage, or the Alterations which on my Remonstrance they concur'd with me in thinking necessary, but have now taken away? Or should I embark, could any thing material be done by People under Circumstances so highly discouraging?

For my own part, my Lord, I am able and willing to put up with as small Accomodations as any Man living can be content with. Six Feet square is more than sufficient for all my personal conveniences, nor are any of my People desirous of a larger Allotment. Tis our great Cabbin which is too small, and that is in reality the Shop where we are all able to work, which, if not sufficiently large, will deprive the Workman of a possibility of following their respective employments, and prevent me from reaping the Fruit earned by voluntarily exposing myself to danger, and incurring a material Expence.

Neither personal hazard nor expence however will I withold when likely to meet with their proper encouragement. Born with an Attachment to a singular pursuit, I have already performed two Voyages, and in the course of them have merited, I hope, some share of Public regard; and though my Services are upon this Occasion refused, I shall always hold myself ready to go upon this, or any undertaking of the same Nature, when ever I shall be furnished with proper Accomodations for myself and my People to exert their full Abilities.

To explore is my Wish, but the Place to which I may be sent almost indifferent to me. Whether the sources of the Nile or the South Pole are to be visited, I am equally ready to embark in the undertaking when ever the Public will furnish me with the means of doing it properly; but to undertake so extensive a pursuit without any prospect but Distress and disappointment is neither consistent with Prudence nor Public Spirit.

As to the position of no other Ship being fit for the Voyage because no other could take the Ground,[6] I cannot omit putting your Lordship in Mind that within these few weeks the Emerald, one of our sharpest Frigates, lay on Shore on the Gun fleet a much longer time than the Endeavour did upon the Coast of New Holland, after which she was got off. Sir John Lindsay[7] also hauled up the Stag, another of our Frigates, at Trincomaly,[8] and shifted her Rudder Irons during the course of his last Voyage. What more, my Lord, did the Endeavour do, or what more could any Ship have done in that particular point on which the opinion of the Navy Board so materially rests?

If these then are capable of taking the ground, how much more so must the Launceston[9] (the Ship for which we have petitioned your Lordship) be, as all Seamen know that the bottoms of that Class of Ships are flatter than any others employed in His Majestys Service. For my own part, I can only say that was your Lordship to think proper to let us have her for our intended Expedition, I would gladly embark on board a Ship in which safety and Accomodation, both of which must be consulted in a Voyage of this kind, are more nearly united than in any other kind of Ship I am acquainted with; and [I] well know that there are many Commanders in His Majesty's Service of undoubted Abilities and experience who would willingly undertake to proceed with her on the intended Expedition, ambitious of shewing the World that the success of such an Undertaking depends more upon the Prudence and perserverance of the Commander than upon any particular Built of the Ship that may be employed.[10]

I cannot dismiss this Letter without thanking your Lordship for the many particular Favors which I have received at your Lordships Hands in the commencement, and during the prosecution of this, my Favorite Undertaking, of which I shall ever retain a most grateful sense. I do not doubt that was not your Lordship prevented by forms of Office I should still continue to receive the same countenance and Assistance, and that if it should be thought proper to alter or enlarge the present Equipment your Lordship would still continue your Protection. As I am not conscious that by any part of my Conduct I have forfieted that claim to it which your Lordships great condescention and goodness originally conferr'd upon me,

I am with the utmost respect,
Your Lordships
most obliged and most
obedient Humble Servant,

Jos: Banks.

[Windsor R.A. 1322, by gracious permission of Her Majesty the Queen; M.L./D.L. Banks Papers Series 06.001; C.C.M.M., Sandwich papers; C.K.S. Stanhope of Chevening MS. U1590 S1/S2 (a copy in Sarah Sophia Banks's hand).]

[1]The expedition set sail under the command of Captain James Cook in HMS *Resolution*, with a second vessel called HMS *Adventure*, 1772–1775. This was Cook's second great voyage of exploration. Banks had organized personnel and materials in the expectation of even greater achievement in the fields of natural history and science than was possible on the *Endeavour* mission. However, the 'principal Ship' was unsuitable for his special needs, and the alterations he proposed were tried then rejected. Banks withdrew in favour of a trip to Iceland when his plea for a different vessel was finally refused. He drafted a carefully worded letter explaining his reasons. The reply to Banks's long explanation exists as a modern copy: Montagu to Banks, 2/6/1772, C.C.M.M., Sandwich Papers.
[2]Banks intended to take a party of 16 people, including himself, Dr. Daniel Solander, four artists, two secretaries, and eight servants and assistants.
[3]The Navy Board argued that changes to HMS *Resolution* were unnecessary, and also that those Banks wanted had caused her to be unstable. Banks felt that the work had been badly executed in order to confound him, and that a chance to increase knowledge was not being firmly grasped: Letter to the printer of the Gazetteer, M.L./D.L. Banks Papers Series 06.002; Montagu to Banks, [June 1772], C.C.M.M., Sandwich Papers; Observations

on Banks's letter above by the Navy Board, 3/6/1772, C.C.M.M., Sandwich Papers, and Windsor R.A. 1323; Memorandum by Sir Hugh Palliser, C.C.M.M., Sandwich Papers.

[4]Lord George Anson (1697–1762): Admiral of the Fleet. In 1740 he was commodore of a squadron of six ships sent to the Pacific, all of which struggled to round Cape Horn in stormy weather, with two being driven back and one wrecked.

[5]Dr. James Lind accompanied Banks on the brig *Sir Lawrence*, along with most of the party Banks had assembled to travel with Cook. Parliament had made a special grant of £4,000 to Lind to sail with Banks on HMS *Resolution*. Instead, Banks chartered a 190-ton brig, which sailed from Gravesend for the Hebrides and Iceland on 12 July 1772, the very same day that Cook took his ships out of Plymouth Sound.

[6]Important in determining whether a ship will ground or not, and, if it does, what possibility there is of it being freed. HMS *Endeavour* had a shallow draw, which probably prevented disaster when she sailed on to Endeavour Reef in June 1770.

[7]Sir John Lindsay (1737–1788): Rear-Admiral RN. From 1769 to 1772 he was Commodore and Commander-in-Chief in the East Indies, with his pennant displayed on HMS *Stag*.

[8]Trincomalee, Sri Lanka. The inlet forms a natural harbour, which was used by the British as a naval station until 1957.

[9]A 44-gun ship. Banks wanted a larger vessel in which more could be done. He also wanted a greater say in the direction of the venture so that opportunities to benefit learning were fully explored.

[10]The plans of the *Resolution* with alterations to accommodate Banks's party are now in C.C.M.M., Mulgrave Papers.

Letter 8

1773

To Thomas Falconer

New Burlington Street
12 January 1773

Dear Sir,

I know not whether I ever thankd you for the useful & instructive letter[1] which you sent me during the time of my preparation for my intended voyage to the South Sea. If during the hurry & confusion occasiond by my disappointment it was forgot, I trust your candour has forgiven already an omission which I now ask pardon for.[2]

The only return I can attempt to make you is to communicate some account of what my last voyage has offerd to my observation. The course I steerd was through the western Islands[3] to Iceland, from whence, after having remaind 8 weeks, I returnd by the Orkneys to Edinburgh, & from thence by land to London.

As the subject matter of the whole would be too long for one letter, I shall divide it. In this I shall confine myself to what happend before we left the western Isles. If that proves amusing to you, you shall in the course of a few weeks receive the rest.

Disappointed in the execution of my favourite project of visiting the Southern Hemisphere, but not destitute of hopes that the next year might revive it, I did not chuse to discharge the draughtsmen that had been retaind for that purpose. So, with the concurrence of the gentlemen who had intended to accompany me, it was resolvd that some short voyage should be undertaken, tho the season was far advancd, in order to shew our freinds that our disappointment had not been owing to any tardiness of our own, or disinclination to undertake voyages. Four months was all we had left. Nothing within the compas[s] of so short a time appeard an object of curiosity equal to Iceland; a countrey which had been peopled by Europeans 800 years ago, who for some ages signalised themselves both in Learning & adventure, but had now been almost unvisited by travelers for many ages. Accordingly, it was determind that that should be our rout, &, having hird a Brig of 150 tons, we saild from London on the 12th of July.

After a passage rather tedious we arrivd at *Ila*[4] at the latter end of the month, from whence we proceeded to *Jura*,[5] thence to *Oransay*.[6] *Ila* is the most fertile of the western Isles that I have seen, tho probably Bute & some of those that lye nearer the main[land] are more so. The general soil is what is there calld whyn stone,[7] such as London is pavd with, but in the middle of the Island is a patch, as it were, of Lime in which are lead mines which are workd to some advantage;[8] also the mineral of Emery[9] & some Copper, tho not iron, [are] workd. The Island in general abounds with a Kind of stone, marl, of which the inhabitants have but just discoverd the use, but from which they will probably in a very short time reap much benefit as it in a very short time reduces Ling land[10] to a good turf by merely being spread over it.

The misfortune of this &, I believe, most of the western Isles seems to be thinness of population in general. Large parcels of Land are leasd out to people who are not capable of improving or even cultivating a tenth part of them. Nor will people be content with this extra quantity of Land, which is usd for breeding of Cattle so that, of necessity, $\frac{9}{10}$ of the countrey remain[s] in much the same state as they were left in after the creation.

Jura is separated from Isla by a straight of less than an English mile in breadth. Here Thurot[11] laid his little fleet, trusting that, by the rapidity of the tides, he might go out at any time in a contrary direction from that in which an Enemy might enter to pursue him. The Scotch speak of him with great affection. He seems to have behavd very civily to them, & to have signified that he had particular orders from his master so to do.

The Island itself is much less fertile than Ila, owing cheifly to the swamps with which [a] great part of it is coverd.

Oransay, a small Island separated from *Colonsay* by a narrow channel dry at half Ebb, is low & rather fertile. Here are the remains of what is thought to be the second place of Religious worship which was founded in these Kingdoms. The Ruin is large & in good preservation, but from the ill contrivance of the different apartments plainly appears to have been built at different times. At the west end is a beautiful cross in good preservation, with several characters upon it which we were not able to read.

From Oransay we proceeded along the shore of the Jura to the northernmost end of that island where, in a straight between that Island & *Scarba*, both the reports of the living & printed accounts had fixd the site of a whirlpool almost as extraordinary & as terrible as that of *Mael Strom* on the coast of Norway.[12]

Willing to examine so extraordinary a Phoenomenon with all possible care & attention, we Landed on the little Island of *Scarba*, &, pitching a tent, waited with patience for the turn of the tide which was to exhibit such wonders, but in vain. Tho [the] tide turnd & ran with more than usual rapidity, indeed [it] did not produce a whirl sufficient to have endangered the smallest boat. The tides indeed were not high, which made the waters less turbulent than they are at the time of spring tides, but upon enquiry at the gentlemans house who lives nearest to it,[13] he told me that it was not at all terrible except at the concurrence of a spring tide & a westerly wind. In his time, however, only one boat had been lost in it, & that a very small one managd by only two men.

From *Scarba*, then, a very small & unfruitful Island, we took our departure, &, passing among the *Luing* or *Slate* Isles, found ourselves among a variety of tides that made it difficult to determine what course to steer. At this rather critical conjuncture it was, I beleive, that we passd the ship of our freind Mr. Pennant,[14] which I had all along carefully lookd out for not without hopes, if I should be fortunate enough to meet him, of tempting him to enlarge his plan, & wishing much to assure him, at all events, that as I had lookd upon myself while among these Islands as treading upon ground which by prior right he had taken possession of, I should communicate to him every observation I could possibly make whenever he thought fit to publish any account of them.

From hence we proceeded through the straights, or as it is calld the Sound of *Mull*, passing between that Island & the Countrey of *Morven*, the supposd residence of Fingall; both sides affording a variety of romatick & beautifull prospects, especially that of Morven. Here, as we were obligd to stop every tide, we landed often, & in one of these excursions I saw the method of making Kelp, an article which raises a great revenue to this country, tho a few years ago totaly neglected or let out at a small price to people who came every year from Ireland for that purpose. The process, as it appeard rather curious to me, I shall here describe.

The crop,[15] which consists, I believe, of every Kind of sea weed except the sea Thongs, is reapd once only in three years (for so long it takes to grow again to maturity after being cut down). It is tedded or made just as hay is, & if it receives much rain during that time becomes totaly unfit for its purpose. When it is about two thirds dry, that is that it will take fire with the addition at first of some hay or dry stuff, it is ready to burn, which is done in [a] Kind of trough built up of turf about 12 or 14 feet in leng[t]h, & 4 or 5 in breadth. When the fire is once well lighted, the weed will burn alone. The people then attend constantly, supplying it with plenty of weed till what they call a floor is formd; that is, till the bottom of the Kiln is coverd with 16 or 18 inches of ashes, which then are as light & powdery as wood ashes falling from the fire. The moment that the weed is intirely consumd, the attendants begin with instruments almost like hoes to stir about the ashes, which by this kind of motion are reducd from a powder into a state maybe of partial fusion in which they resemble much, in consequence, the dough from which bread is made, & from which, when they cool into the hard s[t]onelike form in which we receive them, [are] so hard that in using them we are obligd to apply large hammers to break them. In this manner they go on, raising floor upon floor till the Kiln is full, when they cover it with turf, & let it remain till the sessells [vessels] come to fetch it away. In the stirring alone consists the whole mystery & dificulty of making the Kelp, for the ashes, unless reducd into that stony consistence [consistency], are of little value. I confess I was much surprised at the effect of that operation which, tho it may be well known to the chymists, I was before totaly ignorant of. You would oblige me much by sending me your opinion of it.

The morning after this Highland lecture upon Chymistry, it was our fortune to anchor immediately in the famous Harbour of Tobir more, near a gentlemans house called Drumlin. Mr. M'lean.[16] The owner of it, according to the Hospitable customs of the countrey, came off to breakfast with us, & invited us ashore, which invitation we most readily accepted. At his house we found his son & an English Gentleman, Mr. [Leach], who told us that on a small Island not very distant from thence was to be seen an appearance much resembling the Giants causeway in Ireland.[17] He had, he said, been looking out carefully for Mr. Pennant in order to have given him the same information, but not having been fortunate enough to see him, rejoicd much in the opportunity he had of communicating it to us. As I still lamented the not having had an opportunity of visiting the giants causeway, which I had at first intended, this was an opportunity not to be lost. I instantly orderd the ship to wait my return in the Harbour of Topir more, & set out in the boat accompanied by both the gentlemen by 10 o'clock in the morn. As the distance was full 40 miles, & we had not a breath of wind, we were obligd to work hard at our oars; nor did we arrive at the place of our destination till 10 at night, when it was to[o] dark for us to Judge at all whether or not our curiosity was likely to meet with gratification equal to the pains we had been at.

The small Island of *Staffa*,[18] on which we were, is situated on the west side of Mul[l], about 4 miles from the shore, opposite an inlet calld *Loch na Gaul*, & about 9 miles from the celebrated Island of Y Columb Kil or Iona. It is about an English mile in leng[t]h, & half as broad. It is inhabited by a few young cattle, some sheep, & one family consisting of seven people living in a miserable highland house without a chimny, according to the custom of the countrey, where a stranger might well imagine smoak to be one of the necessaries of life, as he must find much more of that than of any other Kind of provision in every house into which curiosity or necessity may induce him to enter.

In this house, & a small tent which we had carried with us, we slept not very comfortably, you may imagine, so at the earliest dawn of the succeeding day we were up, & prepared to prosecute our undertaking.

We had been taught to expect something resembling the Giants causeway in Ireland, but had not the least Idea that what we were to see was at all to rival that noted curiosity, either in the extent of the whole, or the size of the component parts. Many such things were before known to exist in the Islands of Canna, Sky & Mul[l], but they seemd, from the descriptions I had heard or read of them, to be little more than Efforts of nature towards that regularity, which, in the giants causeway, she had compleated.

Think, then, how agreable was our surprize when, on walking down to the sea beach, we were at once presented with the view of the ranges of Pillars near a mile in Extent, some of which were 57 feet high, supporting the Mass of the Island above them, which, as the ground of it swelld into hills or sunk into valleys, seemd monstrous Pediments. Vast as the Columns were, [the island] seemed to Lean heavy upon them, & claim all their streng[t]h to support.

Below these to the water was that kind of Pavement from whence the Giants causeway has got its name, viz. the tops of the Joints of the Pillars, which have been broken off, & remain shewing their angular figures, which are separated from each other only by small cracks. These were, Exactly as those of the Giants causeway, formd of a hard black stone (possibly the Basaltes of the ancients). Their surfaces were concave & convex, & of all figures from the triangular to heptangular, tho pentagons & hexagons were far the most frequent. The thickest we measurd was $4\frac{1}{2}$ feet in diameter.

At the north End of this range of pillars was a very singular sight. The pillars themselves there, instead of standing upright, lay on their sides & were bent in all possible directions, yet remained whole. The Joints indeed were visible in them, but adheard nearly as fast to Each other as any other part of the pillar.

But curious as this was, or striking as that vast assemblage of Columns above describd, the beauty of a third object made it almost the most desirable sight of the three. I mean a Cave, which had been hollowd out by the sea among these pillars from the first breaking of the causeway; that is from a to b in the sketch which I send.[19] The depth is 371 feet from the point, which is coverd by the arch, that is from c 251. The Arch itself is from the water, 117. The sides consist of Pillars. The roof also is fretted with the bottoms of those which are broken off. Underneath is water on which a boat may go quite to the farther end, by means of which light is reflected so strong that at the farther extremity we could easily read a common printed book.

A cave of these dimensions has always been lookd upon as an extraordinary sight, even without ornament, & the aid of a few stalactites has made people boast of it as a beautiful wonder. What, then, must we esteem this cave to be, ornamented as it is by such a range of columns, & fretted as the roof is, & illuminated by its own light? I confess I cannot conceive that the grotto of Antiparos,[20] the famous cave in the neighbourhood of Naples, can bear a comparison with it. Nor can I say that I have in my life receivd so much pleasure from the contemplation of any artificial arrangement of Pillars as I did from this one.

We had nobody with us who had seen the Giants causeway, but I have since spoken with several, who all agree in the great superiority of this Island over it in curiosity as well as in beauty & extent. The bending pillars, of which you receive sketches, are not those that I have heard of. These I take to be a great illustration of the subject.

After having spent just 12 hours upon this Island, the whole time in a state of satisfaction which none but a traveler can feel, we embarkd for Iona or y Columb Kil,[21] where we arrivd at night, & spent part of the morning in viewing that Island, remarkable only as the first seat of Christianity. The buildings are extensive, & rather pretty considering the time in which they were erected. Many people of high note have been buried here, but cheifly without stones. We saw at least none whose antiquity claimd any degree of respect, tho we were told that some were there in ancient Irish character, but did not meet with any such.

On each of the 4 sides of this Island, which answer the Cardinal points, is a stone in which seamen place great faith, beleiving that if they clean carefully any one of them a wind will arise from its respective quarter. When we were there the stone on the North side was nicely swept, & a northerly wind arising fannd us gently away to our ship, where we arrivd at night.[22]

The harbour of Topir more in which we lay is the best in this part of the highlands capable of containing in safety an immense number of ships. Here tradition tells us that one of the great Spanish Armada[23] was lost flying from our victorious fleet. The name of her having been Spanish makes the people here dream of Dollars by bushels, & wish much that they had a diving bell to search for them; tho in all probability, being an armd ship, she was not chargd with that Kind of Lading.

The next morning we Landed on a small Islet near the main[land] of Scotland in order to hunt the roe [deer], but we were not fortunate enough to see any; tho the scotchmen saw seven, & my servants some.

The Season being now far advancd, we resolvd to spend no more time among these Islands, but to push directly for Iceland after having seen St Kilda. Accordingly, we saild, passing in sight of *Canna*, *Rum*, *Egg*, *Sky*, the *Uists*, *Harris & Lewis* & many more, [and then] arrivd at the But[t] of Lewis. Here we had the wind S.W. with thick weather, a fair wind for Iceland, but a foul one for St. Kilda. So, thinking it absolutely certain that Mr. Pennant, who set out intirely to visit these Islands, must have been upon it, we resolvd for Iceland, where my next shall pilot you if you signify that you have had any entertainment from this.

Beleive me,
your obligd & Faithful...

[Warren R. Dawson, *The Banks Letters: A Calendar of the Manuscript Correspondence*, London (1958), p. 318. Dawson noted: '...formerly in the possession of Dawson Turner, bound in a volume with the original drawings, etc. D.T. Manuscripts Sale, 1859, Lot 26.' Dawson cited 'Hawley Coll.' for this letter, by which he meant: 'Banks papers in the possession of Major Sir David Henry Hawley, 7th Bart. of Mareham-Le-Fen, Lincolnshire', p. xxxviii. A typed transcript of the letter is at the Natural History Museum, London, in the Banks Archive. This was made from a Dawson transcript at B.L. Add. MS. 56031(7), ff. 10–19.]

[1]Falconer to Banks, 16/4/1768, R.B.G. Kew Archives B.C. 1 22, N.H.M. B.L. D.T.C. I 20–22.

[2]Banks's expedition to Iceland largely arose from his disappointment at the arrangements for the voyage of HMS *Resolution*, 1772–1775, which sailed under Captain James Cook. Banks had gathered together equipment and personnel for this mission, and after withdrawing decided to use them for a trip northwards instead. This was made in a 190-ton brig, *Sir Lawrence*, with Captain James Hunter and a crew of twelve. The *Sir Lawrence* sailed from Gravesend on 12 July 1772, the same day that Cook took his ships out of Plymouth Sound. Banks took an impressive team: Dr. Daniel Solander as Banks's companion botanist; Dr. James Lind as a scientific companion and observer; Reverend Uno von Troil (1746–1803) as a travelling companion and scholar; Lieutenant John Gore as an experienced sailing companion; John Clevely Jnr. (1747–1786), James Miller (fl.1772–1782) and his brother John Frederick Miller (fl.1772–1785) as draftsmen; Sigismund Bacstrom (fl.1772–1775) and Frederick Herman Walden (fl.1772) as secretaries; John Riddel as a young seaman and traveller; James Roberts and Peter Briscoe as servants, and eight other servants.

They were all listed at the beginning of Banks's *Journal of a Voyage to the Hebrides begun July 12th 1772 and a small part of the continuing Voyage to Iceland and the Faeroes as far as September 6th (Sunday)*, Blacker-Wood Library, McGill University, Montreal. This account begins with general statements concerning the reasons for the voyage, and follows the journey from 12 July to 5 September, when the text abruptly ends. The notes at the Centre for Kentish Studies refer to the journey to and from Hekla, 17 September to 30 September, and the six days from 16 October to 22 October, ending with the remark, 'Idle tird resolve to go away fair or foul', C.K.S. Stanhope of Chevening MS. U951Z31. By early December Banks was back in London, having travelled down from Edinburgh, which he visited with Solander and Lind. Banks never published his Iceland journals. However, Uno von Troil published: *Bref rörande en resa til Island MDCCLXXII...*, Uppsala (1777). This was translated into English: *Letters on Iceland: containing observations on the civil, literary, ecclesiastical and natural history...*, London (1780).

Banks thought that another chance to go to the Pacific would arise in the next year, perhaps on an Honourable East India Company vessel.

[3]In this letter Banks outlined the course taken through the Hebrides, which are islands off the west coast of Scotland. These islands consist of two principal groups: the Outer Hebrides, and the Inner Hebrides. Banks sailed on to Iceland after touring them. 'Appendix: Maps' includes his route at this stage of the voyage. Some of the islands he mentioned in detail are explained below.

[4]Islay, one of the larger and southerly islands of the Inner Hebrides. Banks was on Islay from 1 to 7 August, during which time he visited the proprietor of Islay, Daniel Campbell, the town of Killara, and a 'dirty nasty' cave at Laggan. Rain and wind drove Banks and his party from the west of the island, and on 5 August they arrived at the Sound of Islay, where Charles Freebairn's house was 'situated in a very romantick spot under an almost overhanging clift, close by the sea side...'. There are a variety of soils on Islay, of which more than half might be tilled and cropped. It is the most fertile of the Hebridean islands.

[5]Jura, a mountainous island of the Inner Hebrides, north-east of Islay, and south of Scarba. Banks crossed to Jura on 6 August, and climbed one of the Paps, four high conical peaks of quartz rock. The most southerly is called *Beinn a' Chaolais*, or the Peak of the Sound. A small cairn was raised at the top before Banks returned with his group to Mr. Freebairn's.

[6]Oronsay, a small island immediately south of Colonsay. There are the remains of religious buildings on these islands. On Oronsay the ruins are of a priory or monastery of Cistercian or St. Augustin monks. The abbey would have been on Colonsay, but it has been destroyed. Banks visited Oronsay by boat on 8 August.

[7]The 'whyn stone' is most likely to be a dolerite, commonly found on Islay. Whin or Whinstone was applied to any hard, dark stone in the north of England and Scotland. Since the eighteenth century the term has been restricted to igneous rocks such as dolerite or basalt. See Letter 17, Note 1.

[8]Lead mining here did not survive. By the early nineteenth century whisky distillation was more profitable.

[9]Emery is a greyish-black variety of corundum ($Al_2 O_3$), which can be crushed and used as an abrasive for polishing tough surfaces. The main uses of corundum are based on its hardness, such as for grinding wheels or emery paper. Banks may be mistaken in calling this substance emery. Dolerites have been used for building and sculpture on Islay, but there are no emery deposits on Islay or Jura.

[10]'Ling land' refers to heath land, which the locals fertilized with 'marl'. Marl is a calcareous clay, created by deposition under marine or lacustrine conditions. It contains 35–65 per cent soft calcium carbonate, the remainder of the sediment being clay. Semi-lithified marl

is called marlstone, and fully lithified marl is an argillaceous limestone. The limestone of Islay is valuable for agricultural uses of the kind Banks described.

[11]François Thurot (1727–1760): French privateer; naval officer. His reputation grew with a string of victories against the British navy. When anchored off Islay, he carefully avoided plundering the islanders to supply his ship and crew, thus earning their affection. Thurot fought the British off the Forth in October 1759, and then sailed for Loch Foyle with a force of 1200 men. However, he was killed in action against three British frigates in 1760. His body was washed ashore, and is buried at Kirkmaiden-in-Furnes,Wigtownshire.

[12]Maelström, Malström or Moskenström, a whirlpool off the north-west coast of Norway, between the islet of Moskenö and the island Moskenäsö. It is dangerous only when the north wind blows against the tide. The whirlpool Banks sought was called Corryvreckan, and is in the Gulf of Corryvreckan between Jura and Scarba.

[13]Banks refers to this person in his journal: 'Mr. MacNeal the principal gentleman in the Island', *Journal of a Voyage*, 9/8/1772, McGill University.

[14]Thomas Pennant was touring Scotland and the Hebrides at the same time that Banks passed through the islands. He was allowed the use of Banks's journal for his edition of *Tour of Scotland and a Voyage to the Hebrides...*, Chester and London (1774–1776), in which Banks's description of Fingal's Cave won praise from such readers as Horace Walpole (1717–1797). Pennant's account also included engravings from the work of John Frederick Miller and John Clevely Jnr., both of whom Banks had employed as draftsmen on this trip. Evidence of Pennant's recent excavations were found by Banks on Iona.

[15]Banks described this process well. He was among the first Englishmen to witness and record what was becoming a significant Highland industry. Kelp making was one of the most profitable ways an island or coastal landlord could extract a profit from his estate in this period, although the methods used did not favour tenants, who were often treated as poorly paid labour. Kelp ash was produced by burning particular seaweeds. It was an essential material for bleaching, soap boiling, glass manufacture and the production of alum: processes which required soda.

The seaweeds harvested for this were the deep sea tangle, sea-oak or black tang, bell-wrack or yellow tang, and serrated seaweed or prickly tang. They were cut with billhooks in November, and carted or carried in creels to a convenient beach to dry. Burning started in March when the dried weed was placed in kilns, which were built or dug on the shore. The kilns were frequently U-shaped, and are now a common archaeological feature in the Western Isles, sometimes being mistaken for burial cists. The labourers would first place peat in the kilns, and spread the seaweed on top. The alkaline material became a glowing liquid mass once burning took place, and cooled in about two days. The result was a solid substance, which was cut out and stacked for shipment south. During the Napoleonic Wars this industry thrived as foreign supplies were frequently blockaded. However, in 1832 the industry collapsed following the removal of import duty on barilla, much of which came from Spain. Later on, alkali production from salt was improved as well, and so this way of life died. Banks gave a more detailed account in his daily record of events: *Journal of a Voyage*, 11/8/1772, McGill University.

[16]Sir Allan Maclean (d.1783): master of Drimnin.

[17]Giant's Causeway, a celebrated formation of prismatic basaltic columns, east-north-east of Portrush, Ireland.

[18]Staffa, an island of the Inner Hebrides, west of Mull. Staffa is composed of volcanic tuff and trap-rock (basalt). Except on the north-east shore, it is almost everywhere surrounded by cliffs. These are hollowed with seven caverns, the most famous of which is Fingal's

Cave. Fingal's Cave is bordered by columnar basalt, with a vaulted roof of amorphous basalt, and a floor of tuff. Banks rowed to Staffa by boat on 12 August, arriving at night with 8 companions. They slept in a tent and at a local dwelling. The magnificent description of Fingal's Cave was the result of one day's energetic observation on 13: 'Compard to this what are the Cathedrals or the palaces built by man ? Mere models or playthings, imitations as diminutive as his works will always be when compard to those of Nature. Where is now the boast of the Architect ? Regularity, the only part in which he fancied himself to exceed his mistress, Nature, is here found in her possession, & here it has been for ages uncounted...How amply does Nature repay those who study her wonderfull works.' *Journal of a Voyage*, 13/8/1772, McGill University. Banks gained control of his emotions following this moving statement, and set about the task of carefully surveying the cave.
[19]The drawings made for Banks in the Hebrides, Iceland and Orkneys in 1772 are at B.L. Add. MSS. 15509 to 15512.
[20]Antiparos, one of the Cyclades, just west of Paros, Greece. It has a famous stalactite cavern.
[21]Icolmkill, I-Columbkill or Iona, is the isle of Columba's retreat, and lies to the south west of Mull. In the sixth century St. Columba (521–597) founded an abbey there. It became a centre of learning in the north. It was also a centre of missionary enterprise undertaken by the Culdees. The ruins of a cathedral or abbey may be found on the east side of the island. The churchyard there contains two fine crosses, and the tombstones of Scottish and Danish kings. After an extraordinary day on Staffa, Banks and his group departed for Iona, arriving at 8 in the evening. They were welcomed by local people, and toured the island the next morning. At 12 midday they left for Tobermory, and the *Sir Lawrence*.
[22]Outside the west door of the main abbey church on Iona is a large block of granite, which is 1.32 m in length, 0.55 m in width at its widest point, and 0.4 m high. A Greek cross is inscribed on it, and a hollow has been made measuring 0.93 m by 0.34 m, with a depth of just 45 mm. The block may have been a trough used by pilgrims to wash their feet. According to local tradition it was called 'the cradle of the north wind' because a favourable north wind could be summoned by its power.
[23]The Spanish Armada sailed from Lisbon in May 1588 to assist in an invasion of England by a Spanish army in the Netherlands. It consisted of 130 ships. The Armada was defeated in a series of engagements with the English navy, and driven northwards. The remains of the fleet suffered further losses when attempting a stormy route home around Scotland to Ireland and then the Atlantic. The *Florida* was sunk near Tobermory, and guns of brass and iron were retrieved, but nothing of greater value was found in the wreck. Only 76 ships of the Armada ever returned to Spain this way.

iii. **Fingal's Cave on the Island of Staffa, 13 August 1772.**
Engraved for Thomas Pennant from a drawing by Frederick Miller, and published in Pennant's *A Tour in Scotland and Voyage to the Hebrides*, Chester (1774), vol. I, p. 263. From Horace Walpole's library copy of the book. The General Library, Natural History Museum, London.

1773

To Sarah Sophia Banks

[New Burlington Street]
[February 1773]

Dear Sister,

If it should be cloudy tomorrow we shall not dine till four.

The motto of the print is taken from a part of Virgils Eneid,[1] where Eneas encouraging his people says:

nunc vino pellite curas
cras ingens iterabimus e[a]quor.[2]

In English:

now drive away your cares with wine
tomorrow we */shall/* will again cross the immense ocean.

The last part only you see I have Chosen,

Adieu,

J.B.

[M.L./D.L. Banks Papers Series 93.07.]

[1]Virgil (Publius Vergilius Maro), (70–19 BC): Roman poet. The *Aeneid* was Virgil's greatest work. It describes the epic adventures of Aeneas, the hero. The first six books concern Aeneas's voyages, and the second six books concern the wars in Latium.

[2]The motto is inscribed on a sheet underneath Banks's left hand in the portrait by Sir Joshua Reynolds (1723–1792), National Portrait Gallery. The source appears to be: Horace, *Odes*, 1, 7, 31–32: 'O fortes peioraque passi/mecum saepe uiri, nunc vino pellite curas; cras ingens iterabimus aequor.'

iv. **Joseph Banks F.R.S., F.S.A.**

Portrait by Sir Joshua Reynolds, c.1772–1773. The National Portrait Gallery, London.

Note: Reynolds commenced work on this portrait before Banks withdrew from the second voyage to the South Pacific under James Cook, 1772–1775. It was completed once Banks returned from journeys he made to the Hebrides, Iceland and the Orkneys, July–November 1772, and afterwards to Holland, February–March 1773. By this time he was an experienced global explorer, and the motto written on a sheaf of papers underneath his left arm suggests further travels yet to be made. This impression is also conveyed by the atmospheric view glimpsed through a window in the backgound. Reynolds captured Banks's open, confident and direct bearing at the age of 29–30. Engravings of this portrait were made in 1773 for sale. It was exhibited at the Royal Academy that year as well.

Letter 10

1773

To Constantine John Phipps F.R.S., 2nd Baron Mulgrave

[New Burlington Street]
[April 1773]

Instructions sent out with Captn. Phipps on His Northern voyage.[1]

A voyage so short as the intended one is not likely to furnish many objects of Natural history. Indeed, if it was, I can have little to say as Mr. Irwin[2] is so well acquainted with the desiderata of Zoology, & Mr. Lions[3] with those of Botany.

You will receive with this a Cask of Double dis[t]illd malt Spirits, a Case with Glass bottles, a ream of Drying paper, a Tin box for insects with Pinns & Snippers for taking & securing them. If any Specimen should be found which is too Large for to be containd in one of the bottles, I shall beg the favour of your purcer to supply a Cagg or Boreca into which it may be put.

Whatever place you may touch at to the Northward of Ferroe[4] must of consequence produce many unknown vegetables. It is therefore desireable that Every different Kind, as far as it is possible to distinguish the least variation, may be preservd. For that purpose the simple method is to take each plant, &, spreading the leaves a little open with the hand, to place it between the leaves of a quire of paper. Between each plant two intermediate leaves should be left, & fresh quires supplied into which the plants should be removd Every three days at least till they are thoroughly dry.

The great but indeed inevitable inconvenience attending this method, is the difficulty of drying the Quires out of which plants have been taken in order to render them fit for receiving the plants again in their turns. This I have commonly Effected by laying them on the booms or quarter deck, with the open side of each Quire to windward, & a boy to attend them whose business it is to turn each to windward again as soon as the wind has blown it over, which, if properly managd, it does leaf by Leaf. This in dry weather is very sufficient, but in wet recourse must be had to the Galley Fire.

Some plants will be dried Enough in two or three times changing. The most succulent you find will probably not require more than ten times. If any should, I would advise that they are taken out of their sheets & laid on a locker, or any place where nothing may be laid upon them, for 8 or 10 hours at a time till they are quite dry, which they cannot fail to be in two or three days.

All Kinds of Animals found in the northern regions will be curious. If it is possible to bring them home alive, a young white Bear[5] I should be particularly glad of as I have great reason to suppose that the white bear is an animal differing even genericaly from the Brown or Black bear.

Of Seals, it is more than probable that you will meet with several Kinds which have not been taken notice of by naturalists. We know indeed but one kind with any degree of precision.[6] It varies very much in Colour. I should therefore advise that if any Kind is seen which differs from the common in any other way, [such] as proportions of its parts to each other, that the skin complete, & head with the teeth of such animal, is preservd. I have heard of one Kind whose hair is long & shaggy like that of a water dog. If any Kind is found to be so after they come to maturity, I should much wish to see them.

Whales are a Kind of animal which Naturalists are almost totaly unaquainted with. Accident might bring you in Company with whale fishers. If so, Foetus's of any species preservd in Spirits would be very acceptable, as would parts [which] might be preservd in spirits, for the doing [of] which Irwin has particular instructions from Hunter.[7]

Birds of the diving Kind, [such] as Puffins, murrs[8] &c., must abound where you mean to go, & probably new species. Lions will easily tell you the few which are Known. Any such that you may meet with, if they should be too large to be preservd whole, might still be brought home in part by taking off their skins with the heads & feet sticking to them, which should be likewise put into spirits.

Observations concerning the species of Known Birds which you may see, especialy in their breeding places, will be of use as there are several whose migrations we are utterly unaquainted with.

Fish of the High Northern latitudes we Know very little of. I should therefore advise that a specimen at least of Every sort, which is not very Evidently the same as some one Known among us, may be put into spirits. How desireable it is to Know from whence the immense shoals of Herrings, Macarels, Capelings,[9] Pilchards, Shads, Salmons &c., which Every year Come from the Northward & are seen to a Certain Southern point, which they never pass, derive their origin. Any observations tending Ever so little to illustrate that subject, I should receive with great pleasure.

The very Blubbers &c. which, in many Shapes & sizes from that of my hat to an almost invisible littleness, swarm in the polar regions are not insignificant as they seem a necessary branch of all perfect nature does. [That] the whale or any other animal use[s] them as food is a question to be answerd. A few individuals of as many species as can be procurd will be very acceptable.

When ever a sounding line or Cable is hove in, a Careful hand attending it is almost sure, especialy if it has been long overboard, to find something adhering to it which may be Curious. Crabs, Star fish, &c. are often [found], & once a most Curious many headed Polype was taken in this way.

Shells &c. may be brought up in the Trap for soundings, the value of which, you Know, is from fashion very high. Consequently, they are desireable to most people. Many species of shell are found adhering to sea weed, & under Stones on the beach. There your boats Crew may Gather whenever they are left to wait without any more essential employment.

Sea weeds of all Kinds may be preservd by merely washing them in fresh water, & laying them three or four days to dry, after which, packd up in Canvas bags, they can receive no Kind of Detriment.

Insects will not abound unless the Countries you are to visit should be as hot as our wild projector has supposd them to be. Such as you find, Lions well Knows how to take & impale with the instruments alotted for that purpose.

Mosses also, & all Kinds of Lichens with which the high latitudes very much abound, may be preservd in the same Kind of bags, in which they may be put while yet a little damp as they are then pliable, but when dry become quite brittle.

Of Plants, must be chose[n] such as have, if possible, flowers & Seed pods, or fruits upon Each plant. If that is impossible, two of each Kind should be taken for each compleat specimen intended to be brought. Of all trees, shrubbs & plants that are sufficiently large, the specimens should be chosen nearly as long as the paper.

So much for my own department. I Know that you will excuse my Blunders, so shall venture to swim a little into other peoples.

The Large trunks of Trees, which are found floating on the Northern seas, seem to afford matter for observations from whence Conclusions might be drawn of the utmost Consequence toward judging whether or no there is a Passage by the Pole to the other side of the globe; whether the currents realy run from the pole as a center in all directions, supplyd again by under currents, as those most urge who suppose the polar waters heavyer than those nearer the Equator on account of their greater saltness, or dencer on account of the greater degree of Cold they are subjected to &c. &c. It might therefore be satisfactory to Know if Possible what:

1. What & how many species of trees are thus found floating on the Sea?
2. Is the Bark yet left on any of those found on the North side of Spitzbergen, or in more northern latitudes?
3. Have they all sufferd material friction against hard substances, which may be Known by the ends of the branches, in what manner they have been broke off?
4. Have any of them been cut off by the hands of men? Are they unrooted or broken off by force of Winds &c. above ground?
5. In what parts of the sea are the greatest quantities of them found?
6. Why may it be that, tho they are in plenty on the North Side of Iceland, they are very seldom seen on our Coast?
7. In what parts of the sea do they most abound?
8. What quantity are worm Eaten, & what proportion free from worm?
9. Is the worm the common ship worm?[10] To what size does it grow? Are the shells with which the holes are lined yet unbroken near the mouths of the

holes? Are any remains of the animals left, or may be in high lat[itudes] are they still alive & working? Specimens of different grains of woods & different barks, also of the works of worms or other insects upon drift wood, would be very desireable.

10. Are the planks or timbers of Ships which have been cast away on this Coast ever worm Eaten?

For the satisfaction of one of my Friends, be so good as to weigh certain musquet Balls against weights of brass both in air & water, &, noting down with all convenient accuracy the weights of Each on the paper in which it is wrappd up, let them be brought home to me with a bottle of the water in which they were weighd carefully corkd up.

God Bless you, & send you to the Herring Hall or the Source of the migration of macerel, & thence home to your Ever affect[ionate], but never Emulating,

J. Banks.

[U.W. Thordarsdon Collection. 3–8a.]

[1]Phipps led an expedition to the Arctic in HM Ships *Racehorse* and *Carcass*. The ships returned on 25 September 1773. They had been prevented from exploring any farther north than 80–81° by ice-fields. The natural history collections made were limited, as Banks had expected. Nevertheless, both Banks and Dr. Daniel Solander classified and described them for Phipps's *A voyage towards the North Pole, undertaken by His Majesty's Command, 1773*, London (1774). The 'Natural History' catalogue, including English names where Phipps knew them, was printed in the 'Appendix', pp. 183–204. Most of the animals, plants and insects recorded were from Spitzbergen.
[2]Dr. Charles Irving (d.1794): marine inventor.
[3]Israel Lyons (1739–1775): botanist; mathematician. In 1764 Banks wanted additional university tuition in botany, and arranged at his own expense for Lyons to come from Cambridge to Oxford to provide it.
[4]Faroe Islands, island group in the North Atlantic Ocean.
[5]The white or polar bear (*Ursus maritimus*). The polar bear is currently *Ursus maritimus*, after Phipps's original scientific name, which makes it generically the same as the brown bear (*Ursus arctos*), and the American black bear (*Ursus americanus*).
[6]Common or harbour seal (*Phoca vitulina*). It is not the most numerous seal in the high Arctic. That is the ringed seal (*Phoca hispida*).
[7]Dr. John Hunter FRS (1728–1793): surgeon and anatomist.
[8]Murr(e) at the time Banks was writing this would probably have been used to refer to both razorbills (genus *Alca*) and guillemots (genus *Uria*). For the British, common guillemot is now the name of two species of birds of the genus *Uria*: *Uria aalge*, which is common guillemot, and *Uria lomvia*, which is Brünnich's guillemot.
[9]'Capelings', capelin or caplin (*Mallotus villosus*), which is a small fish very similar to a smelt. It is found off the coast of Newfoundland, and is much used as bait.
[10]Ship worm (*Teredo navalis*).

Letter 11

1773

To Thomas Falconer

New Burlington Street
2 April 1773

Dear Sir,

I must beg your pardon, &, more particularly, Dr. Haggarths for the imperfection of my account of Kelp burning,[1] but I hope you will both attribute it to the Known difficulty of drawing a description of any fact in which material circumstances only are inserted, & those of little import left out: a thing impossible for any man to perform perfectly till he Knows the system to which his description may be applyd. I am inclind to beleive that Dr. Haggarth is mistaken in his Solution, because I omitted to relate in my description that after the ashes were sufficiently stirred into their doughy consistency, they still remaind glowing hot enough to give fire almost instantly to an additional quantity of sea weed which was thrown upon them; in which state of heat, I apprehend it is impossible they should imbibe a sufficient quantity of atmospherical water to produce the requird Effect. But to our Business.

On the 18th. of August we left the But[t] of Lewis resolvd to steer immediately for Iceland, & on the 29th. came to an anchor at a place calld Besested,[2] lying at the bottom of a deep bay on the S.W. side of the Island. Here, as we found a tolerable harbour, we resolvd to remain. We went ashore therefore, & to our great Joy found that the sides of the Harbour were constituted of the surface of an ancient flow of Lava.

The Governor Stifsamptman,[3] as he is calld, receivd us with great politeness, & allotted to us certain houses, belonging to the Danish merchants who were returnd to Copenhagen, for our residence here. We soon fixd ourselves, not so convenient as in English houses, but more so than we could have done in the ship. To give a detail of our daily proceedings from this time would be tedious & unentertaining. I shall therefore proceed as well as I can to give as good an account as I am able of the countrey & its inhabitants.

The Island of Island [Iceland] contains an area I beleive Larger than that of Ireland. As we were only 6 weeks ashore upon it, it cannot be expected that we could examine so large a space. Indeed we traveld 13 days on horseback, & merely reachd Hecla[4] as the extremity of our Jaunt. Wherever we traveld Lava, & that stone which is Calld by Sr. Wm. Hamilton[5] Tuffa,[6] were the only strata. So much has this Island sufferd by Fire that we were told that since it has been inhabited the whole S.W. promontory, & near or quite a hundred miles inland from it, was on fire at once. Many are the mountains which have occasionaly vomited up fire, but in this land of emptiness, it has often hapnd that the level ground has opend itself into a crater, & thrown out water, stones, Lava &c. over all the adjacent countrey.

We seldom traveld half a day without [encountering] hot baths, whose waters were generaly of a boiling heat. These were constantly surrounded by quantities of that stone which by the mineralists is calld Lebes[7] from its similitude to that formd on the bottoms of Kettles &c. Some of them boild so violently that in the center of them the water was raisd by the agitation of the heat 6 or 7 feet high almost constantly. Others, from the impulse of a power which I confess myself unable to trace, spouted up their water periodicaly to an immense hight; particularly that calld Geiser, situated about [?] miles N. from Scalholt,[8] of which you receive some drawings.[9]

This wonderful volcano of water, if I may use that Expression, which has continued its eruptions ever since the Island was inhabited about 800 years [ago], rises from hundreds of smaller wells [in] a basin 57 feet in diameter, formd like a funnel of the Lebes it deposits. [It] gradualy slopes from the sides to a hole in the center, 19 feet in diameter, out of which the water rises.

We arrived there in the morning at 6 o'clock, about half an hour after the eruption, which generaly happens morning & evening, was over. the Crater was then Empty, & we had an opportunity of examining the central hole, which for about 10 feet down was also free from water; but this was not quite safe, for every now & then a burst would come up, throwing the water some fathoms high without giving us the least warning. This continued till half past 8, the water rising gradualy all the time till it overflowed the edge of the central hole. From this time till half after ten only one eruption hapnd, & that a small one. The Crater however was filld two thirds at half past twelve. Three strokes were heard underground like the firing of a distant Cannon. The water in the crater instantly rose so high as to overflow the Edges, & the ground all around it was shaken very perceptibly. At a quarter after three many distinct blows of the same Kind as the last were heard. The Earth shook much, & the crater overflowd more than before. At 43 minutes after 4 louder noises were heard, which shook the ground above a mile from the crater, & much water ran over. At 49 minutes after 4 still louder noises were heard than before; & at 51 minutes after 4 the great Eruption hapnd. A Column of water, not solid but consisting of many pointed Cones, at once flew up to the hight of 92 feet above the edge of the crater, & continued by Jerks rising & falling 4 minutes. Its thickness we guessd to be about 26 feet, but its hight was measurd with a quadrant by Dr. Lind, who had prepard a base, as accurately I beleive as that operation will admit of.

How this effect is producd is certainly a mystery which nature has Lockd up from the eyes of curious mortals with great care, guarded as she here is by fiery dragons. It is impossible probably for the art of man to gain an opportunity of inspecting her operations. Analogy then is the only resource, & that tells us that as water when expanded by fire into the form of steam occupies 14000 times the space that it did while only water, such an expansion so much more violent than that of the air containd among the grains of Gunpowder seems, & probably is, sufficient to produce even the effects above describd. This goes, however, no farther than a general guess at the Gross cause. Experiment may in time assist me to account for the particularities of this Phaenomenon. We are now Employd in trying some, the results of which, if they are successful, I may one day send to you.

You should have receivd this Letter sooner had not my absence in Holland, through which I have made a short tour,[10] prevented me. I wish I could send you anything from that countrey worth observation. Holland is a great Hydraulick machine, which must continualy work in order to discharge its water. Dutchman, Lazier than any other race of mankind, have from that very principle invented more methods of Easing themselves from the curse of Labour than any other nation. Hence it proceeds that their Engines, Mills &c. are better & more numerous than those of any other nation.

Mr. Pennant is in town. He has had my Journal, & the drawings which I made in the western Isles. I lookd upon them, as I told you, as his right. I, while in that countrey, Lookd for him with assiduity, conceiving myself as no more than a poacher who might get leave of the Lord to shoot upon the mannor, but in return owd at least the offer of whatever he might Kill.

Adieu, dear Sir. I snatch an opportunity of sending this. In the mean time assure yourself that you shall hear form me again the very first Leisure I have.

Beleive me,
Your Affectionate & very Hble Servant...

We are employd in fitting out an expedition in order to penetrate as near to the North Pole as Possible. It consists of two Boom Ketches,[11] chose as the strongest species of Ships, therefore the best to Cope with the Ice. They will sail before the middle of the next month commanded by a good Freind of mine, Captn. Phipps. Your opinion of the Frigid Zone cannot but be usefull to him, & very agreable to me at this Juncture.

[Warren R. Dawson, *The Banks Letters: A Calendar of the Manuscript Correspondence*, London (1958), p. 318. Dawson noted: '...formerly in the possession of Dawson Turner, bound in a volume with the original drawings, etc. D.T. Manuscripts Sale, 1859, Lot 26.' Dawson cited 'Hawley Coll.' for this letter, by which he meant: 'Banks papers in the possession of Major Sir David Henry Hawley, 7th Bart. of Mareham-Le-Fen, Lincolnshire', p. xxxviii. A typed transcript of the letter is at the Natural History Museum, London, in the Banks Archive. This was made from a Dawson transcrpt at B.L. Add. MS. 56301(7), ff. 20–24.]

[1]Letter 8.
[2]Bessastadir.
[3]Lauritz Andreas Thodal (1717–1808): the Norwegian Governor of Iceland, 1770–1785.
[4]The journey to Hekla was made between September 18 and September 28 in 1772. Banks was accompanied by Dr. Daniel Solander, Dr. James Lind, Reverend Uno von Troil, John Clevely Jnr., James Roberts, and two of Thodal's servants. Hekla is a volcano in the south of Iceland, rising some 5,110 ft above sea level. It is composed chiefly of basalt lava, but slag-sand and loose ashes cover a good part of it too. In 1766 it erupted violently, causing widespread damage, and killing much livestock. Indeed, throughout the eighteenth century earthquakes and eruptions devasted large areas of Iceland. Particularly disastrous events were the eruptions in 1727 of Öraefajökull, and in 1755 of Katla. Coming between the two, in 1732, tremors destroyed or damaged over fifty farms. The worst catastrophe followed Banks's visit. In 1783 and 1784 the Skaftár eruptions took place. The Laki fissure expelled basaltic lavas along a 27 kilometre long chain of more than 140 craters for eight months. Earthquakes and vocanic ash ruined over 1000 farmsteads in the south of Iceland. Crops, cattle, sheep and horses subsequently died, as did over 11,000 people, one fifth of Iceland's human population.
[5]Sir William Hamilton's varied interests included vulcanology, which he actively pursued when he was British Envoy in Naples. He regularly reported to the Royal Society on the volcanic activity of Vesuvius and Etna, and his accounts still form a valuable contribution to early scientific investigations of this subject.
[6]Banks apparently meant tuff, which is formed of consolidated volcanic ashes and other erupted material.
[7]This term is not current, but Banks seems to be referring to tufa, a chemical sedimentary rock usually formed by evaporation around the mouth of a hot or cold calcareous spring or seep, or along a stream carrying calcium carbonate in solution.
[8]Skálholt, in the south-west of Iceland, some 55 miles east of Reykjavík, and some $15\frac{1}{2}$ miles directly south west of the Great Geysir.
[9]The journey took Banks and his party past Thingvellir, across the the river Brúará to the hot springs at Laugarvatn, and then on to the Great Geysir at Haukadalur.
[10]Banks visited Holland from 12 February 1773 to 22 March 1773: *Journal of a trip to Holland beginning with the time of leaving London February 12 1773 & ending with the day of returning there again March 22 1773.*
[11]HM Ships *Racehorse* and *Carcass*, which were commanded by Constantine Phipps on his voyage of exploration to the Arctic. 'Boom Ketches' refers to bomb-ketches, which were small, strongly-built vessels originally designed to carry mortars for bombarding. Bomb-ketches were considered robost enough to cope with ice if it was encountered on this mission. See Letter 10.

v. **View of the eruption of Geysir.**
Watercolour with ink, 1772, by John Clevely Jr. The British Library, Add. MS. 15. 511, f. 43.
By permission of the British Library.

Letter 12

1778

To Carl von Linné the younger

Soho Square
5 December 1778

Sir,

With pleasure I receivd your favors,[1] & the 1st Edition of your Learned Father's Systema,[2] for which I return you my thanks. I always had the highest respect for that valuable man, & shall pay every duty to his memory which gratitude can dictate. I have invariably studied by the Rules of his System, under your Learned Friend Dr. Solander, So that the Plants in my intended Publication will be arrangd according to his Strictest rules.[3] Such as are of Genera describd by him will have his names. The new ones, which I think will almost outnumber them, will be named Either in honor of distinguishd Botanists, or, according to the Rules in Philosophia Botanica,[4] by names derivd from the Greek.

Uninterruptedly, however, as I have applied to the work of engraving for near 5 years, I have not yet advancd above half of my intended progress. About 550 plates are engravd, &, I think, if circumstances as yet unexpected do not Oblige me to cut it Short, it will Extend to double that number. Understand by this how impossible it will be for you to Quote it in a work intended for Publication in the course of this year.

The Plants which you so Kindy sent me by Mr. Troille[5] I fear are lost. The Ship has not been heard of, tho more than a year is Elapsd Since she Saild.

In one thing it will be in your Power to Oblige me much, & I shall not Want for gratitude: if you will Kindly undertake to supply me with as good a Collection of Mutis's[6] plants as you can Spare, without damaging your own Collection. A small bit, you Know, is of great use to a Botanist. When you have not a duplicate, a small branch or part, broken from your Specimen, may serve without damaging it too much. I shall be able & willing to make returns in things which you cannot easily Obtain Elsewhere.[7]

I am, with all due attention,
Your Affectionate,
Faithfull Servant,

Jos. Banks, P.R.S.

[L.S. Linnaeus Correspondence I. 330.]

[1]Linnaeus the younger to Banks, 12/6/1778, B.L. Add. MS. 8094, f. 178.

[2]Carl von Linné (Carolus Linnaeus) FRS (1707–1778): Swedish botanist and naturalist. Linnaeus established a system for classifying the products of the natural world, especially plants. His ideas were eagerly adopted, not least in England where naturalists wanted a more scientific basis for their work. Linnaeus's approach was first outlined in *Systema naturae...*, Apud Theodorum Haak: Lugduni Batavorum (1735) and *Fundamenta Botanica...*, Amsteraedami (1736). Later, Erasmus Darwin (1731–1802) edited a translation of Linnaeus's *Systema Vegetabilium...*, Gottingae (1774), and called it *A System of Vegetables...*, Lichfield (1783). Darwin frequently consulted Banks about this work. Moreover, Dr. Daniel Solander was one of Linnaeus's favourite pupils, and one of Banks's closest friends.

[3]Banks intended to publish an illustrated record of his botanical findings from the *Endeavour* voyage — the *Endeavour* Florilegium. He estimated such a work would run to some 14 volumes, and might cost him £10,000. Sadly, Banks never completed his ambitious plan. The drawings, however, may still be seen at The Natural History Museum, London, in the Botany Library: 'Catalogue of the Natural History Drawings commissioned by Joseph Banks on the Endeavour Voyage, 1768–1771, held in the British Museum (Natural History)', J. Diment *et al*, *Bulletin of the British Museum (Natural History)*, *Historical Series*, London (1984), vols. 11, 12, 13. The Natural History Museum, London, published the Florilegium in association with Alecto Historical Editions, 1980–1988. This was done using the original copper plates, employing colour *à la poupée*.

[4]C. Linné, *Philosophia botanica*, Stockholmiae (1751).

[5]The Reverend Uno von Troil was a traveller and scholar, who accompanied Banks on his voyage to Iceland in 1772. He became Archbishop of Uppsala.

[6]José Celestino Mutis (1732–1808): Spanish physician and botanist, who discovered the *Cinchona*.

[7]Note not in Banks's hand reads 'Printed in Linn. Corr. v. 2. 574.' This refers to: J.E. Smith (Ed.), *A selection of the correspondence of Linnaeus and other naturalists, from the original manuscripts*, London (1821), vol. II, pp. 574–5.

vi. **The heath banksia or red honeysuckle (*Banksia ericifolia*).**
This Australian plant was named by the younger Linnaeus in honour of Joseph Banks in 1781 [*Supplementum plantarum*..., Brunsvigae (1781)]. It was gathered at Botany Bay, 28 April-6 May 1770, by Banks and Dr. Solander. Painting by John Frederick Miller, 1773, from field notes and specimens, and after Sydney Parkinson, 1770. Botany Library, Natural History Museum, London, A7/324.

Letter 13

1780

To Dr. Benjamin Franklin F.R.S.

Soho Square
29 March 1780

Sir,

By the hands of Mr. Paradise[1] I have receivd a copy of the instructions for the Protection of Captn. Cooke, which you circulated among the Armd Vessels of your Friends in N. America. I perusd this paper with the Greatest pleasure, for having never doubted myself that the liberal & enlargd sentiments I had always admird in your mind remaind there in full lustre,[2] I could not but rejoice at the triumph which such an indisputable proof afforded me over those who, warp'd by politicks or party, wish'd to entertain a different opinion of your character.

Give me leave, then, as the Friend of disinterested discovery, & of Captn. Cooke, to return you my warmest thanks.

Permit me also as President of the Royal Society to thank you in the name of many of our most valuable members, who, abstracting themselves from all less generous considerations, fix their whole attention upon the great Object of Science, & would take the most publick method of conveying to you their acknowledgements, were they not sensible that such an act might be wilfully misunderstood.

Some Medals are preparing to be struck in memory of our late Friend.[3] The Royal Society will intreat the King of France[4] to accept one at their hands as a testimony of the high sense they entertain of his Generous orders in favor of that excellent navigator. It would give me pleasure to learn that the Congress issued similar orders. If they did, I shall rejoice in the opportunity of transmitting to America the like permanent token of our regard & gratitude.[5]

Adieu, my Dear Sir. Beleive my Mind incapable of being Led astray by the influence of political opinions. I respect you as a Philosopher, & sollicit the continuance of your Friendship in full Persuasion that all your actions are conformable to your most conscientious Ideas of rectitude, whatever my wishes may be as a native & inhabitant of a country with which you are at war.

Your Faithful & affectionate
Hble Servant,

Jos: Banks.

[A.P.S. Franklin Papers — Bache Collection.]

[1]John Paradise FRS (1743–1795): linguist; also interested in collecting minerals.
[2]Franklin was an American minister in Paris, 1776–1785. He persuaded the French Government to exempt Captain James Cook's ships from attack by the French Navy. He also sought safety for British ships engaged in exploration from American vessels. One of Franklin's first official acts as minister plenipotentiary was to write a passport for Cook. This was done on 10 March 1779, when Cook was already dead.
[3]Banks proposed the Royal Society memorial medal for Cook, who was killed in Hawaii early in 1779. The medal was the work of Lewis Pingo (1743–1830), the chief engraver to the Royal Mint, and was struck at the Mint in bronze, silver and gold. Banks later sent a gold medal to Cook's wife, Elizabeth (1742–1835), with a fine tribute to the great navigator: Banks to E. Cook, 12/8/1784, R.S. Cook 23.
[4]Louis XVI, King of France (1754–1793).
[5]Banks later stated that Congress did not issue any special orders in favour of Cook: Banks to Kippis, N.H.M. B.L. D.T.C. IX 282–283.

vii. **Dr. Benjamin Franklin F.R.S.**
Portrait, 1767, painted by David Martin in London. The White House, copyrighted by the White House Historical Association.

Letter 14

1782

To Sir William Hamilton F.R.S., Ambassador at the Court of Naples

Soho Square
13 February 1782

My dear Sir William,

Many thanks for your Favor,[1] & more For the annexd account of the volcanic rain sent you by your correspondent in Sicily.[2] It has been read to the Society, & met with much approbation. Only one thing puzzles us all. He calls himself an inhabitant of the 3d Region. Now all the authors we have read describe the third region to be uninhabited. If your friend, then, is not a Hermit, he leaves us all at a Loss what to call him.

We have read a paper[3] at the Royal Soc[iety] about a week ago, which gave me pleasure as it producd an illustration of your Observation of Filaments of Glass found near your Volcano at the time of its last tremendous eruption.[4] It is what, I beleive, is Calld Lana Philosophorum: small Fibres of Glass which are forcd by the Bellows of a Wind Furnace employd in melting Iron out of the hole into which the Nozzel is inserted. They are not thicker than a Cobweb. Yours, & those of Mauritius, [are] thicker than horse hairs, but scarce any other difference seems to exist between them. No doubt your vast Furnaces of nature have their blast holes! Tho we may be at a loss to Know whence the air comes that blows them. If that should not be allowd, Elastick vapor, which bears the Column of Smoke above the heavens, & Projects stones to such astonishing distances, will produce the same effect as its action must be that of a Fluid.

We are sadly disturbd here with the news Ch[arles] Greville[5] has given us that your observations on the Cult of Priapus[6] have causd the Priests to enquire into & censure that curious remain of antiquity.

We are all at this moment Electrical Mad. A house guarded by 8 sharp pointed conductors has been struck, & Many Electrical heads are laid together to find how that could happen. If they give any verdict without much quarrelling, disputing & protesting I shall wonder much.[7]

Poor Seaforth,[8] you read by the Papers, is dead. He went off suddenly, & was drunk when he died. It would have been extraordinary luck if he had been sober as it was for some time his custom to Keep up a regular fire of Drams day & night whenever he was awake.

May your Observations on the Nobility of Naples succeed. I should much delight to Join in them. They seem so like my Friends in the South Sea, the Lousy rogues whose manners tho are more simple than those of their masters.

Adieu, my dear Sir, beleive me
Your very Faithful,

Jos: Banks.

[B.L. Egerton MS. 2641, ff. 134–135.]

[1]Hamilton to Banks, 17/7/1781, B.L. Add. MS. 34048, ff. 12–14, N.H.M. B.L. D.T.C. II 15–16. Hamilton closed his letter with: 'that your *Great Toe* & your purse may never fail you, is the wish of...' This was a euphemism for the emblem of Priapus.

Banks was elected to the Society of Dilettanti in 1774, and in the same year was briefly Very High Steward of the Society. He acted as Secretary and Treasurer, 1778–1794, and then he served just as Secretary until 1797.

[2]*Philosophical Transactions*, vol. 72 Pt. I (1782), pp. 1–7, and Appendix i–vi, 'Account of a new Kind of Rain. Written by the Count de Gioeni, an Inhabitant of the 3d Region of Mount Etna; communicated by Sir William Hamilton, K.B. F.R.S.' Read 8/11/1781.

[3]*Philosophical Transactions*, vol. 72 Pt. I (1782), pp. 50–52, 'An account of some Scoria from Iron Works, which resemble the vitrified Filaments described by Sir Wi[l]liam Hamilton. In a Letter from Samuel More, Esq. to Sir Joseph Banks, Bart. P.R.S.' Read 17/1/1781 [1782].

[4]Hairs and filaments of glassy volcanic rock are produced in the eruption of many volcanoes, typically in those which produce basaltic, and therefore runny lava. The lava is so fluid when erupted that it forms droplets, which draw long filaments behind. Today this is called 'Pelé's hair' after the Hawiian goddess of volcanoes.

[5]Charles Francis Greville FRS (1749–1809): collector of minerals and precious stones. He was a man of taste, and a member of the Society of Dilettanti, who lived with Emma Hart in Edgware Row until she left in 1786. She went with her mother to Naples, and married Greville's uncle, Sir William Hamilton. Banks travelled to Holland with Greville in 1773.

[6]Hamilton had discovered evidence in the province of Abruzzo of a pagan fertilty cult involving phallus worship. He reported his findings to the Society of Dilettanti, of which Banks was a member. In 1783 a limited number of slim volumes were published: *The Worship of Priapus. An account of the fête of St. Cosmo and Damiano celebrated at Isernia in 1780...*, London. These contained a letter from Hamilton to Banks describing the cult, which was dated 30 December 1781, and a brief essay by Richard Payne Knight (1750–1824). Richard Payne Knight was one of Hamilton's acquaintances, and in 1786 he produced *An Account of the Remains of the Worship of Priapus, lately existing at Isernia, in the Kingdom of Naples...in two letters: one from Sir William Hamilton...to Sir Joseph Banks...*, London. This included Hamilton's letter to Banks with a longer essay, and was condemned for its provocative mixture of sexual and religious ideas. See Letter 27.

[7]*Philosophical Transactions*, vol 72 Pt. II (1782), pp. 355–378, 'Proceedings relative to the Accident by Lightning at Hecklingham.' Read 14/2/1782. Banks perhaps recalled the famous dispute in the 1770s when the Royal Society advised the government that pointed conductors

were best for protecting buildings and their contents against lightning strikes. Benjamin Franklin was foremost among the scientists responsible for reaching this conclusion. However, an ammunition store at Purfleet was struck by lightning despite having pointed conductors. There was a serious explosion, and some people felt that the Royal Society and Franklin must be wrong. This feeling became stronger when the American Colonies revolted with Franklin as a prominent rebel. Loyalists, and George III in particular, favoured blunt lightning conductors. George III tried to make the Royal Society withdraw its resolution recommending pointed conductors, but the President, Sir John Pringle (1707–1772), dryly observed: 'Sire, I cannot reverse the laws and operations of nature.'

[8]Kenneth Mackenzie FRS (d.1781), Lord Ardlive, Viscount Fortross, Earl of Seaforth: officer in Scottish regiments; chief of the clan Mackenzie. The Earl of Seaforth was a member of the Society of Dilettanti with Banks. He sometimes got into financial difficulties. In June 1777 Banks helped Seaforth to escape his creditors in London by fleeing with Seaforth and his mistress to St. Omer. This was the only time Banks visited France.

Letter 15

1782

To Dr. Charles Blagden F.R.S.

Soho Square
9 November 1782

My dear Dr.,

We began the Meetings of the [Royal] Soc[iety] on Thursday before a numerous audience by Mr. Herschell,[1] in a paper written for the purpose, dedicating his new planet to the King under the name of Georgium Sidus, & acknowledging his Majesties Bounty to him; tho neither the name is well conceivd or the paper well written. It was tolerably well receivd. The Astronomer[2] after the meeting got up most unexpectedly to declare his approbation of the name, & thank Mr. Herschell for the benefit he had receivd by his Labors, particularly in the improvement of telescopes, by which said he, 'You was enabled to discover this new planet.' His Observations are to shew that its path is Circular, its dimensions $3\frac{1}{2}$ larger than the Earth, & its period $82\frac{1}{3}$ years. He shewd me a very han[d]some letter from De Lalande[3] in which he is praisd for having had the bonheure & glory of discovering a planet. The Idea of animation in the Moon is founded only on his having some time ago instituted a set of Observations for the end of discovering whether animals inhabit it, but in which he has as yet made no progress.[4]

Magellan[5] is returnd & looks wondrous sown [?]. The good news which Gen. Elliot[6] has sent us will beyond doubt tend to your release. We say here that Lord How[e][7] might have Fought more than he has done.

Your Faithful,

Jos: Banks.

[R.S. BLA. B. 11.]

[1]On 13 March 1781 the astronomer Dr. (later Sir) William Herschel FRS (1738–1822) discovered a seventh planet in the solar system, the first to be found in modern times. He thought it was a comet, but later observations showed otherwise. Herschel was given the privilege of naming the planet. He chose 'Georgium Sidus' as a dedication to George III, who had granted him a pension of £200 a year, and a further £2000 for the improvement of his telescopes.

The announcement of Herschel's decision was made in the *Philosophical Transactions*, vol. 73 (1783), pp. 1–3, 'A Letter from William Herschel, Esq. F.R.S. To Sir Joseph Banks,

Bart. P.R.S.' Banks did not think the name 'well conceived', and it was not generally adopted. Johann Elert Bode (1749–1826), a German mathematician and astronomer, suggested 'Uranus' instead, and this has proved more popular. The next paper published in the *Philosophical Transactions* for this year was also by Herschel: pp. 4–14, 'On the Diameter and Magnitude of the Georgium Sidus; with a Description of the dark and lucid Disk and Periphery Micrometers. By William Herschel, Esq. F.R.S.' Read 7/11/1782 — a Thursday.
[2]Reverend Nevil Maskelyne FRS (1732–1811): Astronomer Royal, 1765–1811.
[3]Joseph Jérome le François de Lalande FRS (1732–1807): astronomer.
[4]Herschel thought he noticed movement on the Moon: Banks to Blagden, [1/11/1782], R.S. BLA. B. 95. Banks doubted the reports, and, like Blagden, awaited proof.
[5]João Jacinto Magalhaens (Magellan) FRS (1722–1790): physician and botanist.
[6]Possibly George Augustus Eliott, 1st Baron Heathfield (1717–1790): general, and defender of Gibraltar against the Spanish, 1779–1783. He was a formidable commander, who successfully resisted every attempt to take the Rock until the sea blockade was broken by Earl Howe. When peace was concluded, and the siege lifted, Eliott returned to England.
[7]Possibly William Howe, 5th Viscount Howe (1729–1814), then lieutenant general of ordnance. He was a controversial British commander in the wars of the American Revolution, but had resigned his post in America in 1778.

William's brother was Richard Howe, Earl Howe (1726–1799), an Admiral in the Royal Navy. In 1782 he was naval commander of the Channel, guarding against a combined French and Spanish fleet. Blagden referred to this in earlier letters: Blagden to Banks, 14/7/1782, N.H.M. B.L. D.T.C. II 157–159. In September Howe led a British fleet from England to relieve Gibraltar, a task he accomplished in October despite attempts by a much larger French and Spanish force to prevent it.

Letter 16

1783

To Dr. Benjamin Franklin F.R.S.

Soho Square[1]
13 September 1783

Dear Sir,

For having it in my power to Answer with precision the numerous questions which are askd me by all sorts of people Concerning the Aerostatique Experiment,[2] which, such as they may be, are suggested by every newspaper now printed here, & considerd as a part of my duty to Answer, is an Obligation for which I am indebted to you; & an Obligation of no small extent I consider it. I Lament that the vacation of the Royal Society will not permit me to Lay your paper before them as a Body immediately, but it shall be the first thing they see when we meet again as the Conciseness & intelligence with which it is drawn up preclude the hopes of any more Satisfactory being receiv'd.[3]

Most agreable are the hopes you give me of Continueing to Communicate on this most interesting subject. I consider the present day, which has opend a road in the Air, as an epoche from whence a rapid increase of the stock of real Knowledge with which the human species is furnishd must take its date; & the more immediate Effect it will have upon the Concerns of mankind greater than any thing since the invention of Shipping, which opend our way upon the face of the water from Land to Land.

If the rough Effort which has now been made meets with the improvement that other Sciences have done, we shall see it usd as a counterpoise to Absolute Gravity, [and] a broad wheeld waggon traveling with 2 only instead of 8 horses; the breed of that Rival animal in some wise diminishd, & the human species increasd in proportion.

I have thought as soon as I return from my Present banishment of constructing one & sending it up for the purpose of an Electrical Kite — a use to which it seem[s] particularly adapted.[4]

Be pleasd to direct your Favors to Soho Square. They are sent to me without delay wherever I am.

Beleive me,
Your Obligd
& Obedient Servant,

Jos: Banks.

[A.P.S. Franklin Papers.]

[1]Banks was in Lincolnshire when this letter was written, although he gave the place of origin as Soho Square. His other letters of the same date confirm that he was not in London, as do his concluding remarks referring to 'banishment' and correspondence being forwarded from Soho Square 'wherever I am'.

[2]Banks means ballooning, which had caused a sensation in England and France. The balloon Banks had in mind was designed by Jacques Alexandre César Charles (1746–1823), and Nicolas Robert (1761–1828). It was filled with hydrogen and released from the Champ de Mars on Wednesday, 27 August.

[3]'Two letters from Dr. Franklin to the President' were read on 6 November: Franklin to Banks, 30/8/1783, R.S. L&P. VIII 36, U.P.L. Benjamin Franklin Papers IX 31; Franklin to Banks, 8/10/1783, R.S. L&P. VIII 37, Benjamin Franklin Papers IX 32.

[4]Franklin flew a kite in a thunderstorm in 1752. He was thereby able to charge a Leyden jar, and produce sparks from the end of the wet string, which he held with a ribbon of insulating silk. Lightning conductors were developed from Franklin's original work on electricity.

Letter 17

1783

To Dr. Charles Blagden F.R.S.

Revesby Abbey
13 October 1783

My dear Dr.,

By the Last Post Mr. Fox,[1] by his Majesties Command, sent to me for my opinion a Memorial from Cassini de Thury[2] recommending a Junction with London of the Triangles which the French have Carried from Paris to Calais, & thence to Dover Castle, for the purpose of ascertaining Triginometricaly the situations of the Observatories, in which, in case the Gentleman of the R[oyal] S[ociety] should decline the execution, he offers his services.

My answer is that I consider the operation as doing honor to our scientific Character, & benefit to Astronomy, & that I have no doubt but we have people Enough in the R.S. able & willing to undertake it. If his Majesty will permit me to ask the advice of the Council of the R.S., I shall be enabled, I hope, in a short time to give in a Plan, estimate & names of proper people for his Majesties information.[3] If I had a Frank I would send you a Copy of the Memorial, but in that case must beg you not to Exhibit it to any one as might be thought wrong, but you may converse as much as you please of the absurd as well as good parts of it. There are both.

Thank you for the Music. The Ladies[4] have receivd it safe. The Music of Gretna Green is of the very best, not a song but has passd the Ordeal of Criticism long ago — some before we were born, such as Jack Lettin &c.[5]

Surely no one who had given the Answer you did to Belcher[6] about London Practice, could ever hesitate what to say to Heberden.[7] If you should hereafter at any [time] undertake practice, I have no doubt of your success, but would never have you think of the Countrey. Your knowledge of Medecine would cut into a Regiment of Leaches to my Knowledge.

Thank you for the trouble you have given yourself in transcribing from Charles' Letter. He seems to have quite as much wit as Philosophy.[8]

Johnson, as I told you in my Last, arrivd on Saturday last, &, after spending Monday & Tuesday in making his Board, began to draw. Never was [an] Animal more out of his Element. His business is to Copy fair the foul scrawls of Architects into pretty plans to tempt gentlemen to build houses. How Greville Could think of sending him I do not guess, unless upon the principle of its being a good job for him. He goes on now very industriously. He talks of being able in the time I stay to make ground plots of the different floors of Tattershall Castle, & one elevation. Dear drawings they will be. All the detail which I most wanted is out of his way, as is perspective, which he never tried.[9]

Out of his Element is my poor Turtle I suppose. If he is at any time in danger of his Life, I beg it may be savd by an incision in his Throat, & his body presented to the Club.[10]

Yours Affectionately,

Jos: Banks.

I have a pretty Long account of the Meteor from Boys[11] of Sandwich in a Letter to Simmons,[12] which I will send when I can get a Frank — may be in this post with the Memorial.[13]

[R.S. BLA. B. 19.]

[1]Charles James Fox (1749–1806): statesman, and the first Foreign Secretary from 27 March 1782.

[2]César François Cassini de Thury FRS (1714–1784): astronomer; Director of the Royal Observatory, Paris. The 'Memorial' concerned was *Memoire sur la jonction de Douvres á Londres. Par M. Cassini de Thury, Directeur de l'Observatoire Royal; de la Société Royale de Londres*, (1783). Permission was sought by the French Government in early October for a joint trigonometrical survey to ascertain the relative positions of the Royal Observatories in Paris and Greenwich. In accordance with the King's wishes, the matter was referred to Banks and the Royal Society.

[3]Banks had already considered the best method by which such a survey could be made: Banks to Armstrong, 26/1/1776, Rylands English MS. 700/3, 'I should not chuse to encourage any Survey which is not grounded upon several large Triangles which can only be accurately measured by Instruments of the best Construction.' Since the proposed survey satisfied these requirements, Banks happily supported a project likely to give 'honor to our scientific character & benefit to Astronomy'.

He was also well acquainted with men able to undertake the task in England, men like William Roy FRS (1726–1790). Roy directed the survey, and the area chosen for the base line was an open, flat expanse in south-west Middlesex called Hounslow Heath. This was near to Spring Grove, so Banks found it easy enough to oversee the initial stages of triangulation, which took place in the spring and summer of 1784. In due course, and with the use of the famous '3-foot theodolite R.S.' made by Jesse Ramsden FRS (1735–1800), the British and French teams completed their work by uniting trigonometrical surveys across the English Channel. In late 1787 they could measure the distance between points in England and France with great accuracy. The origins of modern map-making had been carefully drawn, and can be traced in papers published in the *Philosophical Transactions* for this period.

[4]Banks's wife was Lady Dorothea Banks (*née* Hugessen), (1758–1828). His sister was Sarah Sophia Banks. They are usually referred to as 'the Ladies'.

[5]As promised, Blagden sent music by the Horncastle coach on 6 October: Blagden to Banks, 4/10/1783, F.M.C. Perceval Bequest H. 184, N.H.M B.L. D.T.C. III 106–107, 'The songs in Gretna Green are now published, & I shall send a copy of them by the Horncastle coach next Monday morning, if there be such a thing as the Directory says. Most of them I see are old, whether the new ones have any degree of merit the Ladies must decide. I add the last collection of Vauxhall songs, as the first air in it is said to be a great favourite.' A note facing this letter reads: 'A musical Farce by Stuart brought out this year — the songs alone were printed'.

[6]John Belchier FRS (1706–1785): London surgeon.

[7]Dr. William Heberden, the London physician, had sought Blagden's views on the possibility of moving from London to work in Canterbury as a physician earning £1000 a year. Blagden declined because 'tens of thousands could never make amends for being separated from those friends & those pursuits wherein the comfort of my life is centered': Blagden to Banks, 9/10/1783, F.M.C. Perceval Bequest H. 186, N.H.M. B.L. D.T.C. III 112–115.

[8]Blagden quoted from a letter by Jacques Alexandre César Charles to show how French 'madness' for ballooning was leading to 'a most violent party spirit'. He did so to illustrate that 'Ridicule & invective, verse & prose, are employed without mercy on this occasion': Blagden to Banks, 9/10/1783, F.M.C. Perceval Bequest H. 186, N.H.M. B.L. D.T.C. III 112–115.

[9]Possibly John Johnson (1754–1814) who was a draftsman employed by Banks on the recommendation of Charles Francis Greville. Banks wanted someone to make drawings of Tattershall Castle in Lindsey, Lincolnshire, while he was at Revesby Abbey. Banks explained to Blagden that this was 'to do somewhat in the way of antiquity during my stay here': Banks to Blagden, 21/9/1783, R.S. BLA. B. 13. By 7 October Blagden had completed arrangements for Johnson to come from London. The work cost Banks £25.6s.9d. A number of documents and pencil sketches of the castle by Johnson were sold at Sotheby's on 17 June 1880 in lot 980.

[10]As Banks was in Lincolnshire, Blagden kept him informed about this animal, which 'was not likely to live many days' due to cold weather: Blagden to Banks, 11/10/1783, F.M.C. Perceval Bequest H. 187, N.H.M. B.L. D.T.C. III 116–120. The turtle died and was eaten as soup at the Crown and Anchor on 23 October, a fate less unpleasant perhaps than the dissection Blagden was looking forward to, 'I should have much greater satisfaction from Sheldon's dissecting it': Blagden to Banks, 21/10/1783, F.M.C. Perceval Bequest H. 192, N.H.M. B.L. D.T.C. III 139–142. The 'Club' mentioned here was the Royal Society 'rebellious' dining club, founded and led by Banks, 1775–1784.

[11]William Boys (1735–1803): surgeon; topographer. He was born in Kent, and in 1767 was mayor of Sandwich. He wrote a history of Sandwich.

[12]Samuel Foart Simmons FRS (1750–1813): physician. He was born at Sandwich, Kent.

[13]There was considerable interest in 'meteors' in late 1783, more than one having been witnessed in the skies above Europe. The correspondence hurrying between Blagden in London and Banks at Revesby Abbey contained vivid descriptions of them. Two major events occurred on 18 August and 4 October. The excellent accounts Blagden sent to Banks were used as the basis for a paper to the Royal Society: *Philosophical Transactions*, vol. 74 (1784), pp. 201–232, 'An Account of some late fiery Meteors; with Observations. In a Letter from Charles Blagden, M.D. Physician to the Army, Sec. R.S. to Sir Joseph Banks, Bart. P.R.S.' Read 19/2/1784.

Letter 18

1783

To Dr. Benjamin Franklin F.R.S.

Soho Square
9 December 1783

Dear Sir,

The Friendship which I have experienced from you in your so speedily sending me Accounts of the Progress of the new Art of Flying, which makes such rapid advances in the Countrey you now inhabit, I beg to acknowledge with real gratitude.[1] I wish I had more than gratitude, something to Communicate in return, but times must Come when I shall be able to repay the debt which you have accumulated upon me with so much Friendly perseverence; & beleive me I shall do it with a grateful pleasure.

Charles's[2] Experiment seems decisive, & must be performd here in its full extent. I have hitherto been of Opinion that it is unwise to struggle for the honor of an invention which is about to be Effected. Practical Flying we must allow to our rivals. Theoretical Flying we claim ourselves. Bishop Wilkins[3] in his Mechanical Magic has, I am informd (for I have not yet got the book), a proposal for flying by means of a vessel filld with rare Air; & Mr. Cavendish, when he blew soap bubbles of his Inflammable air,[4] evidently performd the experiment which Carried Charles the memorable flight of the 1st instant. When our Friends on your side of the water are coold a little, however, they shall see that we will visit the repositories of stars & meteors, & try if we cannot derive as much Knowledge by application of Theory to what we find in the Armories of heaven as they can do.

Mr. Mitchell has given us a very curious paper in which he considers light as subject to the power of */attraction/* gravitation like all other bodies.[5] If, says he, should there be any material difference in the magnitude of the Fixd Stars, the light of the Large ones would move more slowly, &, in consequence, be liable to a different refraction from that of the smaller ones, but no such thing can be Observd with our best Telescopes. We have scarce a right to judge them not varying from each other in any immense quantity of magnitude, for was any one to be 100 times larger than another the difference would be discernable.

A miserable Comet made his appearance to Mr. Nathan Pigot[6] in his observatory in Yorkshire on the 19, & the weather has been so hazy in the evenings that it has scarce been Observd since it was on the:

		Right Ascen		North dec
[Nov.] 19 at 11^h: 15′	—	41°–0′–0′′	—	3°–1′
20–0: 54′	—	40°–0′–0′′	—	4°–3′

21 it was seen in the place it was expected but the night was too hazy to observe it.

It appears like a Nebula, with a diameter of About 2 minutes of a degree. The nucleus for it is seen with difficulty when the wires of the instrument are illuminated, but it is not visible with an opera glass.

Mr. Pigot.

Novr. 29. It was seen near the Chin of Aries. It appeard like a nebulous Star. As there was so much moon light it was dificult to find it.

Decr. 1. It was removd near the preceeding Eye of Aries, but conceiving other astronomers who have fixd instruments have noted its place, he has not calculated the distance from any Known.

Mr. Herschell.[7]

We are told that a Man has prepard Wings at a very considerable expence. Indeed, they say £1000,[8] [and] that the models upon which they are constructed have flown, & that the reality now in London, but packd up in a Box, should by a comparative calculation carry 150 lb more than the man. The machine consists of 4 wings, two of which beat while the other two are drawn back. Some people whose opinion in mechanics is lookd upon as Authority have said that they must succeed. Credat Judeaus say I. I must see it before I beleive it.

I am, dear Sir,
with real gratitude
& sincere thanks,
Yours Faithfully,

Jos: Banks.

I open this to thank you for Mr. Faujas's book,[9] which I receivd this moment on my return home from dinner.

[A.P.S. Franklin Papers.]

[1]Franklin To Banks, 30/11/1783, B.L. Add. MS. 8095, f. 177.
[2]Jacques Alexandre César Charles.

[3]John Wilkins FRS (1614–1672): Bishop of Chester; Warden of Wadham College, Oxford; Master of Trinity, Cambridge; founder member and first Secretary of the Royal Society. Banks had in mind a book by Wilkins called *Mathematical Magick, or the wonders that may be performed by Mechanical Geometry*, London (1648). Wilkins speculated on the means by which man might fly in Book II, chapters 7 and 8. Banks may also have recalled another work by Wilkins: *The Discovery of a World in the Moone...*, London (1638). In it Wilkins offered a 'pretty notion', Proposal 14, p. 221: 'So suppose a cup, or wooden vessel, upon the outward borders of this elementary aire, the cavity of which has been filled with fire, or rather aethereall aire, it must necessarily upon the same ground remaine swimming there, and of it selfe can no more fall, than an empty ship can sinke.'
[4]Henry Cavendish FRS (1731–1810): chemist and physicist. In 1766 he demonstrated the existence of a light gas that he called 'inflammable air', subsequently known as hydrogen. In an early paper, one related to even earlier notes, Cavendish referred to its properties: *Philosophical Transactions*, vol. 56 (1766), pp. 141–184, 'Three papers, containing Experiments on factitious Air, by the Hon. Henry Cavendish, F.R.S.' Received 12/5/1766. Part I of his 'Experiments on Factitious Air', pp. 144–159, was entitled 'Continuing Experiments on Inflammable Air', and in it Cavendish measured the specific gravity of hydrogen as compared to air. He showed hydrogen was eleven times lighter than atmospheric air. 'Experiments with bubbles of the kind Banks referred to were described in: Frankland to Banks, 20/11/1784, R.B.G. Kew Archives B.C. I 181, N.H.M. B.L. D.T.C. I 92–93.'
[5]Reverend John Michell FRS (1724–1793) astronomer; geologist. Banks is referring to his paper in the *Philosophical Transactions*, vol. 74 (1784), pp. 35–57, 'On the Means of discovering the Distance, Magnitude, &c. of the Fixed stars, in consequence of the Diminution of the Velocity of their Light, in case such a Diminution of the Velocity should be found to take place in any of them, and such other Data should be procured from Observations, as would be farther necessary for that Purpose. By the Rev. John Michell, B.D. F.R.S. In a Letter to Henry Cavendish, Esq. F.R.S. and A.S.' Read 27/11/1783. This is the famous 'black hole' paper, in which Michell suggested that if a star is massive enough for its escape velocity to exceed the velocity of light, its light will not reach us. Michell later devised apparatus for Henry Cavendish, which was used in work on Sir Isaac Newton's FRS (1642–1727) gravitational constant. Cavendish determined the density and mass of the Earth using this apparatus. For further details: R. McCormmach, *British Journal for the History of Science*, 'John Michell and Henry Cavendish: Weighing the Stars', vol. 4 (1968), pp. 126–155.
[6]Nathaniel Pigott FRS (d.1804): astronomer. *Philosophical Transactions*, vol. 74 (1784), pp. 460–462, 'Observations of the Comet of 1783. In a Letter from Edward Pigott, Esq. to the Rev. Nevil Maskelyne, D.D. F.R.S. and Astronomer Royal.' Read 24/6/1784. The numerical values given by Banks differ slightly from those in Pigott's paper. Units, degrees, minutes and seconds have been added to the letter by the editor.
[7]Dr. (later Sir) William Herschel FRS.
[8]See Letter 19.
[9]Barthélemy Faujas de St. Fond (1741–1819): traveller; savant. He published an account of balloon experiments in France in 1783: *Description des expériences de la machine aerostatique de MM. de Montgolfier...*, Paris.

Letter 19

1783

To Matthew Boulton F.R.S.

Soho Square
9 December 1783

Dear Sir,

We have been Amusd here for Some weeks in inspecting certain Rods of Steel Said to be intended as the supports of wings with which a man is to fly, & very lately certain boxes arrivd said to contain the very wings in Question. Now, as the said Rods & wings are Said to have been made at Birmingham, you will Oblige me not a little if you will Communicate to me any knowledge you may have on a subject which I consider as realy interesting, & must do so when I Estimate the interest the wing makers have in it, who we are told have expended £1000 in the project.[1]

I beg my Compts. to Dr. Priestly,[2] & Should be glad to hear that he has receivd a Letter from me, which, as it requires no answer, may be communicated to me in yours.

Beleive me, sir,
Yours Faithfully,

Jos: Banks.

We have a man here of Establishd Mechanic Knowledge, who declares himself posessed of a discovery to Construct Fire Engines[3] at a very reducd proportional Expence, & in sizes small enough to apply to many of the smaller mechanical Purposes. He Expects profit for its success, & wishes to deal with a good Partner.

[B.C.A., Matthew Boulton Papers, box 272, 8.]

[1]William Whitmore, a jobbing smith, made the rods for the wings. William Thomas Gill made the springs, and George Donisthorpe the wheels and pinions. Direction was provided by Patrick Miller (1731–1815), an inventor who later developed water wheels for steam navigation: Boulton to Banks, 11/12/1783, B.C.A. Matthew Boulton Papers, box 272, 9; Boulton to Banks, 11/12/1783, B.C.A. Matthew Boulton Papers, letter book 148, p. 10.
[2]Dr. Joseph Priestley FRS.

[3]Steam engines, which Boulton and Watt Ltd. produced, drove industrial development in England forward. Banks had a Boulton and Watt machine installed at a mine on his estate at Overton in Derbyshire to replace an older Newcomen engine. It started work in 1781, being used to pump water out of Gregory Mine.

Letter 20

1783

To Dr. Charles Blagden F.R.S.

Spring Grove
27 December [1783]

My dear Dr.,

I cannot consider it as a real matter of Complaint against me that I have put Gentlemen into the Council; nor do I wonder at all when I hear the names of any one of the Gentlemen who you tell me oppose me. They have every one of them had their papers repuls'd, & probably wish to print them in the R[oyal] S[ociety] Transactions.[1]

For the Charge of remaining in the Countrey, I never intend to Canvas for approbation of my Conduct. I stand or fall According to the Estimation in which it is held. I have meant well, & I beleive done well for the Society, & so many of my *Macaroni*[2] friends are convincd. We shall see how many think so.

To hold the Presidency under any tenure of Being the moderater in any Shape whatever of a Debating Club I shall always reject. The Statute explaining the business of the ordinary meetings is in my opinion sufficiently explicit. Under that I accepted, & by that only will hold my seat.

It is a matter of Glory to me to be President of the Royal Society, but I am too independent even for a Salary to accept the place of Chairman among debaters.

In Short, if I am supported I shall do well, & if I have deservd support I shall be supported. If I have not, I shall resign without a repentant thought, for as I have always done what I thought for the good of the Society, I shall allways continue the same Conduct if I remain; & upon that Conduct I must be judgd as I cannot honestly adopt any other.

Cool headed people will consider the real advantage of the Society. If it is judgd for the advantage of it that I remain, I shall. If not, I shall retire to my plough without a sigh.[3]

Beleive me,

Yours Faithfully,

Jos Banks.

[R.S. BLA. B. 98.]

[1]Banks had been unhappy with the Society's Foreign Secretary, Dr. Charles Hutton FRS (1737–1823), a Professor of Mathematics at the Royal Military Academy, Woolwich. Hutton was effectively deprived of his post late in 1783 when it was decided that it should be filled by a person resident in London alone. Another mathematician, the Reverend Samuel Horsley FRS (1733–1806), rallied some Fellows against Banks, and acrimonious disputes followed. There was a feeling that Hutton had been treated harshly, and that the mathematical sciences were ill-served by a naturalist President. Further, Banks was accused of judging candidates for fellowship more on their social standing than their academic qualifications. January 1784 saw a motion of confidence in Banks carried by 119 to 42. Subsequent votes and resignations reinforced Banks's position, and his opponents withdrew.

[2]An ironic reference to the way Banks and his supporters had been described by their opponents. 'Macaroni' was a disparaging term for men who had travelled, and acted in an affected or foppish way. It was also used to refer to amateur collectors of no scientific standing. Banks had in mind words used in: Blagden to Banks, 27/12/1783, F.M.C. Perceval Bequest H. 201, N.H.M. B.L. D.T.C. IV III 180–181.

[3]See also Letter 55.

Letter 21

1784

To Dr. Charles Blagden F.R.S.

[Soho Square]
[2 March 1784]

My dear Dr.,

I sincerely Condole with you on a Loss, which, howsoever it might be Expected in the Course of nature, must be felt severely by a son of the disposition which I have always found in you.[1]

I shall most certainly Call upon you tomorrow morning, & if possible see & talk to Ld Mulgrave[2] before I come,

Yours Sincerely,

Jos: Banks.

[R.S. BLA. B. 99.]

[1]Royal Society dissensions had occupied both Banks and Blagden when Blagden's mother died: Blagden to Banks, 1/3/1784, R.B.G. Kew Archives B.C. 1 157, N.H.M. B.L. D.T.C. IV 13. Blagden suggested that Mulgrave be consulted on Royal Society matters.
[2]Constantine John Phipps FRS, 2nd Baron Mulgrave.

Letter 22

1784

To Johann Heinrich Merck

Soho Square
29 June 1784

Sir,

I sat down with great pleasure to return you many thanks for your Obliging & Polite Present of your dissertations[1] in the Fossill Bones of your Part of Germany, which arrivd safe to my hands.

I was not so fortunate in your former present of the first Letter, never having receivd it, & having hitherto applied for it to my Bookseller in Germany in vain. Consequently, your present had to me a double value derivd from the excellence of its Contents, as well as the real inclination I felt to Profit by reading it.

Few if any subjects of natural history appear to me equaly interesting to us inhabitants of this planet with that you, sir, have chosen: the investigation of the Antiquities of Creation, whose traces, impressd on the most durable materials, seem left for the purpose of informing us that at some distant period of remote antiquity many */species of/* Animals were to be met with on its surface whose species are now extinct. What their power of Body or of mind may have been, we must not enquire after till a new life opens our Faculties for the reception of more Knowledge than in our present state of Existence we are capable of; but we may Guess that different Periods of the increase of human Knowledge, or rather of the development of reason, requird different Companions for him here. An Early inhabitant of this Globe, tho he had not Learnd from his Ancestors the use of the various */wiles/* Stratygems with which he overcomes beasts of Prey, or the Assistance he derives from other animals in overtaking the swift, or The arts of Cultivation by which he secures a succession of Food, might easily overtake an Enormous Ruminating Animal, &, having destroyd him, might give food to a Family for some time.

I wish the Custom of the Royal Society would Permit me to demand of them an opinion of the subject you have chosen, as well as of the mode in which you have executed it, as I am confident both would meet with the Eloge they so much merit; but an ancient resolution of that body, contrivd cheifly to demonstrate that independence on which an Englishman so particularly values himself, Forbids them to give an opinion as a Body on any subject. Permit me, however, to assure you from myself that I have receivd both Pleasure & instruction from your Labours, which I sincerely hope you will continue for the advancement of science.

I am, Sir, Your Obedient
& Obligd Servant,

Jos: Banks.

[F.M.C. Perceval Bequest H. 115.]

[1]Merck's last letter to Banks: Merck to Banks, 18/6/1784, B.L. Add. MS. 8095, ff. 309–310. Merck published a number of pieces between 1780 and 1789, including a letter in August 1782, another in May 1784, and a third in May 1786. Each explained his findings and theories. In date order they were: *Lettre a Monsieur de Cruse...sur les os fossiles d'eléphans et de rhinocéros qui se trouvent dans le pays de Hesse-Darmstadt*, Darmstadt 1782; *Seconde Lettre a Monsieur de Cruse...sur les os fossiles d'eléphans et de rhinocéros qui se trouvent en allemagne et particulierement dans le pays de Hesse-Darmstadt*, Darmstadt 1784; *Troisieme Lettre sur les os fossiles d'eléphans et de rhinocéros qui se trouvent en allemagne et particulierement dans le pays de Hesse-Darmstadt, Addresee a Monsieur Forster...*, Darmstadt 1786.

Letter 23

1784

To Johan Alströmer F.R.S.[1]

[Soho Square]
16 November 1784

I am ashamed at the long delay in answering your letter, which requested a personal account of Solander's life. I have hastily described the greater part of it, leaving out only those details we both know well.

It seems Solander came to England because John Ellis,[2] the Author of *Essay on Corallines*, and Peter Collinson[3] asked Linnaeus[4] to send an able young scholar from Uppsala who might promote the science of Natural History, then much neglected in England. Soon after his arrival, Solander became acquainted with Carteret Webb,[5] who held an important Judicial office, and whose advice the Government often sought. Carteret soon introduced Solander to the Lord Chancellor of England, Lord Northington.[6] He was charmed by Solander's open and friendly nature, and invited him to his Hampshire estate, even though Solander had been in the country but a week. It was not long before the Lord returned to London on business. However, he left Solander in the care of his wife and daughter, who were to teach Solander to speak English — by using that language and no other. Solander submitted happily to this, and was instructed daily by two of our most lively and beautiful ladies. No one was surprised at the rapid progress Solander made, and when he returned to London after six weeks he spoke English quite well. Indeed, he spoke it correctly, clearly and with a pleasant pronunciation throughout his life.[7]

The post of Assistant Librarian at the British Museum was chosen for him.[8] His salary did not amount to more than 60 pounds a year, but there was the promise of a promotion in the future. Moreover, his friend, Webb, obtained an annual income of 100 pounds from the Trustees of the Museum. Solander was instructed to start to catalogue the natural history collections of the British Museum on 26 February 1763.[9] He was also elected a member of the Royal Society in London. Solander worked diligently on the catalogue, and produced fine copies of each description. These became the property of the Museum, which possesses a large number. No doubt they will be published one day, and would certainly be much appreciated by the public.

In his spare time, Solander helped his friends, many of whom sought his assistance in all aspects of Natural History. His judgements in this science were soon held to be above appeal. He identified numerous plants in Peter Collinson's herbarium, aided his friend Ellis in his investigations of corallines, and helped his countryman

Gustaf Brander[10] to finish the *Fossilia Hantoniensia*. Solander even designed and built the first conservatory in England along with John Ellis, which was completed for Carteret Webb.[11]

At that time, or perhaps in 1764 when I was studying at Oxford University,[12] I met Solander for the first time. Our acquaintanceship became a firm friendship, the end of which caused me much grief. His loss is irreplaceable. Even were I to meet such a learned and noble man as he was, my old heart could no longer receive the impression which twenty years ago it took as effortlessly as wax, one which will not disolve until my heart does.

Solander became the Assistant Librarian of the British Museum on 19 July 1765. When I sailed to Newfoundland in 1766, his botanical instructions were of great use to me, and on my return we together catalogued those plants I had not seen before.[13] As we did, I noticed my understanding of botany grew considerably. At this time we met Lady Anne Monson,[14] after whom the genus *Monsonia* was named, Mr. Lee[15] and others, who all studied botany and entomology enthusiastically. They profited a great deal from Solander's teaching.

Commodore Wallis[16] returned from the South Pacific in 1768, and the Government decided that a ship should be sent to observe the Transit of Venus, and afterwards explore more widely. When I was told of this, I immediately applied to join. Once I had permission to go, I started to prepare for the voyage.

I informed Dr. Solander, who was very excited by my plans, and immediately offered to furnish me with information on every part of natural history which might be encountered on such an ambitious and unparalleled mission. Days later, when we were guests dining at Lady Monson's table, we discussed the excellent opportunity I now had to improve science and achieve fame. Solander suddenly became passionate, and sprang from his chair with an eager look asking, 'Would you like a companion?' To which I replied, 'Someone like you would be a constant benefit and pleasure to me!' 'Indeed so', said he, 'I want to go with you'. From that moment all was decided. The following day I asked Sir Edward Hawke, First Lord of the Admiralty,[17] for permission to take a fellow traveller with me. His Lordship refused me at first, being more a sailor than a Philosopher, but he was soon persuaded to agree.

Our voyage lasted three years, during which time I can say of Solander that he combined a placid temperament with diligence and such a sharp awareness that nothing was ever left incomplete. We never exchanged an angry word. However, we disagreed often on all manner of subjects, but always ended as we had begun, in good spirit. One of us usually accepted the other's reasoning once it had been properly discussed.

We were well supplied with books on the natural history of the Indies. Storms were seldom strong enough to interrupt our study time, which lasted each day from about 8 a.m. to 2 p.m., and after the smell of food had disappeared, from 4 or 5 p.m. until dark. We worked at the great table in the cabin with our draftsman opposite. We directed his drawing, and made rapid descriptions of our natural history specimens while they were still fresh. When we had been at sea a long time, and they were exhausted, we completed our descriptions, and added synonyms using our library.

These were entered directly into books in the form of flora for each country we had visited. Before we arrived home those of Madeira, Brazil, Tierra del Fuego, the Islands of the South Pacific, and New Zealand were in the presses. The descriptions of the little island of Saon,[18] and the interesting Island of St. Helena were also finished. The death of our secretary[19] prevented fair copies being made before we landed.

We both know how Solander lived in England. In the brightest part of the day he devoted himself to Botany, but his fondness for company never allowed him to spend the evening in the Museum. Indeed his friends would have been very upset if he ever tried such a thing. You and your countrymen know well how graciously he was received among the eminent people of England. Lords and Ladies chose Solander to explain the collections at the British Museum because of the taste with with which he described them. His remarks were so enlightening and engaging that the King himself spoke with Solander whenever the two met.

Solander spent one day in every week with the Duchess of Portland,[20] whose conchoid collection he arranged. His descriptions of the collection were given to the Duchess, and we expect shortly to see them published, exactly as Solander left them, by Mr. Lightfoot,[21] the author of *Flora Scotica* and the Duchess's chaplain.

The botanical work I have in progress is nearing an end. Solander's name will appear on the title page along with mine since everything was done jointly. Hardly a sentence was written while he lived to which he did not contribute. As all the descriptions were made when the plants were still fresh, little remains but to refine those Drawings which are not quite complete, and to add synonyms from books which we did not take with us to sea, or which have since then been published. In a few months such minor additions should be made, if only the engravers can put the final touches to it.

You have already heard of the circumstances of Solander's death.[22] I was away from London on a visit when a messenger came with news of his perilous condition. I travelled all night and arrived in the morning, but all hope was gone. He had been well on the morning of his illness, and was among friends when he had a stroke. No doctor of note in London failed to attend him with an opinion, but all agreed that there was definitely no hope. Two years before he died Solander had a copious nosebleed, which seems to have been a critical discharge, because when his head was opened, a haemorrhage in the brain was found to be the cause of death.

Thus, to my lasting grief, my friend died on 12 March 1782.[23] He was buried on the 19th. Every Swede in London was present, and although many Englishmen wished to come, I was the only one allowed to accompany Solander to his grave. There were more than enough Swedes to do him this last honour, and the English who were not present wept in their hearts no less.

This too early loss of a friend I loved during my mature years, and who I shall always miss, forces me to draw a veil over his passing immediately I cease to speak of it. I can never think of it without feeling such acute pain as makes man shudder. If honour, justice, moderation, kindness, diligence, if ever such natural gifts and accomplishments deserve a place in a better world, nothing but my own failings will prevent us from meeting again.

[Johan Alströmer added a comment to this letter.]

Who could express more clearly the memory of a mourned friend than Banks has Solander? The same spirit of public service, and the same zeal for science makes one man recognizable in the other. Therefore, I have published the important aspects of this most tender friendship. It will serve as a fitting memorial to Solander.

[The original has not been identified. This letter appeared as a translation in Swedish in *Upfostrings Salskapets Tidningar*, No. 14, pp. 105–110, Stockholm d. 21 February 1785. It was then translated into German in *Berlinische Monatschrift*, September 1785. It appeared in English as a translation in an article by Roy Rauschenberg, *Isis*, vol. 55 (1964), pp. 62–67. Another English version was recently published along with the Swedish one of 1785: E. Duyker and P. Tingbrand (Eds.), *Daniel Solander Collected Correspondence 1753–1782*, Stockholm (1995), pp. 410–419. The current edited translation was made by N. Chambers, and Walter and Karin Eyles from the Swedish.]

[1]After Solander's death in 1782, Alströmer, who was President of the Royal Swedish Academy of Sciences, asked Banks to provide an account of Solander's life and work in England. Banks's reply, dated 16 November 1784, was published in *Upfostrings-Sälskapets Tidningar*, 21 February 1785. This was then translated into German and published in the *Berlinische Monatschrift*, September 1785.

[2]John Ellis FRS (c.1710–1776): naturalist; King's Agent for West Florida, 1764, and for Dominica, 1770. His works include *An Essay Towards a Natural History of the Corallines Commonly Found on the Coasts of Great Britain and Ireland*, London (1755). In 1768 he received the Copley medal of the the Royal Society for his research in this area. After Ellis's death Banks arranged for publication of *The Natural History of many uncommon Zoophytes collected by John Ellis, arranged and described by D.C. Solander*, London (1786), Martha Watt (Ed.). This was dedicated to Banks by the editor, Martha Watt (1755–1795), who was Ellis's daughter: Watt to Banks, 1/9/1782, B.L. Add. MS. 33977, f. 173; Watt to Banks, undated, B.L. Add. MS. 33982, f. 243.

[3]Peter Collinson FRS (1694–1768): naturalist; antiquary; merchant. He maintained strong trading connections with America, and established a botanic garden at Mill Hill.

[4]Carl von Linné (Carolus Linnaeus).

[5]Philip Carteret Webb FRS (1700–1770): antiquary; MP for Haslemere, 1754–1761 and 1761–1768; joint-solicitor of the Treasury, 1756–1765.

[6]Robert Henley (c.1708–1772), 1st Earl of Northington: lawyer; MP for Bath, 1747–1757; Lord Keeper of the Great Seal, 1757–1761; afterwards Lord Chancellor. He married Jane Hubard (d.1787) in 1743, and they had eight children. Three were sons, and five were daughters.

[7]There is general agreement that this was not how Solander learned English. Solander's early letters make no mention of it. Indeed, according to Ellis, Solander did not meet Lord

Northington until 1764: 'I have lately introduced him [Solander] to our Lord Chancellor...', Ellis to Linnaeus, 1/1/1765, J.E. Smith (Ed.), *A selection of the correspondence of Linnaeus and other naturalists, from the original manuscripts*, London (1821), vol. I , pp. 162–164.
[8]Solander was employed in March 1763 to catalogue the natural history collections at the British Museum. In 1768 Solander gained special permission from the Trustees to join Banks on his scientific voyages to the South Pacific and Iceland. He returned to work in the Museum in 1773.
[9]Solander's understanding of the Linnean system enabled him to bring scientific order to the natural history collections at The British Museum. He was appointed in 1760 with this in view, and started a detailed 'slip catalogue', which was never completed. His notes were later amalgamated, possibly with some earlier museum records, to form the Solander Slip Catalogue. This was bound into volumes in the nineteenth century, of which there are 27 in the Zoology Library, and a further 24 in the Botany Library, The Natural History Museum, London.
[10]Gustavus Brander FRS (1720–1787): Swedish merchant, who donated his collection of fossils to the British Museum. Solander described them in *Fossilia Hantoniensia collecta et in Musaeo Britannico deposita a Gustavo Brander*, London (1766).
[11]Ellis appears to have been mainly responsible for this design, and Solander assisted him.
[12]Banks was at Christ Church, Oxford, 1760–1765.
[13]Banks sailed to Newfoundland and Labrador in HMS *Niger* under Captain Sir Thomas Adams RN (1738–1770) in 1766. This was his first important voyage of exploration.
[14]Lady Anne Monson (c.1714–1776): amateur botanist. She collected at the Cape of Good Hope and in Bengal.
[15]James Lee (1715–1795): nurseryman at The Vineyard, Hammersmith. He translated parts of Linnaeus's work into English to form *An Introduction to Botany...Extracted From the Works of Dr. Linnaeus...*, London (1760). Lady Monson assisted Lee with this work. Both were commemorated by Linnaeus in the names of plants.
[16]Samuel Wallis.
[17]Sir Edward Hawke. See Letter 1.
[18]'Saon' may be Sawu or Savu, which is an island situated between Timor and Sumba.
[19]Herman Dietrich Spöring died on 23 January 1771 of tertian malaria and dysentery.
[20]Margaret Bentinck Cavendish (c.1714–1785), Dowager Duchess of Portland: collector, especially of shells, who took a keen interest in natural history.
[21]Reverend John Lightfoot FRS (1735–1788): botanist; librarian and chaplain to the Duchess of Portland. Lightfoot arranged and catalogued the Duchess's collections. He received help from, among others, Banks and Solander with *Flora Scotica...*, London (1777).
[22]Banks was not present when Solander became unwell during breakfast at Soho Square on 8 May. Solander had been recounting to Charles Blagden the story of the blizzard on Tierra del Fuego in 1769 when two of Banks's servants died of exposure. Blagden wrote to Banks immediately Solander became seriously ill: Blagden to Banks, 8/5/1782, F.M.C. Perceval Bequest H. 160, N.H.M. B.L. D.T.C. II 126. Solander was treated by a number of physicians, including Blagden, Dr. John Hunter, Dr. William Heberden and Dr. William Pitcairn FRS (1711–1791). However, he died at 9.30 pm on 13 May. Dr. John Hunter performed an autopsy on Solander's body the next day, and discovered evidence of a haemorrhage in the brain.
[23]This is not when Solander died.

viii. **Dr. Daniel Solander F.R.S.**
Oil on canvas, c.1776, by John [Johann] Zoffany. The Linnean Society, London.

Letter 24

1787

To an Unknown Correspondent[1]

[Soho Square]
[c. February 1787]

As the sole object of Government[2] in Chartering this Vessel[3] in our Service at a very considerable expence is to furnish the West Indian Islands with the Bread-Fruit & other valuable productions of the East, the Master & Crew of her must not think it a grievance to give up the best part of her accommodations for that purpose. The difficulty of carrying plants by sea is very great. A small sprinkling of Salt-water, or of the Salt-dew which fills the air even in a moderate gale, will inevitably destroy them if not immediately washed off with fresh water.

It is necessary, therefore, that the Cabin be appropriated to the sole purpose of making a kind of Greenhouse, & the key of it given to the Custody of the Gardiner; and that, in case of Cold weather in going round the Cape, a Stove be provided by which it may be kept in a temperature equal to that of the intertropical Countries.

The fittest vessels for containing the plants, that can easily be obtained, I conceive to be Casks,[4] saw'd down to a proper height, & properly pierced in their bottoms to let the water have a passage. In both which articles the Gardiner's directions must be followed. Of such half tubs properly secured to the floor as near to each other as they can stand, a considerable number may find room in the Cabin, each of which will hold several Plants; & these I consider as a stock which cannot be damaged or destroyed, but by some extraordinary misfortune. /As these Tubs, which will be very heavy, must be frequently brought upon deck for the benefit of the sun, the Crew must assist in moving them, as indeed they must assist the gardiner on all occassions in which he stands in need of their help./[5]

Besides these, must be provided Tubs so deep that the tops of the plants placed in them will not reach to their edges. These must be lashed all round the Quarter deck, along the Boom, & in every place where room can possibly be found for them; & for each a cover of Canvass must be made to fit it, which covers it will be the duty of the Gardiner to put on and take off as he judges fitting. No one else must interfere with him in so doing on any account whatever.

As the plants will frequently want to be washed from the Salt dampness which the Sea air will deposit upon them, besides allowance of water, a considerable provision must be made for that purpose; but, as the Vessel will have no Cargo whatever but the plants on board, there will be abundant room for water Casks, of which she must be supplied with as large a quantity as possible that the Gardiner may never be refused the quantity of water he may have occassion to demand.

No Dogs, Cats, Monkies, Parrots, Goats, or, indeed, any animals whatever must be allowed on board, except Hogs & Fowls for the Company's use; & they must be carefully confined to their Coops.

Every precaution must be taken to prevent or destroy the Rats as often as convenient.

A Boat with Green boughs should be laid along side with a gangway of Green boughs laid from the hold to her, & a Drum kept going below in the Vessel for one or more nights; &, as poison will constantly be used to destroy them & [the] Cockroaches, the Crew must not complain if some of them who may die in the ceiling make an unpleasant smell.

As it is likely from analogy that the Easterly winds will prevail to the South of the line from the month of March to that of September, it [is] to be hoped that the Vessels will be fitted out with as much dispatch as is convenient with a view of her not losing a year, which will be the case if she loses the first monsoon.

Her first destination will be New Zealand, where she is to take on board 2 tubs of Flax plants.[6]

From thence she is to proceed to the Society Isles, where she must stay till the Gardiner has procured a full stock of Bread-Fruit trees; & if Otaheite, which will be probably visited first, should not supply a sufficient number of such as are of a proper age for transplanting, she must proceed to Imao, Maitea, Hudheime, Ulietea & Bolabole, and stay till enough are procured.[7]

She is next to proceed towards the Endeavours Streights,[8] which seperate New Holland from New Guinea; &, if she wants water in her passage, she may put into the Friendly Isles[9] in making the Streights which lie in Lat 10. 40. The Master must not be surprized if he falls in with a reef[10] as he may be assured that with a little attention he may explore a passage through it. In these Streights he must find some harbour in which he may fill water, which, as they are full of shelter, there cannot be any difficulty in performing.

From hence to Prince's Island[11] in the Streights of Sunda[12] will be the next run; &, if water should be wanted in the passage, it may be procured at Java where the Endeavour watered. At Prince's Island the Gardiner will have some Trees to get on board, which may make it necessary to spend some time there.

From thence to the Isle de France[13] will be an easy run. From thence round the Cape, at which place the Ship must not touch unless there is absolute necessity, they must proceed to St. Helena,[14] where she will receive orders from England pointing out the places in the West Indies at which she is to touch & deliver her Cargo.

[N.H.M. B.L. D.T.C. V. 210–216; M.L./D.L. Banks Papers Series 45.03.]

[1]The date and recipient of this statement are both unclear. In the Dawson Turner Copy, volume V, it follows a letter from Thomas Townshend (1733–1800), 1st Viscount Sydney, to Banks, 15/8/1787, pp. 208–209. Townshend requested instructions from Banks for a

voyage by HMS *Bounty* in that letter. However, the document above could not have been Banks's reply since it referred to an earlier plan.

[2]The vessel was for a mission to transport the bread-fruit plant from the East to the West Indies, a mission officially approved by William Pitt the younger on 13 February.

William Dampier (1652–1715) was one of the earliest European explorers to discover the value of the bread-fruit as a food source. Both Banks and James Cook were impressed by it when they visited Tahiti in 1769 as well. Soon after the return of HMS *Endeavour* in 1771, Valentine Morris (1727–1789), who owned plantations on St. Vincent and became governor there, wrote to Banks to suggest transplanting the bread-fruit from the Pacific to the West Indies: Morris to Banks, 17/4/1772, B.L. Add. MS. 33977, f. 18. In 1775 John Ellis produced an excellent account of the bread-fruit, noting a number of other useful plants which might be grown in the West Indies: *A Description of the Mangostan and the Bread-fruit*..., London. The idea was still being discussed in August 1786 when a Jamaican planter called Hinton East (d.1792) urged it on a visit to Banks at Spring Grove.

It has been suggested that Banks intended the statement above for Arthur Phillip (1737–1814) as the first Governor of New South Wales. Phillip was to detail a ship from the fleet at Botany Bay to make the voyage, but this was not attempted. Banks noted accordingly: 'Instructions For the Vessell From Botany Bay the Scheme was Changd when on the Eve of Execution', M.L./D.L. Banks Papers Series 45.03. HMS *Bounty* was sent from Britain instead.

[3]A note from the margin reads: 'Charts must be provided'.

[4]A note from the margin reads: 'This was altered; & pots substituted by the Special advice of the Gardiners; but I doubt whether it was not the best'.

[5]A note from the margin, which can easily be inserted.

[6]New Zealand flax, also New Zealand hemp and New Zealand cloth plant, (*Phormium tenax*). Banks had been excited by this plant on the *Endeavour* voyage. For instance: *Endeavour Journal*, vol. II, pp. 10–11. It could be used to manufacture canvas and cable.

[7]These islands are all part of the Society Islands in the South Pacific. The spelling of their names tends to vary. In sequence they are: 'Imao', Moorea; 'Maitea', Matiea or Mehetia; 'Hudheime', Huahine; 'Ulietea' was possibly Raiatea. The island's older name was Ioretea; 'Bolabole', Bolabola or Borabora.

[8]Endeavour Strait, the strait immediately north of Cape York Peninsula, Australia.

[9]Friendly Islands, which are the Tonga Islands.

[10]The Great Barrier Reef.

[11]Prince's Island, the Malay Archipelago, at the west entrance to the Sunda Strait.

[12]The Sunda Strait lies between Sumatra and Java.

[13]Ile de France, which is Mauritius, Indian Ocean.

[14]St. Helena, Atlantic Ocean.

Letter 25

1787

[*To Charles Jenkinson, 1st Baron Hawkesbury, and 1st Earl of Liverpool*]

[Soho Square]
30 March 1787

My Lord,

It is fully my opinion that the plan of sending out a Vessel from England for the sole purpose of bringing the Bread Fruit to the West Indian Islands is more likely to be successful than that of dispatching one of the transports from Botany Bay,[1] & I am inclined to believe it will be at least as economical. In the Botany Bay scheme, besides the delay of re-equipment, the vessel will have 480 degrees of Longitude to sail through, while that sent on purpose [from England] will have only 420, which two differences taken together I cannot estimate at less than 4 or possibly five months demurrage.

The business of fitting her for so unusual a purpose as that of carrying a cargo of Plants may also certainly be done better & cheaper here than abroad, where a large number of valuable casks must have been sacrificed; & people may also readily be found [here who are] more capable of conducting her through a voyage in which difficulties are likely to occur than the master[2] & crew of a Merchant Vessel would probably have been able to do.

For such a service, a Brig of less than 200 Tons Burthen would be fully sufficient, & she may be navigated by:

1 Master	1 Sergeant
1 Mate	6 Marines
20 Seamen	

In all, 29 men of her crew, which, with the Gardiner,[3] will amount to no more than 30 Souls. [There should be a] 64 pounder, & close Quarters.[4] Her track should be round Cape Horn to the[5] Society Isles, through Endeavours Streights[6] to Princes Island.[7] /At Princes Island such Bread Fruit trees as have died may be replacd by Mangosteens, Jacks, Durions[8] &c. &c., & dry rice may also be procured./ Thence round the Cape of Good Hope to the West Indies, where she may deposit one half of her Cargo at his Majesties Garden at St. Vincents for the benefit of the Windward Isles, & carry the remaining half down to Jamaica.

As from analogy it appears that if the Monsoons blow in the Sea between New Holland & New Guinea, the Easterly one will commence in March or April, it will be proper that She should sail from hence in the month of July.

The name of the person intended to take charge of the Plants is *David Nelson*,[9] & the Terms I propose for him are: £25 as an outfit to purchase clothes & necessaries; a Salary of £50 a year, with his mess on board; and, as he stands engaged to the present undertaking on those terms, & has left his place to accept them, I hope I shall not be thought unreasonable in proposing that his pay shall be continued from the time he engage[d], & that he be allowed board wages till he is shipped at the rate of ten shillings a week.

He sailed with Captain Cook on his third voyage round the world in my service for the purpose of collecting plants & seeds, & was eminently successful in the object of his mission. He had been regularly educated as a Gardiner, & learned there the art of taking care of Plants at Sea, and guarding against the many accidents to which they are liable, which Few people but himself have had any opportunity to know practically. He learned also how to conduct himself on board a ship, & made an acquaintance with inhabitants of the South Sea Islands & their Language, which will in all probability facilitate his obtaining the number of Plants wanted, a matter in which, as the Indians have never been accustomed to sell them, & a large number will be wanted, difficulties may arise.

[N.H.M. B.L. D.T.C. V. 143–146; R.S. M.M. 6. 60. In addition, this letter was sent to Constantine John Phipps F.R.S., 2nd Baron Mulgrave, P.R.O. HO. 42/11]

[1]A plan to send a ship from the fleet at Botany Bay to transport the bread-fruit plant from the East to the West Indies had previously been considered, but then rejected: see Letter 24. A vessel called HMS *Bounty* was dispatched from Britain instead. It was commanded by William Bligh. Bligh entered the English Channel on 23 December 1787 after some delay, and proceeded to Tenerife for supplies. Following courageous attempts to round a stormy Cape Horn, Bligh turned towards the Cape of Good Hope on 17 April. Repairs to his battered ship were made at False Bay at the Cape in May and June of 1788. A course was then set for Adventure Bay, Tasmania, where Bligh anchored on 21 August. New Zealand was passed on its southern side before HMS *Bounty* sailed northwards into the Pacific. Matavai Bay, Tahiti, was reached on the 26 October. The stay at Tahiti lasted for 23 weeks while the task of acquiring the bread-fruit was undertaken (1015 were collected).

Afterwards there were some brief stops, and on 11 April 1789 Aitutaki was discovered. Annamooka was the last island to be visited on 23 April. On 28 April a number of the crew mutinied under the leadership of Fletcher Christian (1764–1794?). They set Bligh and 18 others adrift in a 23-foot-long open boat. Bligh made his great voyage to Timor in this vessel, sailing some 3,700 nautical miles on the way. The bread-fruit were destroyed, with William Brown, the assistant gardener, joining the mutineers. Thus, HMS Bounty never reached the West Indies. It was stripped and burned off Pitcairn Island in 1790.

[2]William Bligh commanded this mission. He also led the subsequent bread-fruit voyage, made in HM ships *Providence* and *Assistant*, 1791–1793.

[3]Banks chose David Nelson (d.1789) and William Brown as gardener and assistant gardener for the voyage.

[4]In the event, 46 men sailed on HMS *Bounty*. She had been purchased as the *Bethia* late in May 1787, and listed as an armed vessel. She carried 4 four-pounders and 10 swivel guns, but weighed only 215 tons.

[5]The text makes better sense with this marginal insertion as a footnote: 'In the Society Isles the Bread Fruit is in the highest perfection & abundance.'

[6]Endeavour Strait, Australia.

[7]Prince's Island, the Malay Archipelago.

[8]Mangosteen (*Garcinia mangostana*); 'Jacks', jackfruit (*Artocarpus heterophyllus*); 'Durions', durian (*Durio zibethinus*), which smells foul, but has delicious creamy arils.

[9]David Nelson proved himself an able collector and a reliable crew member on HMS *Discovery* on James Cook's third voyage, 1776–1780. Banks employed Nelson as a collector, 1776–1780. This faithful collector died at Kupang on the journey with Bligh back to England.

Letter 26

1787

To Sir George Yonge, F.R.S., Secretary for War

Soho Square
15 May 1787

Sir George,

It gives me the greatest satisfaction to find, by the dispatches you have done me the honor to communicate to me, that your patriotic wishes relative to the establishment of a Botanic Garden at Calcutta have been anticipated by the Governor & Council there; & that your designs of settling a correspondence between that & his Majesty's Garden at St. Vincents is also received with the warmth which a plan of such general benevolence deserves.[1] Ceres[2] was deified for introducing wheat among a barbarous people. Surely, then, the natives of the two great Continents, who, in the prosecution of this excellent work, will mutually receive from each other numerous products of the earth as valuable as wheat, will look up with veneration to the monarch who protected, & the minister who carried into execution, a plan the benefits of which are above appreciation to the present generation, & will extend their benificent influence to the latest posterity of those who receive them.

Col[onel] Kyd, on whose intelligence & activity the whole execution of the plan seems at present to rest, has made some just remarks on the botanical dispatch with which Ld. Cornwallis[3] was furnished at the time of his Lordship's sailing from hence. In truth, we here are little acquainted with the produce of Bengal. The Coast of Malabar[4] has been well investigated by the Dutch, who have published in the Hortus Malabaricus[5] a full account of its vegetable productions; & the labors of the late Dr. Koenig,[6] who, had he been a little more encouraged, would have done much more than he has for science & public utility, have made me, to whom he bequeathed his papers, tolerably acquainted with that of Coromandel.[7] But Bengal & its dependencies remain a vast blank in the book of information, which no one hitherto has attempted to fill up. For this reason no one has materials sufficient to enable him to form a just demand. Much praise indeed on the whole is due, for, by affording to the Gentlemen in Bengal an opportunity of displaying readiness & activity in executing his Majesties Commands, which has done them infinite credit, it will infallibly inspire a similar alacrity on this side of the water, & by these means give his Majesty the inexpressible pleasure of knowing that his paternal & benevolent intentions toward his Asiatic & American subjects will be speedily carried into effectual execution.

We expect with impatience the packages of plants which Col. Kyd's Letter announces. His mode of packing young trees is quite unknown to me, & that of sending seeds certainly the best we are acquainted with, notwithstanding the many experiments which have been made in hopes to improve it.

Our politic neighbours, the French, have preceded us several years in the execution of similar projects; & from the result of their experiments we learn with certainty that his Majesty's Ideas will, if carried into full execution, be attended with the most ample success. Mr. Paire,[8] Intendant of L'Isle de France, sent two expeditions in the years [blank] & [blank], each consisiting of two Vessels in search of the nobler species, & obtained by each a number of Plants of nutmegs & cloves. Had not dissentions in the government of L'Isle de France delayed considerably the increase of these Cargoes, they would long ago have staid with the Dutch the spice trade. At present we learn that they have so large a quantity of both Clove & nutmeg trees actually planted out that, from the state of them & their period of growth, with which M'Céré[9] the director of their Garden is perfectly acquainted, they have no doubt of being able before the end of the present Century to supply as large a quantity of these spices as the whole consumption of Europe will demand. Nor have they stoped here. They have now nutmegs & cloves flourishing in their West India Islands, &, as a prelude to the commencement of their commerce, they have this year published an analysis of their Cloves, by which it appears that they are fully equal to those of the Dutch in every quality applicable either to the purpose of food, medecine or perfume.

With such an example before us, it is clearly demonstrable that nothing but activity is wanting to carry his M[ajesty']s Commands into execution on our part, and that a large portion of that virtue will be necessary to enable us to retrieve the advantage which our active neighbors have obtained over us in point of time. To recommend activity to Col. Kyd would be superarrogation; nor need any Spur be given to Mr. Anderson,[10] the industrious and active superintendant of his Majesty's Garden at St. Vincent's. We may all remember, however, in our united efforts to serve this interesting cause, which is so eminently honored with his Majesty's protection, that we have nobler prospects in view than the mere attempt of Filching from another country its commercial advantages.

To exchange between the East & West Indies the productions of nature usefull for the support of mankind that are at present confined to one or the other of them, to increase by adding this variety the real Quantity of the produce of both Countrys, & by that means their population, furnishing at the same time to the inhabitants new resources against the dreadful effects of Hurricanes & droughts, to one or the other of which all intertropical countries are subject, are the more immediate objects of his Majesty's present intentions, the disinterested humanity of which seems alone sufficient to inspire diligence & activity into the minds of all those who may be fortunate enough to be allowed a share in the honor necessarily consequent in having carried ideas of such exalted benevolence into Execution.

Future Correspondence will by degrees enable both partys to know with accuracy the things each wants & the other possess; and there is great reason to hope that Government will by the next Ships send out an able botanist to Bengal who has resided Several years in the West Indies, who will at once point out the things proper to be sent by each to the other. Little, therefore, need be demanded at present, & the more so as the Cargo Col. Kyd has already dispatched is to be expected by the first ships. In hopes, therefore, of being speedily relieved from the pressure of our present ignorance of the Produce of Bengal, I shall hazard only the demand of the few following articles.

Mangifera Indica[11]
Mango

As those which have been imported into the West Indies are of a very inferior sort, the best kinds would be a most acceptable addition. At Goa[12] it has been the ancient custom, & it has been practised with success at madrass, to perform upon the trees the operation of circumcision, which is done by passing a branch through a box or basket filled with rich soil, & cutting off the bark quite round the part which rests in the middle of the box, which must be firmly fixed in its place. This operation should commence with the rainy season, & we are told that, before the end of it, some of the branches, for it does not constantly succeed, will have formed roots sufficient to maintain themselves. These branches being sawed off become Trees which, we are certain, will produce fruit of exactly the same flavor & quality as those produced by the mother tree.

Averrhoa Carambola
Caumrunga

There are said to be three species of Averrhoa in Bengal, all of which produce a very acid fruit, usefull for many purposes. No species of that genus is known in the West Indies.

Eugenia Malaccensis
Jamboo Malacca

I am not certain of the name by which this fruit is known in Bengal. It is similar in general appearance to the Rose apple, but larger, higher coloured & without the smell of rose water; and a pleasant, cooling & wholsome fruit.

Annona squamosa
Atta: Shereefae

A kind of this fruit is said to grow in some gardens near Calcutta, & particularly in that belonging to Col Kyd. [It is] much superior both in size & flavor to any that are produced in the West Indies.

Sitodium Cauliflorum
Jack

A most valuable fruit, affording abundant nourishment in a variety of preparations contrived to suit its different degrees of ripeness, in which respect it approaches, & possibly equals, the bread fruit of the South Sea Islands. A plant or two are said to have been taken on board a French ship in the last war, & to be growing in Jamaica; but whether of a good sort we do not know as they are not yet old enough to bear Fruit.

Licheè

A fruit which the late Governor General, Mr. Hastings,[13] is said to have imported from China, and found to be well suited to the Climate of Bengal. It is both pleasant and wholesome, and consequently would be highly acceptable in the West Indies.

Baúnsmúttee Rice

Is described as a sort much superior in excellence to all the kinds cultivated in the Western World.

Naugurbussee Bamboo

Is said also to be a most excellent sort. There are in Jamaica Bamboos, which were procured from the Spaniards on the main, who brought them originally from Menilha;[14] but as they are seldom used for economical purposes, it is probable they are of a bad kind.

Moth seed — Calla Jeera

The small black seed with which shauls are allways packed. If it is found to possess the virtue of driving away the insects that destroy Furs in Europe, as it does those that attack woollen goods in India. Its value will be great in Russia, & other countries where high prized Furs are used. If a quantity is sent over fresh for the purpose, the experiment may be fairly tried.

Chĕh Root

With which, as we are told, Cottons are died red. A quantity of it, sufficient to try experiments here, would be very acceptable as we are ignorant what it is.

Cajir Gautch

A species of Palm Tree found in abundance on the delta near the mouths of the Ganges, which furnishes abundance of Palm-wine, applicable either to immediate use as a pleasant & wholesome beverage, to the distillation of arrack, in which it is an ingredient, or to the making of a course kind of sugar.

These are all the useful plants which the present state of our information relative to the productions of Bengal allows me to ascertain as known there, & not hitherto introduced into the West Indies. I have no doubt that the Western world will be able to repay the eastern with interest for these, & whatever else a farther pursuit of this interesting subject may make us acquainted with. The following Fruits &c. occur at once as not likely to be known in the East, tho' abundant in the West Indies.

1 C/h/rysophyllum, star apple.
1 Passiflora, water lemon.
1 Spondias.
1 Melicocca.
1 Achras.
2 Winterani/*e*/a c/*o*/anell/*e*/a alb/*e*/a.
1 Cactus, Prickly Pear.
2 Myrtus, Pim/*o*/ento.
1 Memmea.
1 A/*rnene*/nnona.
1 Hibiscus.

Besides which, exchanges may be made to the advantage of both of the different sorts of the following Plants, of which it is probable that each country possesses excellent & useful varieties with which the other is unacquainted.

Roots.

Convulvulus sweet Potatoes.
Arum Cocos.
Diosc/*e*/orea, Yams.

Grains.

Oryza, Rice.
Panicum, Millet.
Zee Indian Wheat.
Holcus, Guinea Corn.

Fruits.

Bromelia — /*Rice*/ Pine apple.
Musa, — Plantain & Bannana.
Citrus, — Orange, Lemon, Shaddock, Lime, Forbidden Fruit.
Cucumis, — Cucumbers.
Cucurbita, — Melons, Pumpkins, Squashes, &c.
Carica — Papaws.

These, Sir, are the few observations that occur to me at present, and, as I hope to have future opportunities of manifesting my respect for your character & situation as well as the uninterrupted zeal with which you have promoted this excellent undertaking, I hope they will be found sufficient. I have therefore only to add, that I shall ever think myself happy when I have the honor to be called upon to throw in my small assistance towards perfecting a plan so highly fraught with disinterested benevolence as this in which you, Sir, have done me the favor to engage me.

[N.H.M. B.L. D.T.C. V. 159–166; P.R.O. 30/11/12; R.A.S.]

[1]Lieutenant Colonel Robert Kyd (d.1793) had eloquently proposed the idea for a botanic garden at Calcutta to the acting Governor-General, Sir John Macpherson (1745–1821): Kyd to Macpherson, 15/4/1786, N.H.M. B.L. D.T.C. V 29–32 and VII 35, R.A.S., S.L. Banks MS. BG 2:1. In Kyd's opinion such an establishment might serve a wide range of horticultural purposes, especially the introduction of plants to be cultivated as food sources in India and beyond.

This view was warmly endorsed when considered by the relevant authorities in London, including Banks. Following his letter above to Yonge, the Secretary of State for War, Banks wrote a similar appraisal to Henry Dundas at the Board of Control: Banks to Dundas, 15/6/1787, N.H.M. B.L. D.T.C. V 184–191, Windsor R.A. 6262 to 6264, U.Y.S.L. Banks MS., R.A.S. Kyd became superintendent of the intended garden in May 1787, by which time the site of an old Moghul fort called Muggah Tanna on the western bank of the Hooghly had been chosen and partially cleared. The Dawson Turner volume VII, Botany Library, The Natural History Museum, London, contains Kyd correspondence for 1787 on this subject.

[2]Ceres was an Italian grain-goddess, also known as Demeter to the Greeks. She was the Earth Mother or patroness of fertility, and was worshipped alone or with the god Cerus.

[3]Charles Cornwallis (1738–1805), 1st Marquis and 2nd Earl Cornwallis: Governor-General and Commander-in-Chief of India, 1786–1793.

[4]Malabar, the maritime districts in the south-west of India.

[5]Hendrik Adriaan van Rheede tot Draakestein, *Hortus Indicus Malabaricus*..., Amstelodami (1678–1703).

[6]Dr. Johan Gerhard Koenig (1728–1785): botanist; traveller. For the last five years of his life Koenig was a botanist in the Honourable East India Company. He worked from Madras with Dr. William Roxburgh (1751–1815). Koenig's collections and manuscripts were bequeathed to Banks when Koenig died of dysentery in June 1785.

[7]Coromandel, the east coast of India where Madras is situated.

[8]This name is spelled differently in other versions of the letter above. The form 'Mr. Le Poivre' is also used. Banks apparently meant: Pierre Poivre, author of such works as *Voyages d'un philosophe, ou observations sur les moeurs et les arts des peuples de l'Afrique, de l'Asie et de l'Amerique*, Yverdon (1768). The English translation of this account appeared in 1769.

[9]Jean Céré (d.1808): Until his death Director of the Botanic Garden, Mauritius. Succeded in that position by his son, Auguste Céré (fl.1808–1820).

[10]Dr. Alexander Anderson (d.1811): botanist; superintendent of the Botanic Garden, St. Vincent. In Guiana, 1791.

[11]Not all of these plants require explanation. However, those which do are: starfruit (*Averrhoa carambola*); 'Eugenia Malaccensis' (*Syzygium malaccense*). This is the Malay apple, which was a Polynesian introduction to the Pacific from the Malay Archipelago; 'Annona squamosa', sweetsop (*Annona squamosa*); 'Sitodium Cauliflorum' is the jack (*Artocarpus heterophyllus*); 'Licheé' is the litchi (*Litchi chinensis*); 'Baúnsmúttee Rice' is *baus-moti*, one of the most favoured 'scented' rices; 'Naugurbussee Bamboo' is unclear; 'Moth seed' is kalanji (*Nigella sativa*); 'Chéh Root' is the chay or Indian madder (*Oldenlandia umbellata*); 'Cajir Gautch' (*Phoenix sylvestris*) is a kind of wild date-palm, which is an important source of toddy and palm sugar; 'Chrysophyllum' is the star apple (*Chrysophyllum cainito*); 'Passiflora' is a kind of passion-fruit (*Passiflora*); 'Spondias' are hog plums (*Spondias* spp.); 'Melicocca' is the mamoncoillo, which is a fruit tree (*Melicoccus bijugatus*); 'Achras' is the sapodilla (*Manilkara zapota*); 'Winterania canella alba' is wild cinnamon (*Canella winterana*); 'Cactus' (*Opuntia* spp.); 'Myrtus, Pimento' is allspice (*Pimenta dioica*); 'Memmea' is the mammee

apple (*Mammea americana*); 'Convolvulus' is the sweet potato (*Ipomoea batatas*); 'Arum — Cocos' is cocoyams, *Colocasia esculenta* (*Araceae*); 'Dioscorea' is the yam (*Dioscorea* spp.); 'Oryza' is rice (*Oryza sativa*); 'Panicum' is millet (probably *Panicum miliaceum*); 'Zee' is maize or Indian corn (*Zea mays*); 'Holcus' is Guinea corn (*Sorghum* spp.); 'Bromelia' is the pineapple (*Ananas comosus*). Banks seems to have drawn partly on his experience on the *Endeavour* expedition for these plants. For instance, he recorded some of them in his journal: *Endeavour Journal*, vol. II, pp. 206–217.

[12]Goa, Daman and Diu, union territory in West India. It is bounded by the Arabian Sea in the west, and landward by Gujarat state.

[13]Warren Hastings (1732–1818): Governor-General of India, 1772–1785; impeached on grounds of corruption and cruelty in his Indian administration, begun 1788, and concluded 1795.

[14]Manila, Manila Bay, Luzon Island.

Letter 27

1787

To Sir William Hamilton F.R.S., Ambassador at the Court of Naples

Soho Square
24 July 1787

My dear Sir,

Last week Chas. Greville[1] took Charge of Seven Copies of the Priapëia,[2] one of which is your right as a member of our Illustrious Society. The other Six you are desird to accept of in gratitude For the Communication which Forms the basis of it. Some disputes arose in the Soc[iety] about the mode in which it should be publishd, & an Idea, in my mind illiberal, that by Keeping it up [in price] people, for the sake of possessing it, would become members was Started by the Duke of Norfolk,[3] & had almost prevaild. We contrivd, however, to pass a law for disposing of the Copies, which is that any member who wishes his Friend to be possessd of one shall move the Society for that purpose, Specifying the name of his Friend. This motion, being Seconded, shall, if 12 members are present at a regular meeting, be put to the ballot, & if $\frac{2}{3}$ of the ballotters are for the motion it shall be Carried. In this manner about 30 have already been disposd of, & if you want any for any Friend of yours I shall be happy to move for you.

This season of the year, which is our vacation,[4] has not Furnishd one Matter of Philosophy worth mentioning. Herschels Volcanoes[5] are accepted & allowd by all I have heard speak on the subject, which, for a matter so liable to Controversy in itself, is not a little remarkable, & shews the high degree of Credit to which, by a constant adherence to truth and a carefull investigation of Facts previous to announcing them, he has raisd himself.

I hope your garden grows, & that the Gardiner will soon grow rich enough to repay what I advancd, without which assistance he could not have budgd as he will tell you; but, indeed, if he does not Find his Royal Pay more puncutal than I have done, he will not soon be out of debt.[6] I take for granted that their Neapolitan Majesties[7] conceive that I have chargd a living Profit upon all I have done for them; in which case I am sure they are not in any shape blameable as they must have supposd that, before I undertook the Job, I was aquainted with the usual Course of pay, & charged Profit accordingly.

You have in your Town a Certain Nicolaus Pacifico,[8] a Letter of whose lays before me. He has a Certain Book, the Posthumous remains of P. Cupani, which I much wish to acquire. He says he will Fish, as he calls it, in Sicily for another

Copy, which he hopes to buy for 50 neapolitan ducats on my account. Now, if you could persuade him to bait well, or let me have the book he has already Caught, I should be much delighted, even tho it Cost more. You Can be so good as to pay on my Account. Be so good as to tell him that the Ship, The Friends Peter Brook, which brings Dr. Smiths[9] Chest addressd to me, is not yet arrivd, but that I will give him the Earliest answer when I have got the box.

Thank Emma for her Letter. I have scrawled a Short Answer. In Short I am not a good womans Correspondent, & I fear she will not again lift up her pen for my encouragement.

Beleive me, my dear Sr. Wm,
Most Faithfully Yours,

Jos: Banks.

[B.L. Egerton MS. 2641, ff. 136–137.]

[1]Charles Francis Greville. See Letter 14, Note 5.
[2]Hamilton expressed pleasure when the Society of Dilettanti published 'the Cult of Priapus': Hamilton to Banks, 29/5/1787, B.L. Add. MS. 34048, f. 36–37, N.H.M. B.L. D.T.C. V 157–158. He referred to Richard Payne Knight's *An Account of the Remains of the Worship of Priapus, lately existing at Isernia, in the Kingdom of Naples…in two letters: one from Sir William Hamilton…to Sir Joseph Banks…*, London (1786). This included a letter from Hamilton to Banks on the cult, which was dated 30 December 1781. Knight had enlarged on Hamilton's earlier work: *The Worship of Priapus. An account of the fête of St. Cosmo and Damiano celebrated at Isernia in 1780…*, London (1783). The Society distributed these publications with care because of their explicit content: See Letter 14. Hamilton also enquired about Dr. (later Sir) William Herschel's discovery of 'Lava in the Moon'.
[3]Charles Howard, FRS (1746–1815), 11th Duke of Norfolk: whig; President of the Society of Arts, 1794.
[4]The Royal Society.
[5]Banks corresponded with a number of people about volcanoes on the Moon. In a letter to Josiah Wedgwood he reported that 'Dr. Herschell, who saw three Volcanoes on the moon in the Last Lunation, sees one of them Still burning. It has increasd in size nearly twice since he first saw it a month ago, & then he calculated the Circle to be 3 miles broad', Banks to Wedgwood, 22/5/1787, U.L.K. Wedgwood MS. Etruria 30491–30. Herschel described this in: *Philosphical Transactions*, vol. 77 (1787) Pt. I, pp. 229–232, 'An Account of Three Volcanoes in the Moon. By William Herschel, LL.D. F.R.S.; communicated by Sir Joseph Banks, Bart. P.R.S.' Read 26/4/1787. Craters on the Moon are sometimes brightly illuminated by earthshine, and this is probably what Herschel mistook for volcanoes.
[6]Johann Graefer (d.1803): nurseryman; gardener to the Queen of Naples, 1786–1799. Hamilton asked Banks to procure someone to go to Naples as a gardener and nurseryman. Banks arranged for Graefer to take up what seemed a promising offer. However, Graefer was treated badly by servants in the Court of Naples who were jealous of his good fortune. Moreover, he found himself in debt. Graefer worked hard, and with Hamilton's help gained royal favour.

Banks was kept waiting some time for payment from Naples for his expenses. He eventually waived the sum Graefer had borrowed.

[7]Ferdinand I, King of Naples and Sicily (1751–1825). Maria Carolina, Queen of Naples and Sicily (1752–1814).

[8]Marchese Niccolo Pacifico (1734–1799): naturalist and patriot. The letter Banks referred to was: Pacifico to Banks, 12/6/1787, B.L. Add. MS. 8096, f. 496. In it, Pacifico promised to enquire after a copy of Francisco Cupani's (1657–1710) *Panphyton siculum, siue Historia naturalis de animalibus, stirpibus et fossilibus quae in Sicilia*..., Panorami (1713).

[9]Dr. James Edward Smith visited Italy on his tour of Europe. Banks sent letters of introduction to the Court of Naples for him: Banks to Hamilton, 24/9/1786, Lady Smith (Ed.), *Memoir and Correspondence of the late Sir James Edward Smith M.D.*, London (1832), vol. I, p. 207. Pacifico confirmed that a case of books Smith wanted had been sent, first in the letter above Note 6, and on the same day in another: Pacifico to Banks, 12/6/1787, B.L. Add. MS. 8096, f. 495.

ix. **The Society of Dilettanti (1).**
Painting, 1777–1779, by Sir Joshua Reynolds P.R.A. Left to right: Constantine John Phipps, 2nd Baron Mulgrave; Lord Dundas; Lord Seaforth; Hon. Charles Greville; John Charles Crowle; Lord Carmarthen; Joseph Banks. Witt Library, Courtauld Institute, London.

x. **The Society of Dilettanti (2).**
Painting, 1777–1779, by Sir Joshua Reynolds P.R.A. Left to right: Sir William Watkins Wyn; Sir John Taylor; Stephen Payne Gallway; Sir William Hamilton; Richard Thompson; Walter Spencer-Stanhope; John Smyth. Witt Library, Courtauld Institute, London.

Letter 28

1787

To Sir George Yonge, F.R.S., Secretary for War

Boston
9 September 1787

Dear Sir,

I am not a little alarmed at the rec[eip]t of yours'.[1] If Capt. B.[2] is not sufficiently instructed, & the purpose of the voyage consequently in danger of being lost, blame must lay somewhere, & no one is so likely to support the burthen of it as your Hble servant.[3] I have therefore enclosed your letter to Mr. Nepean,[4] & requested him to signify to you that before I left town I asked most particularly whether any more was wanting from me toward the instruction for the Bounty, & was told there was not, but, if any thing new should arise, I should be written to on the subject.

They have in the Sec. of States' office the original plan of the Voyage, which I gave in to Mr. Pitt, & the instructions[5] intended for Governor Philip,[6] which were to Guide him in forming instructions for the Gardiner, & the master of the Vessel originally intended to have been dispatched from Botany Bay, which at the time were considered as fully sufficient.

That Capt. Bligh should say he knew nothing of going to the East or West Indies is to me incomprehensible. I had several long conversations with him on the subject of the voyage in which I did my best to explain to him the purpose of Government in incurring the expence of his outfit, & had no doubt he fully understood me; & he had from me, & kept some days in his possession, a Copy of the original plan of the Voyage, & the instructions intended for Governor Philip. I rather apprehend you have mistaken the civil answers of a man, who felt himself in the presence of a superior, for a want of information, & that, if he was to read your letter, he would be shock'd to think what a fool you must have mistaken him for.

As for his attempting to learn any part of the Gardiner's trade, I most heartily wish he may forbear the attempt. I have seen so much mischief happen from dablers in a science (who generally think they know more of it than professors) giving orders, that I dread the Idea. A Capt[ain] never can take directions from a Gardiner, but he may order him to do his duty in Planting &c., as he orders the Boatswain's mate to do his in punishing; but I heartily hope Capt. Bligh will never interfere in directing the manner in which a tree is to be planted, or a Cat of nine tails constructed.

Believe me, dear Sir,
Your faithful &
most humble Servant,

J.B.

[N.H.M. B.L. D.T.C. V. 247–248; M.L./D.L. Banks Papers Series 45.08.]

[1]Yonge to Banks, 7/9/1787, M.L./D.L. Banks Papers Series 45.08, N.H.M. B.L. D.T.C. V 245–246.
[2]William Bligh.
[3]The letter above shows that the one sometimes attributed 'Banks to Yonge', 7/9/1787, M.L./D.L Banks Papers Series 45.08, N.H.M. B.L. D.T.C. V 245–246, is in fact Yonge to Banks. The promise to make a 'botanist of Bligh' sounds like Banks, and the M.L./D.L. version is copied in his hand, but Banks made this copy to be sent to Nepean with Yonge described as a 'dabler': Banks to Nepean, 9/9/1787, M.L./D.L. Banks Papers Series 45.08, N.H.M. B.L. D.T.C. V 249. Banks did not want Bligh to learn more of gardening, and suggested that 'he may forbear the attempt'.
[4]Evan Nepean, under-secretary of state at the Home Office.
[5]The plan for ships from Botany Bay to undertake the mission to transplant bread-fruit was proposed to and accepted by William Pitt the younger. However, it was dropped in favour of sending a ship from England. See Letters 24 and 25.
[6]Arthur Phillip, who was the first Governor of New South Wales, 1788–1792.

1787

To James Hutchinson, Secretary of the American Philosophical Society

Soho Square
4 December 1787

Sir,

Allow me to request the Favor of you to deliver to the American Philosophical Society my most sincere thanks for the honor they have Done me of admitting me a member of their Laudable & Eminent body. Few things, Sir, Could have given me a more lively satisfaction than this Event, which gives me an opportunity of being personaly interested in the Progress which Science is making in a part of the world where She has much to investigate which has never yet been submitted to the Observation of those who have Studied in these Eastern Regions; & much she will, I am firmly convincd, bring to Light there which, cooperating with what these Climates afford, will add real riches to Literature which could not by any other means have been attaind.

Permit me also, worthy Sir, to return you my thanks for the Polite & Friendly manner in which you have been pleasd to Communicate to me the good news. Be assurd, Sir, that your Correspondence will always be regarded as an honor to me, & that your Commands on all subjects which come within the Limits of my abilities [will be] Chearfully Obeyd.

I have not, since I came to London Last, seen Mr. Workman[1] or heard any more of his Plan, which I am somewhat surprizd at; but as I have been confind to my Room the whole time by a Fit of the Gout, it is more than probable that I Shall as Soon as I am releasd.

Nothing has this year been presented to the Royal Society of moment except an account, with specimens, of a vast Lump of Malleable Iron which lies on the surface of the ground in the great Deserts to the N.W. of the Rio del Plate in South America, unconnected with any Strata whatever. The metal is white, tolerably Pure & very malleable.[2]

I am, Sir,
Your Most Obedient
& Most Hble Servant,

Jos: Banks.

[A.P.S. Hutchinson Papers.]

[1]Hutchinson to Banks, 2/8/1787, H.S. Penn Dreer Collection — Physicians etc., N.H.M. B.L. D.T.C. V 202–203: Hutchinson informed Banks of his election and referred to Benjamin Workman's invention of 'an instrument for ascertaining with precision the variation of the Compass'.
[2]*Philosophical Transactions*, vol. 78 (1788), pp. 37–42 and 183–189, 'An Account of a Mass of native Iron, found in South America. By Don Michael Rubin de Celis. Communicated by Sir Joseph Banks, Bart. P.R.S.' Read on 22/11/1787. The paper included a suggestion that the iron mass might have been 'conveyed hither by human art, or cast hither by some operation of nature.'

Letter 30

1788

To Charles Louis L'Héritier de Brutelle F.R.S.

Soho Square
[10] June 1788

Sir,

The melancholy news of the death of our esteemed friend, Mr. Lightfoot,[1] has no doubt come to your ears long before now. He died very suddenly by a gouty spasm in his stomach. His Botanic Collections, which were very respectable, have been purchased by the Queen as her majesty has lately applied herself considerably to the study of Botany. They will, I hope, have due honour done to them.[2]

Our lamented friend acquainted you, as he informed me before his death, that you were elected a foreign member of our Royal Society. The Diploma, which, for a variety of reasons, is not yet signed, will probably be dispatched with this. If not, you will [receive it] as soon as it is finished.

With this you receive a packet of Mosses from Mrs. Dixon.[3] It ought to have been sent half a year ago, but I did not know any means except a traveller. By your Letter to him, I find it may go by the Hanover Messenger. By that means therefore it is now dispatched.

I am in your debt, Sir, for the last parcel of Books.[4] I was in hopes to have been able to pay you by the means of Mr. Lightfoot, but as he, poor man, is no more, I must beg the favour of you to point out some way if you can conveniently do, otherwise, if you will acquaint me with the amount, I will remit it in a bill upon Leipzig.

[N.H.M. B.L. D.T.C. VI. 31-32.]

[1]The Reverend John Lightfoot died on 20 February. He suffered a fit, and then lay speechless in his bed for a few hours before expiring. Banks knew Lightfoot well. For instance, they had passed seven weeks in the summer of 1773 botanizing together after Banks had invited Lightfoot on a group trip through Wales and Staffordshire.

[2]George III paid 100 guineas for Lightfoot's herbarium, which was kept at Frogmore in Windsor Great Park. It was eagerly studied by Queen Charlotte Sophia (1744–1818), and the Princesses Augusta (1768–1840) and Elizabeth (1770–1840). Dr. James Edward Smith was chosen to instruct them, and he also cared for the collection. Most of the Lightfoot Herbarium is at the Herbarium of the Royal Botanic Gardens, Kew. However, the algal specimens were transferred to The Natural History Museum, London.

[3]The surname here is unclear, but it may have been 'Dickson', which was frequently spelled 'Dixon', and the sex also seems to be incorrect.
[4]Probably L'Héritier to Banks, 30/4/1788, B.L. Add. MS. 8097, f. 75. However, the last letter sent was: L'Héritier to Banks, 2/6/1788, L'Héritier Correspondence 5. The exchange of letters, books and plants between Banks and L'Héritier was a long-standing arrangement. It dated from their first acquaintance in 1783, and survived some awkward incidents.

Letter 31

1788

To Antoine Laurent de Jussieu F.R.S.

Soho Square
29 June 1788

Sir,

It is undoubtedly an unpleasant thing to be refused admittance into a Society to which one has been presented as a Candidate, but, in your case, I consider it as a matter which ought not in any degree to vex or trouble you. I, Sir, who have had the misfortune to be rejected by the Academy of Paris am now a member of it. Why then may not the same thing happen to you as I give an example of?[1]

Your literary character, I can assure you, never came into the contemplation of those who voted against you. That is highly respected here, as much as you could wish or expect. As far as I know, your enemies were guided by an antipathy to the doctrines of Mesmer,[2] which, especially since Dr. Franklin gave his opinion against them, is here very prevelant;[3] &, good Sir, difference of opinion in matters of speculative nature have not the least influence on the respect which literary men owe to each other.

I hope that, notwithstanding your disappointment, our correspondence will continue. Mine with my friends at Paris did not cease a moment on a similar occasion.

The Son of Dr. Hope,[4] the late professor of Botany, will be shortly with you at Paris. He brings you two Plates,[5] which are intended for the new publication of Hortus Kewensis,[6] coloured in the best style. They are intended as furniture, & [to] pay Compts. to the Queen,[7] who studies botany intensely & realy reads with perserverance Elementary books; & Lady Tankerville,[8] whose Lord has a very fine Botanic Garden, & who knows plants well, & paints them exquisitely. Accept these as a mark of my homage. Be assured of the continuation of that respect you have so justly inspired me with, & believe me at all times.

Your most
faithful & obedt

&c. &c. &c.

[N.H.M. B.L. D.T.C. VI. 40.]

[1]Banks was elected a Foreign Associate of the Paris Académie des Sciences in August 1787: Marquis de Condorcet to Banks, 14/8/1787, B.L. Add. MS. 8096, f. 432. However, Jussieu waited until 1829 before being elected a Fellow of the Royal Society.
[2]Franz Anton Mesmer (1734–1815): Austrian physician. He studied medicine at Vienna. Mesmer developed a system by which a hypnotic state could be induced, causing a patient to become insensible. The term 'Mesmerism' described his theory.
[3]In 1785 the Royal Society appointed a commission to report on Mesmer's theory, which concluded it was unacceptable. Benjamin Franklin was a member of this commission.
[4]Professor John Hope, FRS (1725–1786): professor of botany at the University of Edinburgh. His son was a chemist: Dr. Thomas Charles Hope (1766–1844).
[5]The plates were of *Strelitzia reginae* and *Phaius tankervilliae* (*as Limodorum tancervilliae*): E. Smith, *The Life of Sir Joseph Banks President of the Royal Society with some notices of his friends and contemporaries*, John Lane (1911), pp. 82–83. The former is the beautiful bird-of-paradise flower, and the latter is an orchid, named in honour of Lady Emma Tankerville.
[6]W. Aiton, *Hortus Kewensis, or a catalogue of the plants cultivated in the Royal Botanic Garden at Kew*, London (1789).
[7]H.M. Queen Charlotte, of Mecklenburgh-Strelitz. In 1761 she became Queen to George III. *The Strelitzia reginae* was named after her by Banks.
[8]Lady Emma, Tankerville (d.1836): wife of Charles, 4th Earl of Tankerville (1743–1822).

Letter 32

1788

To John Lloyd F.R.S.

Revesby Abbey
23 August 1788

My dear Lloyd,

I thank you heartily for 2 brace of Grouse, which indeed My Friends in London have Eat, but that is the same as if they had fallen into my own Share.

I am glad you had so much amusement in your Tour [of] Artoun in the Isle of Mull.[1] I never Saw, nor did I ever hear of, Basaltic Pillars resting on Coal, which, if certain, appears to me a very Curious discovery, & very interesting to those Who search to discover the mode of their Chrystallisation. I shall be particularly glad to see an account of them presented to the R.S., who Know the Abilities of your Companion so well as to wish to hear from him whenever he has leisure to Communicate his Observations to them.

I Observd the Whyn Dykes, & have some good Drawings of them, which shall be at your Service if you have any thing to say on the subject. Your magnetic mountain is not a very unusual thing. In my Travels I have met with many.

The Scouring out of the Bed of the Witham[2] here has producd a most Extraordinary lot of Danish & Roman Antiquities. If I had you here to ride about, & assist to Collect them from the various hands into which they have fallen, I think they would make a most Curious Collection. I got last year above 20 articles, many of them Quite unique, & this year a larger Quantity of River has been Cleard out, but my harvest has not begun.

How do you go on with Your Digging on the Duke of Ancaster's[3] Estate? He always asks me when I see him.

The Ladies have all desired their best Compts to you, as does Mr. Wheler,[4] who is with me. I do beleive you would like to hunt in our Antiquities, among which are some, I have no doubt, of first rate Curiosity.

Yours Faithfully,

Jos: Banks.

[N.L.W. MS. 12415C, letter 17.]

[1]Lloyd made a tour of Ireland and the Hebridean islands of Staffa, Iona, Islay, Mull and Jura: Lloyd to Banks, 13/8/1788, NHM BL DTC VI 61–62. This journey was partly inspired by Dr. Samuel Johnson's (1709–1784) tour of the Hebrides and West Highlands in 1773, and, of course, Banks's own excursions. Lloyd promised an account of his geological observations. Some time later two letters from his 'Companion' appeared: *Philosophical Transactions*, vol. 80 (1790), pp. 73–101, 'Some Account of the Strata and Volcanic Appearances in the North of Ireland and Western Islands of Scotland. In two Letters from Abraham Mills, Esq. to John Lloyd, Esq. F.R.S.' Read 21/1/1790. The 'Whyn Dykes' are described in detail as 'veins of lava'.

[2]The River Witham, Lincolnshire. Banks's interest in waterways included developing canal systems, and bringing Lincolnshire fen land into agricultural use by draining and enclosing it. However, he was also fascinated by local history, and excavations for artifacts of most kinds. These concerns were combined in another paper, which drew on the scouring of the Witham: *Philosophical Transactions*, vol. 86 (1796), pp. 395–451, 'Observations on some ancient metallic Arms and Utensils; with Experiments to determine their Composition. By George Pearson M.D. F.R.S.' Read 9/6/1796.

The Witham offered the opportunity for social gatherings at this time of year as well. Banks presided over large fishing parties along the river, using seine nets drawn by horses on each bank to catch fish to eat. Such a party would last from late morning or noon to ten p.m. This was repeated for about four days, with numerous members of the Lincolnshire gentry invited to dine and converse on boats which followed the nets. For more details of this kind of activity, 1784–1796: 'Minute of the annual Fishery on the River Witham began MDCCLXXXIV', Yale Centre for British Art, Connecticut.

[3]Brownlow Bertie (1729–1809), 5th Duke of Ancaster and Kesteven: Lord Lieutenant of Lincolnshire.

[4]Reverend John Wheler (Wheeler), (d.1818): clergyman in Lincolnshire.

Letter 33

1788

To Thomas March

[Soho Square][1]
30 September 1788

Sir,

In answer to your very Obliging & Friendly Letter of Augst. 18, I can only say that I most fully approve of your Plan of Obtaining Sheep,[2] & most heartily thank you for your very Obliging & friendly assistance in a matter which occupies my mind most Fully. I have only to request that whenever you disburse any money on my account you will do me the Favor either to draw upon me or order me to pay it to your Correspondent as soon as is Convenient, as I am much Convincd of the truth of the old adage which says that short accounts make long Friends.

I know, sir, that by the means necessarily usd in this business we are Precluded from chusing which of the flocks we please, & must get from those that happen to be most Conveniently situated, but if by accident the Flocks of Patrimonio, Perales or Lasteri should be met with, they will be very acceptable as they are much respected for the fineness of their Fleeces here.[3]

Your Friendship on this occasion to a person unknown to you I shall never forget, & I hope the Country will never forget that you are the man to whom they are obligd. If we succeed, of which I have no kind of doubt, it will be a national advantage of a magnitude such as the nation has rarely receivd from individuals.

To depend upon a Country naturaly unkindly to you for the Raw material of the finest branch [of] your Principle manufacture, & to be in hourly danger of the privilege of Obtaining it being resumd, is a humiliating Consideration to a great nation. To put her in possession of that Raw material, then, is an act deserving the best Kind of Gratitude.

Beleive me, Sir,
Your much Obligd
& most Obedient Servt,

JB.

[(S.C.) Banks Collection I. 21. 39–40; S.L. Banks MSS., 'Wool' by box and folder.]

[1]The source appears to indicate Soho Square as the origin of this letter, but other evidence suggests that Banks was at Revesby Abbey.

[2]March to Banks, 18/8/1788, (S.C.) Banks Collection I. 21. 38–39, S.L. Banks MSS. 'Wool' by box and folder.

[3]The plan under discussion was to smuggle sheep with fine wool out of Spain and through Portugal, sheep like the merino. These were used to enrich British wool production. Banks suggested some flocks: Patrimonio (Escorial), Perales, or Lasteri.

Letter 34

1788

To an Unknown Correspondent

Soho Square
20 November 1788

My dear Sir,[1]

Botany goes on well here, but as Mr. Dryander[2] writes answers to all your Questions so much more regularly than I can do, I shall say little about it. Indeed, at this moment Even I am occupied in Politics, as the Singular & unexpected situation of this Countrey makes it necessary every one should be.

It is now a Fortnight since this Crisis of a Fever, which had been coming on some time on the mind as well as the body, reducd the King to a State of Insanity which hitherto has shewn very little signs of Abatement.[3]

The Parliament summond to sit yesterday, tho under an intention of farther adjournment, was obligd, under the impossibility of the Kings signing an instrument to enable Commissioners to Adjourn it, to proceed to business. The Point, I Suppose, as important a one to both parties as Possible, which was whether or not the Prince of Wales, who had long & publicly favord the minority, should be declard Regent.

The Matter, however, passd without a Single word said. For the honor of general decency the minister movd an adjournment for a Fortnight, & the whole House, which seldom has been seen so full, unanimously Acceeded to it.

Few nations will ever produce such a sight as Parliament sitting without Legal authority, for without the third Estate that is the Case, & not the Smallest stir or bustle in the Streets. Every one did his business as if Government had its usual Legality & regularity, & yet not a Tradesman who was not fully acquainted with the real state of things.

I am myself firmly persuaded that the King will recover, but at the same time I may be mistaken. If he is not materialy better before this day, [in a] fortnight debates of the highest importance must take place, but we shall still be just as Quiet as we now are.

Give, I beg of you, my particular remembrances to Dr. Broussonet[4] & my other Freinds, &, if you see Mr. Greville,[5] tell him he ought to be at home at such a moment.

Beleive me always,
Your Obedient
& Most Faithfull Servant,

Jos: Banks.

I put this under Cover as I should not chuse it to be read.

[N.M.M. GAB/2.]

[1]The recipient is unknown, but this letter was sent to France. It was endorsed 'R. le II Xbre'.
[2]Jonas Dryander (1748–1810): bibliographer; botanist; botanical curator and librarian to Sir Joseph Banks, 1777–1810; librarian of the Royal Society, 1810; founding member of the Linnean Society in 1788, and its librarian and vice-president. As part of his work in Soho Square, Dryander published an impressive catalogue of Banks's library in five volumes *Catalogus Bibliothecae Historico-Naturalis Josephi Banks...*, London (1796–1800). He was also a competent botanist, who gave extensive help to the Aitons, father and son, in compiling the editions of *Hortus Kewensis...*, 1789 and then 1810–1813. Indeed, he may be regarded as an author who collaborated in each edition, drawing on the library and herbarium at Soho Square for his scientific research. Dryander died in 1810, and so worked only on the first two volumes of *Hortus Kewensis...*, (1810–1813). See Letter 114.
[3]George III became unwell in June, when he complained of painful abdominal spasms. His health deteriorated, and by late November he appeared mad. Doctors failed to diagnose his mysterious illness, which is now called porphyria. Parliament could not act properly without royal authority, and if the King had died or become incurable the Prince of Wales (1762–1830) would have assumed his place. This prospect weakened Pitt's administration, because the Prince favoured the Whigs under Charles James Fox. The King was moved to the White House at Kew on 29 November, and he spent a cold December there while his condition grew worse. By the time parliament opened on 5 December, the crisis had grown acute enough to cause the Regency debates.
[4]Pierre Marie Auguste Broussonet FRS (1761–1807): botanist.
[5]Charles Francis Greville was in France at this time: Greville to Banks, [3]/10/1788, U.Y.S.L. Banks MS.

Letter 35

1788

To William Devaynes

Soho Square
27 December 1788

Sir,

In obedience to your wishes, I readily undertake to give my opinion relative to the possibility of Tea becoming an object of cultivation & manufacture in the possessions of the East India Company, & the probable means of effecting that very desirable object.[1]

Some Teas are certainly cultivated in moderate Climates. Kempfer,[2] an Author of unimpeached credit, thinks that the best in the world are grown near Miaco, the Capital of Japan, in Lat: 35 N., & he recommends Italy & Sicily as the parts of Europe best suited for that purpose.

At present, however, it is universally allowd that the Chinese Trees are superior in flavour to those of Japan, & in China[3] they are seldom if ever cultivated in so high a latitude. Du Halde[4] tells us that in Quantong, a Province laying between 35 & 38 degrees of Lat., in consequence of Tea being scarce, a sort of Moss & the leaves of various other Trees are sold by that name, & substituted in its room; but this difference may arise from a greater degree of mildness in the climate of Japan very likely to be the consequence of its insular situation.

Singlo Tea takes its name from a mountain in the Province of Tcha-Kiang, laying in Lat. *30. 38. 40.*, which seems to be the climate most favourable to the production of *Green Teas*. Bohea, again, is called so from the mountain Vou y chan in the Province of Fo Kien in Lat. *27. 37. 48*, where the *Black Teas* are said to be most excellent. Both sorts appear to be generally cultivated on the slopes of hills but, from a variety of evidence, it is plain they succeed quite as well in level land, but such land is generally reserved by the Chinese for the produce of Bread Corn, which Purpose the slopes of the hills will not so well suit.

By an examination of the boundaries of those provinces enumerated by Du Halde from the authority of Chinese writers as producing Tea, it appears probable that all merchantable Teas are grown between the *26th.* & the *35th.* degrees of Lat; & that the space between *26 & 30* is most favourable for the culture of Black, & that between *30 & 34* for *Green Teas.*

In the Province of Yu nan a sort of Tea is produced in Lat. *23*, but as the growers of it make it up into cakes, & it is said to have an unpleasant flavour, it cannot be fit for the European Market. Tonquin & Siam have been said by various writers to produce Tea, but Voyagers who have been in those Countries declare that the Rich are the only people in either of them who can afford to drink the Chinese Tea which is imported. The poor substitute the leaves of the other plants found in their Country in its room, still however calling the infusion by the name of Tea.

To search in the territories of the East India Company for all the varieties of climate necessary for the production of the various sorts of Tea would be useless as they do not extend over a sufficient number of degrees of Latitude; nor was it practicable, would it be adviseable, to attempt at once to dispossess the Chinese of the whole Tea Trade.

All undertakings of new manufacture should commence with articles of inferior quality; they being less difficult in preparation, and more certain (as they fall into the hands of the lower orders of people) of being admitted into immediate use than higher prized commodities intended for the consumption of those who have more distinguishing palates, & fewer reasons for being economical in their purchases. Moreover, as the profit they bring is derived from the extent of sale that cheapness always commands, tho little upon each bargain, it ultimately becomes an object of great magnitude & proportional importance.

Black Teas are exactly in this predicament, & they may certainly be cultivated with success in the Northern parts of the Province of Bahar, Rungpoor & Coos beyhar for instance, where the latitude & the cooling influence of the neighbouring mountains of Boutan give every reason to expect a climate eminently similar to the parts of China in which good Black Teas are at present manufactured.[5]

The mountains of Bartan afford in a short distance all the climates that are found in the cooler parts of the Empire of China, & consequently every variety necessary for the production of the Green Teas. If, then, the culture of the Black Teas is once established in the neighbouring provinces of Bahar, & the inhabitants of Bou/*ar*/tan are invited by proper inducements, they will certainly undertake that of the Green; & thus, by a gradual change, the whole of the Tea trade will be transferred into that quarter.

Printed as well as oral authority enables me to assert that the real Tea-shrub is cultivated in the neighbourhood of Canton, & particularly on the Island of Ho nan, immediately opposite the Town, where also, tho' the leaves are usually exposed to sale in their raw state, they are occasionally rolled up as other Boheas are. These Teas are certainly of an inferior quality, but nevertheless there is every reason to believe that large quantities of them are constantly mixed with the cheaper Boheas purchased by our Super-cargoes for the home investment. The reason why the Supercargoes overlooked this circumstance in their letter to the Governor & Council of Bengal, inserted in Bengal public consultations Jany. 23. 1788., it is needless at present for me to enlarge upon.

Mr. James Cunningham,[6] a sensible & intelligent Surgeon, who resided for some time in the Company's service at Amoy where Tea grows, asserts in a Paper printed in the Philosophical Transactions that all the sorts of Tea imported into Europe are gathered from the same species of Shrub, and the most attentive Botanists who have examined specimens brought from China agree with him in that opinion. He says that the differences between them arise from the soil in which the Shrub is planted, & the Season in which the leaves are gathered; to which I have no doubt he ought to have added the climate also.

This being admitted as a fact, & we certainly have no reason to doubt it, the deduction easily follows, which is that the inferiority of the Ho nan Teas is derived solely from their being situated in a soil & climate unfit for their cultivation, & that these very Teas would, on being transplanted into a more proper situation, recover the flavour they were deprived of by the tropical heats of Canton. But supposing they should not recover the whole flavour of the Souchongs, we cannot doubt that a large amount of them would be found marketable as a substitute for the Teas of which they at present make a considerable part by the artifice of the Chinese before alluded to.

The inhabitants of Canton are now in the habit of Shipping themselves on board our India-men whenever hands are wanted. We may therefore with safety conclude that their neighbours at Ho nan may be induced by the offer of liberal terms to follow their example, &, moreover, to embark their Tea shrubs & all their tools of culture & manufacture, & migrate with them to Calcutta, where they will find the Botanic Garden ready to receive them, 20 acres of which might at least be allotted to their immediate reception, which is already cleared & prepared for similar purposes; &, lying under very nearly the same latitude as Canton, could not fail to suit in every particular this infant adventure. Here the Shrubs might be increased under the direction of the able & indefatigable superintendent, & the natives taught the culture & manufacture of the article, who, when fully instructed, might be detached with a proper supply of tools & Shrubs to the places ultimately destined for the permanent Establishment of the manufacture.

Presuming you, Sir, & the Court of directors to agree in the propriety of attempting the Execution of this plan, two things only remain to insure its success, which are: to find out the proper means of getting access to the cultivators & manufacturers of Tea in Ho nan; & to induce them to remove with their shrubs & their tools to Calcutta. To intrust the execution of this business to the Company's Supercargoes at Canton would be imprudent. They have an interest in its failure, & by their influence on the spot could no doubt, if they chose it, prevent the success not of the present only, but of all future attempts of the same nature.

The undertaking ought therefore to commence with the Court of directors here. There must be in the Companie's service officers capable of executing commissions of every degree of delicacy, & you, Sir, or your brethren in the direction, cannot fail of being acquainted with them. To some one of these, whose secrecy in the execution of it may be relied upon, the business ought to be intrusted, &, for my part, if my advice or assistance in preparing the detail of the plan should be thought necessary, I shall be at all times ready to come forward, convinced that the object is of real importance, not to the Company only, but to the Country at large.

I consider it as incumbent on me to use my utmost assiduity in bringing it about, &, sensible of the obligations I have received from the Court of directors, I shall at all times think it my duty to obey with alacrity any commands with which they may be pleased to honor me.

I have the honor to be, Sir,
Your most obedient
& most Hble. Servant,

Jos. Banks.

Note appended to the preced[in]g letter.

At a time like the present, when the slightest degree of political sagacity is able to foresee the near approach of that time when the policy of this nation, and the clamours of its manufactures, will lay it under the necessity of favouring the rising fabrics of Cotton, as it has already favoured those of Silk, by laying restrictions on the importation of rival manufactures from whatever quarter they may be brought, it behoves the East India Company to look forward & consider of the means of providing a proper home investment for the time when that period shall actually arrive.

If they examine into the state of their Territorial possessions, they will find them spread over a variety of climates, from the burning heat of the Tropic to the wholesome coolness imparted by the vicinity of the mountains of Boutan. [They are] fertile in the extreme from one end to the other, peopled by a numerous race of frugal, industrious & intelligent natives, accustomed to labour for lower wages possibly than any other set of men, & intersected in a multitude of directions by navigable Rivers, placed as if on purpose to bring the fruits of their industry at a cheap rate to the centre of commerce, wheresoever it may be that convenience has caused it to be established.

A Colony like this, blessed with advantages of Soil, Climate & Population so eminently above its mother Country, seems by nature intended for the purpose of supplying her fabrics with raw materials, & it must be allowed that a Colony yielding that kind of tribute binds itself to the Mother Country by the strongest & most indissoluble of human ties, that of common interest & mutual advantage.

Yet, evidently as this reasoning points to the discovery of the real & permanent interests of the East India Company, few if any steps have hitherto been taken to establish it on that foundation, & no public interference whatever has been exerted toward encouraging any one new species of culture that has not been marked by the most rigid economy. Individuals, indeed, have cultivated Indigo with profit, & demonstrated by that means the certainty of success attending a rivalship in culture between the East & West Indies; but, notwithstanding this demonstration, the expence of public interference & encouragement, which alone can insure to such undertakings a rapid advance to maturity, has been witheld, probably without its having been once compared with the ultimate advantage to be derived from the object, and the project has hitherto advanced by slow & deliberate steps.

As the commerce of Indigo is on the point of doing, so must those of Coffee, Chocolate, Vanilla, Cochineal, Cotton & even Sugar itself, whenever the rivalship is established, be transferred to the East (tho' the latter, indeed, as the staple commodity of the West India Islands, will probably be settled on them by the Legislature); and it is self evident that every article now received from the West may, on account of the comparative cheapness of labour, be furnished cheaper from the East, notwithstanding the difference of freight */which I am of the opinion may be furnished by the Hindoos/*.

Having thought much on these & similar subjects, & being urged by the request of my worthy friend, the Chairman, I have taken some pains in tracing out the means by which articles not likely to produce rivalship with the Mother Country may be transferred to the Company's possessions, beginning with Tea as an article of the greatest national importance. My thoughts on that subject are annexed to this. If you, Sir, are pleased to call for them, those on Cochineal will speedily follow, provided, however, that the necessary secrecy be granted on subjects which, if they once become the subject of public conversation, must necessarily fail; & that, if the probability of success is admitted, the ultimate importance of each article be taken fairly into consideration, & the Court of directors agree not to relinquish the object on account of any hesitation on the subject of necessary expenditure which, however, is not in any instance expected to arise to a heavy or in any degree an unjustifiable extent.[7]

[N.H.M. B.L. D.T.C. VI. 103–111.]

[1]The trade in tea from China had been growing throughout the seventeenth and eighteenth centuries in Europe. The British thirst for tea had increased in particular. Most of the supplies obtained for the British market were under the monopoly of the Honourable East India Company, whose directors Banks was quite familiar with. Relations with the Chinese were not always good though, and the large quantities of silver which the Far East trade required caused financial concern. Banks therefore responded to a growing interest in the possibility of transferring the tea trade from Chinese control to British possessions. For instance: Hawkesbury to Banks, 1788, S.L. Banks MS. TE 1:2; Banks to Hawkesbury, 1788, S.L. Banks MS. TE 1:3; Banks to Hawkesbury, 29/9/1788, B.L. Add. MS. 38223, f. 201–202.

His choice of the Botanic Garden at Calcutta as suitable for receiving tea plants was prompted by its founder, the 'indefatigable' Robert Kyd, whose ideas Banks had considered: Morton to Banks, 12/11/1788, N.H.M. B.L. D.T.C. VI 88; Memoranda, reports, notes and tables, 1788, N.H.M. B.L. D.T.C. VII 31–71. In a typically thorough way, and one which revealed how much importance he attached to the success of this plan, Banks read widely to supplement what he already knew of *Camellia sinensis*, and spoke to merchants and travellers with experience of the Far East trade: Notes, 1788, S.L. Banks MS. TE 28–30 and various TE tables. His conclusion was that tea plants might grow with advantage in the north-east region of India near Darjeeling and in Assam. However, it was not until 1834 that this experiment was properly tried. Only then was a Committee of Tea Culture formed in Calcutta to accomplish tea cultivation in India. Another typical aspect of the letter above is the reasoned way Banks enlarged his proposal to include other commoditities with commercial value.

[2]Engelbert Kaempfer (1651–1716): German naturalist; chief surgeon, Dutch East India Company Fleet; traveller in the Far East, who visited Japan, 1690–1692. Kaempfer published *Amoenitatum exoticarum politico-physico-medicarum fasciculi V: quibus continentur variae relationes, observationes & descriptiones rerum Persicarum et ulterioris Asiae...*, Lemgoviae (1712). It should be noted that in 1791 Banks privately published *Icones selectae Plantarum quas in Japonia collegit et delineavit Engelbertus Kaempfer...*, and the engravings in this volume were of Japanese plants taken from original drawings by Kaempfer under the catalogue title of 'Delineationes et descriptiones plantarum Japonicarum et Periscarum, 1685, 1690', B.L. Sloane MSS. 2914, 2915, 2917 A, B.

[3]The cultivation of tea in China is ancient, and still thrives. The country produces the widest range of fine quality teas in the world. In Japan approximately 600,000 families produce some 110,231 tons of tea from 148,263 acres of farmland. All Japanese teas are green. Banks referred to a number of places in the Far East. In sequence: 'Miaco', which means 'capital' in Japanese; 'Quantong', which, at latitude 35–38N, is Shandong, east China, (Kwangtung or Guangdong, a province of south-east China, produces more tea than Shandong); 'Singlo' is Songluo, a mountain in Anhui Province, near the border with Zhejiang; 'Tcha-Kiang', Zhejiang, south-east China; 'Bohea' is a local dialect pronunciation of 'Wuji', the well-known tea growing area in Fujian Province, south-east China; 'Vou y Chan', Wuyi Shan (Wuji Mountains), Fujian Province; 'Fo Kien', Fujian Province, south-east China; 'Yu nan', Yunnan, a province of south China; 'Tonquin', Tongking or Tonkin, now North Vietnam; 'Siam', now Thailand; 'Amoy', Xiamen or Hsia-men, a town in Fujian province; 'Canton', capital of Guangdong Province, south China; 'Ho nan', Henan, an island province opposite Canton. It means 'south of the [Pearl] river.'

[4]Jean Baptiste Du Halde (1674–1743): geographer. Du Halde published *Description géographique...et physique de l'Empire de la Chine et de la Tartarie Chinoise...*, Paris (1735).

[5]India is now one of the largest producers of tea, with more than 13,000 gardens, and a total workforce of over two million people. Banks was unaware that a variety of tea plant, then called *Camellia assamica*, already grew on the sub-continent. He referred to a number of places in the north-east area of India. In sequence: 'Bahar', Bihar or Behar, a state in east India; 'Rungpoor', Rangpur, a region in north Bangladesh; 'Coos beyhar', possibly Koch Bihar, a district in west Bengal, India; 'Boutan', Kingdom of Bhután, a state in the east Himalayas.

[6]James Cunningham (d.1709?): botanist; surgeon for the Honourable East India Company in China. He published: *Philosophical Transactions*, vol. 23 (1702–1703), pp. 1201–1207, 'Part of two Letters to the Publisher from Mr. James Cunningham, F.R.S. and Physician to the English at Chusan in China, giving an account of his Voyage Hither, of the Island of Chusan, of the several sorts of Tea, of the Fishing, Agiculture of the Chinese, &c. with several observations not hitherto taken notice of.'

[7]This letter is entitled 'Memorial Sent to the Deputy Chairman'. William Devaynes was made deputy chairman of the Honourable East India Company on 2 December. The chairman in 1788 was Nathaniel Smith (1721–1794).

Letter 36

1789

[To Dr. James Lind F.R.S.]

Soho Square
23 February 1789

My dear Sir,

I congratulate you sincerely on the recovery of his Majesty,[1] to which I can bear the most ample testimony, having long had the honour to be consulted by him on the subjects of Gardening and Farming. I was sent for on Saturday as usual, and attended in the Gardens and Farm for three hours, during all which time he gave his orders as usual, and talked to me on a variety of subjects without once uttering a weak or a foolish sentence.

In bodily health he is certainly improved - he is lighter by about 15Ib. than he was — he is more agile, and walks as firm as ever he did. We did not walk less than four miles in the garden and adjoining country.

I have no doubt that he is able at this moment to resume the reins of Government, but hope he will not do it for some time, lest too much exertion of mind might endanger a relapse.

Believe me, my dear Sir,
Most faithfully yours,

Jos. Banks.

[*The Gentleman's Magazine*, XC, vol. II (August, 1820), p. 99; J. Nichols and J. B. Nichols, *Illustrations of the Literary History of the Eighteenth Century*, London (1817–1858), vol. 4, p. 696.]

[1]As he recovered from serious illness, George III took to strolling, usually in Kew and Richmond Gardens. He often requested Banks's company, and the two men discussed a variety of subjects including gardening, agriculture and sheep.

[2]Robert Fulke Greville described the day: *The Diaries of Robert Fulke Greville*, F. M. Bladon (Ed.), London (1930), pp. 239–241: "I saw Sir Joseph after this Walk, who was surprised & pleased with all he had seen & heard during this Interview, which He said, 'was very near right.' Anxious to know where he might have discover'd a hitch, I found H[is] M[ajest]y had dwelt on going to Hanover. Sir Joseph said He hoped not, & to this He made no reply." p. 240.

Letter 37

1789

To Dr. Charles Blagden F.R.S.

Soho Square
28 March 1789

My Dear Dr.,

Till you have informed me whether the complaints in your letter[1] are leveled at the world in general or at me in particular, & whether or not the reflexions upon our connexion which you gave me the expectation of seeing hereafter are intended to be reflexions on my character & conduct, it is impossible I should be able to answer it.[2]

I have considered you as my friend for many years, &, to the best of my judgement, treated you as such. If I have done otherwise, it has been owing to inadvertency, but I declare I have no consciousness upon the subject. Here then let the matter rest, for if a friend declares himself an enemy, which, mortifying as the idea is, I have reason from the terms of your letter to fear is the case, profession may be considered as deprecation of what I confess I look forward to without the slightest degree of apprehension.

Believe me with sincerity resolved to be as long as you will permit me,

Faithfully yours,

JB.

[N.H.M. B.L. D.T.C. VI. 147; B.L. Add. MS. 33272, f. 58.]

[1]Banks inadvertently upset Blagden. Blagden was Secretary of the Royal Society and a close friend, so this was an unpleasant situation for both men: Blagden to Banks, 30/1/1789, B.L. Add. MS. 33272, f. 54, N.H.M. B.L. D.T.C. VI 128–129; Blagden to Banks, 27/3/1789, B.L. Add. MS. 33272, ff. 56–57, N.H.M. B.L. D.T.C. VI 145–146; Blagden to Banks, 28/3/1789, B.L. Add. MS. 33272, f. 59, N.H.M. B.L. D.T.C. VI 148.

[2]Banks visited the Prince of Wales, and Blagden may have been offended at not being invited. He certainly looked to Banks for advancement, and had been unhappy at the quantity of Royal Society work he had undertaken without pay. However, Banks refused to be drawn into an argument. Instead, he responded warmly to Blagden's cold letters, in which the formal third person was used. Blagden relented in the end, but he never adequately explained such strange, emotional behaviour.

Letter 38

1790

To William Pitt the younger, Prime Minister

Soho Square
23 March 1790

Sir,[1]

the Anziety of My Friend[2] on Account of my not having heard from you is become excessive. His whole mind is certainly Absorbd in the business that has brought him here.[3]

For his integrity, his honor & the superiority of his Abilities, I, who for some years have had an intimate acquaintance with him, undertook to be reponsible when I had the honor of seeing you. If, however, the Numbers[4] which were the subject of my Last Letter have not been Transmitted, he is Certainly without credentials whatever may be the Reason.

Even in that Case, he Wishes as anziously as Ever to have a conference[5] with you. He allows that your treating with him or not must depend wholly upon yourself. He wishes, therefore, only to have an opportunity of Stating to you the Present situation of the business which brought him over more than 1200 miles. It Cannot but be right for you to be acquainted with, as you may have that information on your own Terms.

More than that I can say, which is that I have every reason to beleive you will receive from him information which will give Lights of the utmost importance to the state of the present negociations.

[S.L. Banks MS. HUN 1:3; P.R.O. 30/8/92.]

[1]Marked by Banks: 'No. 3.' This is the third letter he wrote to Pitt in connection with a mysterious Hungarian Baron. Banks kept brief diary notes on this affair. He commented on each meeting in them, and recorded the outgoing correspondence with William Pitt in a numbered sequence as well.

Banks included the Baron's private remarks in the diary, but did not make copies of all the letters sent to Pitt. Some were written and sealed in the Baron's presence, and so there was no opportunity to copy them. Others were drafted, and some drafts remain. Banks's file documents are mainly held by the Sutro Library, California. They may be found together under the title 'Matters relative to Hungary 1790', and provide background details for what follows.

[2]Baron Miklo Vay de Vaja (1756–1824): Hungarian nobleman; engineer. Banks avoided using the Baron's name, and marked many of these letters 'Secret' or 'Private'.
[3]The Baron arrived in Soho Square on 20 March in order to gain secret access to Pitt through Banks. He sought British support for a plot, as he put it to Banks on the 28 March, 'to take H.[ungary] fully from the House of Austria & to Borrow 1,000,000 here for which Landed security would be given. All he said was fully concerted with the K.[ing] of P.[russia] who was to march his army in the beginning of June into H.' The Baron claimed to represent powerful Hungarian interests in which the British could have faith. The means by which Pitt could verify this was a cypher, a series of numbers the Baron would give to prove he was the chosen representative. Vay de Vaja travelled under the cover of an eye complaint, which needed treatment in London.
[4]The cypher given was 179009710971.
[5]Banks wrote to Pitt immediately the Baron arrived, and also visited the Prime Minister in Downing Street on the evening of the same day. Pitt delayed a meeting with Vay de Vaja until 31 March though.

1790

To William Pitt the younger, Prime Minister

Soho Square
28 March 1790

As[1] 9 days have Elapsd since I had the honor of seeing you on a subject, the importance of which depends very much upon the State of Affairs in the present moment, you will, I hope, excuse me if I request the Favor of a second hearing as soon as is practicable to your convenience. I am the more anxious on the subject as the state of suspence in which things are has so strong an effect on the health of my Friend, which unfortunately is in a Critical State,[2] that I suffer no small uneasiness on his Account.

The notification of [the Emperor's] death[3] is an epoch of much importance he tells me, and he much hopes he may be heard before the answer to it is given. He is prepard to put the utmost confidence in you, & to demand in return nothing but a promise of secrecy, & does not wish to be considerd in any but a private Capacity till you shall chuse to receive him in a more confidential one.

[S.L. Banks MS. HUN 1:16; C.U.L. MS. Add. 6958, ff. 783–784; P.R.O. 30/8/90.]

[1]Marked by Banks: 'No 4'.
[2]It seemed that Baron Vay de Vaja would lose his eye. He was unwilling to be treated until his political mission was complete though. Indeed, he became more and more agitated as meetings with Pitt were delayed, fearing that the cypher he had given was incorrect, or that he had been betrayed. He waited two more days before seeing Pitt at Banks's home in Soho Square, and until 9 May for a final encounter at the East India House, Leadenhall Street. Meanwhile the Baron revealed more of his secret plans to Banks. He explained the benefits to Britain of trade with Hungary and Turkey, asked to speak directly with the King, and even produced a document in Latin offering the Hungarian crown to the Duke of York (1763–1827). Towards the end of May, Vay de Vaja passed on the last messages to be sent to the continent hidden in letters and packages arranged by Banks.
[3]Joseph II (1741–1790), Holy Roman Emperor. He was succeeded by Leopold II (1747–1792). The dissidents claimed 'That the new King Leopold was judgd by his Subjects to have the Same arbitrary Ideas as the Emperor had, but to be more Cunning'. They did not want further religious, administrative and social reforms of the kind Joseph II had introduced. They preferred instead 'their old Consitution as it stood 300 years ago'. However, Leopold II's short reign saw the status quo in Hungary largely restored, control over his own dominions strengthened, and Austrian relations with other European nations improved.

1790

To William Pitt the younger, Prime Minister

Soho Square
29 March 1790

Sir,[1]

I have read your Letter[2] to my Friend, & have made him fully comprehend the Contents. He thankfully agrees to the interview tomorrow on your own Terms. I will take Care that he is ready in Time, &, Sir, as he has reason to Fear that he is watched,[3] if it is not inconvenient to you to Come on Foot, I will meet you exactly at 12 o'clock at the Front of St. Anne's Church in Dean Street, & Conduct you thru a back door into my house. If this Plan meets your Approbation, I shall want no Further instructions. If not, I shall beg for a Line of direction from you, which I shall Obey with exactness.[4]

[S.L. Banks MS. HUN 1: 5; C.U.L. MS. Add. 6958, ff. 783–784; P.R.O. 30/8/94.]

[1]Marked by Banks: 'No. 6.'
[2]Pitt to Banks, 29/3/1790, S.L. Banks MS. HUN 1:5: 'I shall be very ready to see Him [Baron Vay de Vaja] as a private Man, and to receive any Information He may wish to give me…it [is] impossible for me to enter into any Ministerial Explanation…'
[3]While at the opera on the 23 March, Vay de Vaja saw a man 'he thinks was a Spy formerly Employd by the present King at Vienna who regarded him with particular attention': 24/3/1790, Diary note, 'Matters relative to Hungary 1790', S.L. Banks MSS.
[4]Nothing came of this political intrigue, and the Baron spent the next few months in England visiting the industrial works of such manufacturers as Matthew Boulton.

Letter 41

1790

To Dr. Joseph Priestley F.R.S.

[Soho Square]
26 April 1790

Sir,

In return for the openness of your conduct in your letter of yesterday,[1] in which you tell me, among other things, that you conceive I had no concern in the rejection of Mr. Cooper,[2] I beg leave to treat you with the same candour by declaring that I am one of those who was not, at the time of the ballott, sufficiently acquainted with that gentleman's merits to be justified in my own opinion in giving him my vote. At the same time, Sir, I assure you with the utmost sincerity that religious prejudice had no influence whatever on my conduct in that respect, for tho' I am convinced that the Majority governed have a right to insist that the Magistrates who govern them do profess the religious tenets which they believe to be the only true ones, & conform to the rites by which the sincerity of their religious profession can alone be put to the test, I never felt the least difficulty to associate in Philosophical disquisition with any person on account of what his creed might or might not be.[3]

So much, Sir, for myself. On the part of the Royal Society, whose Chair you do me the honour to say I have not disgraced, I cannot help being astonished at the sentiment in your letter which expresses a disgust at your recommendation to them being slighted, & still more at your conceiving the rejection of Mr. C. as an intended affront on you.

Be assured, Sir, I shall ever maintain that each individual of the Royal Society is intitled to the utmost freedom of choice in matters of election, & that no person, however high he may rank in Science, can have a right to claim the slightest influence over his conduct beyond what he may chuse voluntarily to bestow. The other, Sir, I conclude you will on a moments reflexion abandon. To hint even that the R.S. were capable of combining together in an act of injustice in order to pass an affront upon you, is what they, as a body, must feel as a serious charge on their character & reputation, & is one, I boldly assert, which is void of the slightest foundation in fact.

With you, Sir, I have hitherto lived in friendship, & have ever set, I trust, a proper value on the influence your discoveries have had in the advancement of science. Your friendship I wish still to retain, & I hope the frankness of my letter will not alter the sentiments towards me which are expressed in yours.[4] I shall therefore in confidence add that, whatever Mr. C's scientific merit may be, no token of which he has hitherto brought forward to the Society, there are, I am firmly convinced, reasons wholly independent of his religious opinions, which, however the partial eye of your friendship may have overlooked, do fully justify the late conduct of the Society, & will, if any appeal is made to the public, acquit them wholly of the charges you are inclined to bring against them.

J. Banks.

[N.H.M. B.L. D.T.C. VII. 108–109; R.B.G. Kew Archives B.C. 2. 9.]

[1]Priestley had proposed Cooper for election to the Royal Society, and canvassed on his behalf. Cooper was rejected, and Priestley complained at what he thought was a personal affront to himself, believing that political or religious considerations had weighed against Cooper: Priestley to Banks, 25/4/1790, S.I.L. Banks MSS. 1257B, N.H.M. B.L. D.T.C. VII 104–105.

[2]Dr. Thomas Cooper (1759–1840): Natural philosopher; lawyer; politician. He accompanied James Watt Junior (1769–1848) on a deputation sent by the democratic clubs of England to France, and was known for his pro-republican views.

[3]Apparently a defence of the Test and Corporation Acts, which denied non-Anglicans full civil and religious rights.

[4]Priestley remained angry despite Banks's firm but cordial explanation: Priestley to Banks, 27/4/1790, S.I.L. Banks MSS. 1257B, N.H.M. B.L. D.T.C. VII 110–111.

Letter 42

1791

To Archibald Menzies

Soho Square
22 February 1791

Sir,

The business[1] on which you are employed being of an extensive nature, as it includes an investigation of the whole of the Natural History of the Countries you are to visit, as well as an enquiry into the present state & comparative degree of civilization of the Inhabitants you will meet with, the utmost degree of diligence and perseverance on your part will be necessary to enable you to do justice to your Employers and gain credit to yourself.

The following Instructions you will consider as a guide to the outline of your conduct, but, as many particulars will doubtless occur in the investigation of unknown Countries that are not noticed in them, all such are left to your discretion & good sense;[2] and you are hereby directed to act in them as you judge most likely to promote the interest of Science, & contribute to the increase of human knowledge.

In all places where the Ship in which you are embarked shall touch, and the Commander shall make sufficient stay, you are to pay a particular regard to the nature of the soil, & to note down its quality, whether Clay, Sand, Gravel, Loam &c. &c., and how it is circumstanced in regard to water. You are to remark particularly [on] the size of the Trees that grow upon it, whether they stand in thick close Groves, or seperate and distinct from each other. You are to consider also, as far as you are enabled to do by the productions, the probable Climate, and whether, should it any time hereafter be deemed expedient to send out settlers from England, the Grains, Pulse and Fruits cultivated in Europe [that] are likely to thrive, and if not what kind of produce would in your opinion be the most suitable.

As far as you find yourself able, you are to enumerate all the Trees, Shrubs, Plants, Grasses, Ferns and Mosses you shall meet with in each Country you visit by their scientific names as well those used in the language of the Natives, noting particularly the places where each is found, especially those that are new or particularly curious. You are also to dry specimens of all such as you shall judge worthy of being brought home, particularly those of which you shall procure either living Plants or Seeds, in order that the Persons who are employed in examining the Plants you furnish to his Majesty's Gardens at Kew may be assisted in ascertaining their names and qualities.

Whenever you meet with ripe seeds of Plants, you are carefully to collect them, and, having dried them properly, to put them up in paper packages, writing on the outside, or in a corresponding List, such particulars relative to the soil and climate where each was found, and the mode of culture in your opinion likely to succeed with it as you may think necessary to be communicated to His Majesty's Gardeners; and you are to forward these packages directed to me for His Majesty's use by every convenient opportunity that shall occur, dividing them for safety's sake into duplicates as often as you shall judge needful.

When you meet with curious or valuable Plants which you do not think likely to be propagated from seeds in His Majesty's Garden, you are to dig up proper Specimens of them, plant them in the Glass Frame[3] provided for that purpose, and use your utmost endeavours to preserve them alive till your return. You are to consider every one of them, as well as all Seeds of Plants which you shall collect during the voyage, as wholly and entirely the property of His Majesty, and on no account whatever to part with any of them, or any cuttings, slips, or parts of any of them for any purpose whatever but for His Majesty's use.

As soon as you shall have provided yourself with living plants, and planted them in the Glass frame before mentioned, you are at all times, when the Ship shall be watered, to acquaint the Commanding Officer what quantity of water you judge necessary for their support and preservation, by the week or month, in order that he may be enabled to make a competent provision of that article for their future maintenance & nourishment.

In all your excursions on shore, you are to examine with care and attention the Beds of Brooks and Torrents, the steep sides of Cliffs, and all other places where the interior Strata of the Earth are laid bare by water, or otherwise to remark [on] the nature of the Earth and Stones of which they are composed; & if among them you discover Ores or Metals, or any Mineral substances which bear a resemblance to such things, or any Beds of Coal, Limestones, or other matters likely in your opinion to prove useful to mankind, you are to collect and preserve specimens of them, carefully noting the exact spot on which each was found. You are also to examine the Pebbles and Sand brought down by the Rivers and Brooks from the Inland Country, and to collect and bring home samples of such as you suspect to contain Mineral substances, even though so minute as not to be discoverable but by a Microscope.

At each place where you land, you are to inform yourself as well as you are able what sort of Beasts, Birds and Fishes likely to prove useful either for food or in Commerce are to be found; and pay particular attention to the various ways of catching them in Traps, or otherways used by the Natives. You are to pay particular attention to every part of the natural History of the Sea Otter, and to learn all you are able concerning wild Sheep said to be found in the Inland Countries, and, if in your power, to procure a Skin of one of them for your Employers. You are also to note particularly all places where Whales or Seals are found in abundance.

At all places where a friendly intercourse with the Natives is established, you are to make diligent inquiry into their manners, Customs, Language and Religion, & to obtain all the information in your power concerning their Manufactures, particularly the Art of dying, in which Savages have frequently been found to excel; and if any part of their conduct, civil or religious, should appear to you so unreasonable as not to be likely to meet with credit when related in Europe,[4] you are, if you can do it with safety and propriety, to make yourself an eye witness of it in order that the fact of it's existence may be established on as firm a basis as the nature of the Enquiry will permit. You are to keep a regular journal[5] of all occurrences that happen in the execution of the several Duties you are entrusted to perform, and enter in it all the observations you shall make on every subject you are employed to investigate; which journal, together with a complete collection of specimens of Animals, Vegetables & Minerals you shall have obtained, as well as such curious articles of the Cloths, Arms, Implements and manufactures of the Indians as you shall deem worthy of particular notice, you are on your return to deliver to His Majesty's Secretary of State for the Home Department,[6] or to such person as he shall appoint to receive them from you.

I am, Sir,
Your most obedient
humble Servant,

Jos: Banks.

[N.H.M. B.L. D.T.C. VII. 197–201; B.L. Add. MS. 33979, ff. 75–78; M.L./D.L. Banks Papers Series 61.03 and 61.04.]

[1]This was the voyage to take possession of Friendly Cove in Nootka Sound from the Spanish, and to complete the charting of the west coast of North America. It sailed under Captain George Vancouver RN (1758–1798) in HM Ships *Discovery II* and *Chatham*, 1791–1795. Vancouver's relationship with Menzies deteriorated, and by the end of the mission Menzies was under arrest. Many of the living plants he collected therefore died. For Banks's view of these events: Notes, N.H.M. B.L. D.T.C. 10(1) 83–86; Banks to unknown correspondent, undated, N.L.A. MS. 9/31.

[2]These instructions seem partly to have been based on the Admiralty instructions, specifically the additional ones, issued for the *Endeavour* voyage, dated 30 July 1768, P.R.O. Adm 2/1332, 160.

[3]A glazed plant cabin. These were used to keep living plants in during a voyage, and were situated on the deck.

[4]Banks meant activities like cannibalism, which he encountered in New Zealand when on the HMS *Endeavour* voyage.

[5]Vancouver demanded that Menzies surrender his journal. Menzies refused saying it was to be given to the Secretary of State, as Banks had instructed. Afterwards, Menzies failed to publish his account, even though he worked hard to prepare a 'clean copy' for publication. He sent the portions he had completed to Banks, who therefore possessed at least three journal parts up to the date 16 February 1794. These were sold at Sotheby's in 1886.

Three volumes were purchased by the British Museum, and are now at the Manuscript Department, British Library: B.L. Add. MSS. 32641. The Linnean Society, London, has a copy of them. Menzies kept the final part, which was sold in London in 1875. It was then sent to the National Library of Australia.

For published versions of the Menzies journals: Menzies in the Pacific, and at Tahiti — D. Shineberg (Ed.), *Bulletin de la Société des Études Océaniennes*, 'Le Discovery á Rapa et á Tahiti 1791–1792: *Journal d'Archibald Menzies*', tome XVIII, no. 3, pp. 789–826, Papeete (1981); Menzies in British Columbia — C.F. Newcombe (Ed.), *Menzies' Journal of Vancouver's Voyage, April to October, 1792...*, Victoria (1923); J. Gorsline (Ed.), *Rainshadow: Archibald Menzies and the botanical exploration of the Olympic Peninsula*, Port Townsend, Washington State (1992); Menzies in California — A.D. Eastwood (Ed.), *California Historical Society Quarterly*, 'Menzies California Journal', California (1924), vol. 2, part 4, pp. 265–340; Menzies in Alaska — W.M. Olson (Ed.), *The Alaska Travel Journal of Archibald Menzies, 1793–1794*, Fairbanks, Alaska (1993); Menzies in the Hawaiian Islands — W.F. Wilson (Ed.), *Hawaii Nei. 128 Years ago by Archibald Menzies*, Honolulu (1920). The journals provide a good record of how Menzies executed his instructions.

[6]The Home Secretary in 1791 was Henry Dundas.

1791

To Edmund Malone

[Soho Square]
27 March 1791

My Dear Sir,

I am required by our Colleagues in the Comm[itte]e appointed to provide money for erecting a monument to the memory of Johnson,[1] to request your opinion respecting the most proper place for it, Westminster Abbey or St. Pauls. If you favour us with your sentiments, I have no doubt they will have great weight in our decision, which we hope to make on the 9th. day of April next.[2]

Sr. Joshua, who prefers St. Paul's, says that the honour as well as interest of the arts are materially at stake, & will receive great advantage if we set the example of a monument in a Church which has hitherto lain fallow for the harvest of the Chisel; that Westminster is already so crowded it would be a deadly sin against taste to increase the squeeze of Tombs there; & that St. Paul's is the most honourable station for the monument of a great man. Burke says waggishly this is borrowing from Peter to give to Paul, but he supports Sr. Joshua fully & firmly.

On the other hand, the supporters of St. Peters say they are engaged to the body of Johnson, to the public, & to the Dean & Chapter to erect it there; that they ardently wish to make an end of a business that has been kept in suspense so very long; & lastly that they fear their funds, between 7 & 800 pounds, will be very short of the necessary expences of such a monument as the Dean & Chapter of St. Paul's will require, which they understand is to be of Colossal dimensions. To their last argument, however, Sr. Joshua has answered by declaring that if, by further solicitations in which he expects the Comme. to co-operate, sufficient money cannot be raised, he will himself furnish it. Thus stood the matter when we parted last night. Pray let us hear from you in time. You will essentially oblige us all, & none more than,

Your Faithful Servt,

J. B.

Present at the Comme
Sr. Joshua Reynolds
Mr. Burke [Edmund]

" Wymondham [William Windham]
" Metcalf [Phillip]
" Boswell [James]
Sr. Wm. Scott [William]
Sr. Jos. Banks[3]

[N.H.M. B.L. D.T.C. VII. 205–206; B.L. Add. MS. 22549, f. 12; M.L./D.L. Banks Papers Series 78.03.]

[1]Banks was a member of the Literary Club, which Dr. Samuel Johnson and Sir Joshua Reynolds presided over at the *Turk's Head* in Gerrard Street. When Banks joined, Johnson wrote to Bennet Langton (1737–1801) to say that he would be 'a very honourable accession', 31/10/1778, a view he repeated to James Boswell, 'he will be a reputable member', 21/10/1778.

Johnson died in December 1784, and Banks was one of his pallbearers. Committees were established to arrange a monument to Johnson. Banks argued that it should be placed in Westminster Abbey with Johnson's grave, which is at Poet's Corner.

[2]Banks was a member of a special committee set up on 5 January 1790 to conclude matters. He withdrew when St. Paul's Cathedral was chosen, and this is where the statue by John Bacon RA (1740–1799) can be found.

[3]Edmund Burke (1729–1797): statesman; William Windham; Phillip Metcalf FRS (1733–1818); James Boswell (1740–1795): Johnson's companion and biographer; Sir William Scott (1745–1836), 1st Baron Stowell: lawyer and Privy Councillor.

Letter 44

1791

To James Wiles

[Soho Square]
25 June 1791

Instructions for Mr. James Wiles, appointed to proceed with Capt. Bligh in his Majesties Ship, the Providence, to the Society Island in the Pacific Ocean, for the purpose of collecting Bread fruit trees & other usefull productions of the Islands in these Seas, to be conveyed to the West Indies, which instructions are to be carried into execution by his assistant Mr. Christopher Smith,[1] in case he himself should be prevented by any unexpected event from executing that Service.

The first duty to be inculcated into the mind of a man who undertakes to Serve his Majesty is obedience to the orders of those his Majesty is pleased to put in command over him. You are therefore to remember on all occasions while you belong to his Majesty's Ship Providence that you obey the orders of the commanding officer, & that wherever you can make yourself usefull, without neglecting the care of your Plants, by performing any extra duty, you cheerfully & readily undertake to give assistance.[2]

The Reason of Government incurring the expence of the Equipment of the Ships with which you will Sail, & the Salary of yourself & your assistant, is with a view of furnishing the West Indian Islands with some of the most usefull productions of the East; &, as the whole of the Success of this undertaking depends ultimately on your diligence & care, it cannot be too strongly recommended to you to guard yourself against all temptations of Idleness or Liquor, as the inattention of a Single day, or, in particular circumstances, of half an hour, may render the great expence incurr'd by Government in this humane & liberal undertaking wholly abortive.

The Ship is ordered to be fitted in the best manner for stowing of the trees & plants, & for preventing as much as possible their receiving Injury; & you will give your attendance on board her as often as you may think it necessary, & suggest any improvement or alteration in the plan proposed which may appear to you calculated for their greater security and preservation.

On your arrival at Otaheite [Tahiti], which is the Island where your principal duty will commence, Capt. Bligh will procure from the natives as many small bread-fruit trees as can be planted in the Pots, Tubs, Cases &c. which will be provided for their reception. To these you will pay the utmost attention, digging them up with all possible care, & planting them in the places intended for them, watering & shading them with the most unremitting attention, till by the circumstance of their growing you are Sure they have taken firm root, of which you will from time to time acquaint the Captain. You will also take charge of Some plants of the Sweet Plantain, called Meia, the apple or Aoee, the root called Peah, of which the Islanders make a kind of Pudding, & the very large Yam, which is of a better kind than any in the West Indies, and all such other Fruits or Plants as Capt. Bligh shall direct you to receive.[3]

On your departure from this Island, the most difficult part of your duty will begin, & must be attended to with the most unremitting assiduity. As you have been bred a Gardener, & know the proper proportion of air, water, light, shade, warmth, Shelter &c. that Plants require, I shall pass over all such matters, & confine myself to the particulars in which a garden on board a Ship requires a different treatment from that of a Garden on shore.

The principle enemy to the health of Plants in Long Voyages is the Spray of the Sea, which Seldom fails to rise into the air at such times as the wind is high enough to turn over the tops of the waves into what the seamen call white caps, & to fall in the form of a salt dew on every thing it meets with. If the Leaves of a Plant are wetted with this dew, the salt contained in it Crystallises upon their Surface as they become dry, &, unless it is Speedily washed off with fresh water, the Plant infallibly perishes.

The most constant attention must therefore be hourly paid to this circumstance, & the greatest care taken that, whenever this salt dew appears in the air in the slightest degree, the windward & top scuttles & ports of the cabin be shut, & the Lee ones also as soon as it becomes sufficiently prevalent to come in by that road; & if, by mischance, it is found to have fix'd itself on any of the Leaves, which may sometimes happen on the Lee side when the Plants are in much want of air, & is easily known by their tasting [of] salt when the tongue is applied to them, the commanding officer must be immediately apprised of the circumstance, & requested to supply a sufficient quantity of fresh water to wash it off, which must be done by sluicing them from the rose of a watering pot, & rubbing them with Linen rag or fine pick'd oakham.

As watering with Rain water is observed to be more conducive to the health of Plants than with that which has been long kept in Casks, you must, whenever it rains in moderate weather, set open the top gratings & Lights, that the plants immediately under them may receive the rain. You must also give your assistance, under the direction of the Commanding Officer, to catch rain water in proper vessels to water the rest, & procure a supply for future use, as rain water, tho it may taste of Tar, will be always better for the Plants than any other that can be procured.

Monkeys, Goats, Dogs, Cats &, in short, every animal that is not confined, is to be dreaded as the smell of the earth is very attractive even to those that do not eat vegetables, & tempts them to Scratch. It is to be hoped that none of the two first will be permitted on board as, howsoever cautiously they are look'd after, they may, if they escape, destroy the whole garden in half an hour. The mischief of most other animals, however, if they are properly confined, may with due attention be guarded against, if they are not suffered to be numerous.

Rats, Mice & cockroaches are the natural enemies of Plants on board a Ship. They destroy them by gnawing the bark & young buds, which you must constantly examine in order to be aware of their mischief before any great progress is made. The utmost attention will naturally be paid by the commanding officer to keep the two former from ever becoming numerous. The latter may be easily kept under by mixing fine powder of white hellebore root with bread dust, also finely pulverized, & a little Sugar. This, strewed upon the deck where they are, will destroy them by millions, & you will be furnished with a proper quantity for that purpose.

As often as the Ship Stops for the purpose of procuring a supply of fresh water, the Plants should be thoroughly sluic'd with it that all remnants of Salt may be wash'd from their Leaves, & also that, if any has got into the mould, it may be dissolved & run out. Tho this may Sometimes consume half a boat load of water, the Commanding Officer will not object to the expence as it can be immediately replac'd without affecting the general stock. Some of the least healthy plants may also be taken on shore, where they must be plac'd in the shade; but I doubt whether they will receive any material benefit unless the Ship is detained a week at least.

Whenever you Shall meet with plants in your opinion particularly beautifull or curious, you are to acquaint the commanding officer, who, if he thinks proper, will give you leave to take on board one or two of each Sort for the use of his Majesty's Botanic Garden at Kew; provided, however, that the stock of bread fruit Trees & usefull Plants is never diminished by the admission of curious ones, which are on no account to be planted, except in Such pots or cases in which the bread fruit & other usefull plants have died.

On your arrival at Timor, or such other places in the East Indies inhabited by Europeans as Capt. Bligh shall chuse to touch at, he will procure the Plants & Fruits which are used for food or otherwise by the natives, among which are the Lansa, Mangostan, Durion, Jamboo of several sorts, Nanca, Tchampadha, Blimbing, Jambolan, Boabidarra, Sal/*o*/ac, Black & Long pepper &c. All these you will plant in the pots, Tubs & cases that are vacant by the Bread fruit or other usefull plants they contained having died. He will also take on board Some bushels of the Mountain Rice, which is cultivated without being overflow'd with water; & you are, as far as you can, to make yourself acquainted with the mode of managing it, that you may communicate the Same to the inhabitants of the West Indies.[4]

As the climate of the Cape of Good Hope is not proper for the bread fruit plant, it is hop'd the ship will not be under the necessity of touching there, lest great loss of Plants should be the consequence. In your Passage round it, however, you are certain to meet with Cold boisterous weather. All possible precautions must therefore be taken, as many pots, Tubs & cases [will be] crowded into the cabbin as can possibly be stowed there, & a fire kept as often as you shall judge it necessary. The Trees upon the Deck must also be covered up as snugly as possible with old sails, & whatever other materials can be Supplied for that purpose. It is hoped, however, that Capt. Bligh will give the Cape a good berth, & use the utmost diligence & dispatch in releasing you from this dangerous situation, in which, if you meet with one night's frost, the whole Success of the voyage may possibly be frustrated.

On your arrival at St. Helena, the Commanding officer will, in obedience to his instructions, direct you to deliver to the Governor & Council there Such proportion of Bread fruit, & other usefull Trees & Plants, as can be Spar'd with propriety for the use of the East India Company, which will be replaced by such trees & Plants as the Governor & Council may think fit to send from the Companies Garden for the use of the West India Islands, & of his Majesties Botanic Garden at Kew, for which it is but too probable that the losses you will have Sustained in passing the Cape will have made ample room. If that should be the case, & any spare room be left after having received all that is prepared for you, you are to fill it up with such wild plants of the Island as are not yet introduced into English Gardens, & particularly to dig up one or two Fern trees with large balls of earth adhering to their roots, & plant them in large baskets for his Majesties use.

From hence the ship will proceed to the Island of St. Vincent's in the West Indies, at which place one half of your cargo is to be deposited in the care of the Superintendant of his Majesties Botanic Garden[5] for the use of the Windward Islands. From him you will receive in return plants growing in pots, which have been prepared in conformity to a list sent from the Royal Botanic Garden at Kew, which are to be put into the Places from whence the bread fruit & other plants have been taken, & attended with the same care & exactness.

You are, during your Stay at St. Vincents, to instruct the Superintendant in the best manner you are able in the manner of cultivating the Plants & seeds that shall be delivered to him, & also in the modes of preparing their produce for food used in the Countries from whence they have been brought.

The Commanding Officer will next proceed to Jamaica, where the remainder of your Cargo intended for the benefit of the Leeward Islands is to be deposited in the hands of such person[s] as the Governor & Council Shall please to direct; except that in all cases where the number of any Species of plant is Sufficiently large to Secure its Success in the Island, one or two individuals are Spared to be reserv'd for his Majesty's Botanic Garden at Kew.

There is good reason to believe that, in gratitude for his Majesties Bounty, Such gentlemen in that Island as are conversant in the Study of Botany will have prepared plants growing in pots to be sent home as a present to him, in which case you are to receive them, lodge them in the places from whence those you leave behind have been taken, & pay all possible attention to them on your passage home.

You are here also to use your best endeavours to instruct Such person or persons as the Governor & Council shall direct to attend you for that purpose in the manner of cultivating the Plants you have brought from the East Indies, & of preparing their produce for food, or such other purposes as they are intended to answer.

If application is made to engage you to stay on the Island & undertake their culture yourself on Such terms of emolument as you approve, you are at liberty to accept them, & will have leave to quit the Ship, in which case your Salary will be paid to Such [a] person as you Shall authorize to receive it in England, up to the time of your leaving her & entering into your new service; &, in case of your refusal of Such offers, your assistant is to have the Same option, provided, however, that one of you remains with the Ship to take care of his Majesties Plants during the Passage home.[6]

On the ship's arrival in English seas, which ever of you is on board must take the earliest opportunity to acquaint me by letter of Her return,[7] & furnish me with a list of the Plants brought home for his Majesty, distinguishing the number of each species, & the kind of Pots or tubs in which they are planted, in order that proper boats may be provided to carry them to Kew, which will be Sent to meet you with as much dispatch as possible, especially if the season should unfortunately [be] cold.

On the arrival of these boats, immediate measures must be taken, with such assistance as the commanding officer can Spare, to embark all the Plants in their respective pots & tubs, & stow them away to the best advantage; which done, you are to embark with them both of you, if both of you return, & never quit them till you have delivered them to his Majesties Botanic Gardener at Kew,[8] who will be ready at Kew bridge to receive them; & you are particularly to take notice that no plant, cutting, layer, Sucker, or part of Plant be on any condition whatever taken away by any other person, but that the whole be Safely & carefully delivered to his Majesties use.

[N.H.M. B.L. D.T.C. VII. 218–226A.]

[1]Christopher Smith (d.1807): botanic collector; assistant collector on HMS *Providence*, 1791–1793; assistant, Botanic Garden, Calcutta, 1794; went to Moluccas, 1796; Superintendent, Botanic Garden, Penang, 1805–1806.

[2]This paragraph, and the remarkable detail of the instructions which follow, suggest the care Banks took to avoid some of the problems that beset the *Bounty* expedition. It is evidence of his determination, this being the second attempt to transport bread-fruit to the West Indies. A newly built West Indiaman, HMS *Providence*, and a brig, HMS *Assistant*, were selected for the task. They left Spithead on 3 August 1791 under the overall command of William Bligh in the *Providence*, with Nathaniel Portlock (c.1748–1817) in charge of HMS *Assistant*. Bligh was unwell at Tenerife, but the ships arrived at Table Bay on 6 November. They then made their way from the Cape of Good Hope into the Indian Ocean and towards Adventure Bay, which was reached on 9 February 1792. After Adventure Bay the ships took a course to the south of New Zealand, and swung north. They arrived at Matavai Bay, Tahiti, on 10 April. Bligh reckoned 2126 bread-fruit were collected at Tahiti, along with 508 other plants.

The ships departed on 20 July. Bligh meandered as he travelled to Timor, charting many new islands, especially among the Fiji Group. The complicated journey through the Torres Strait took 19 days, and the ships anchored at the Dutch settlement of Kupang. The Cape of Good Hope was the next stop. St. Helena was reached on 17 December, where Bligh had been instructed to deposit 10 bread-fruit plants with the governor. St. Vincent in the West Indies then received some 544 plants, and Bligh was given 465 pots and 2 tubs of various items for return to England. He was delayed in Jamaica from 5 February to 15 June 1793 due to the outbreak of war between Britain and France, but the mission to transplant bread-fruit to the West Indies had succeeded. Bligh eventually reached Deptford on 7 August, with a consignment of valuable West Indian Plants for The Royal Botanic Gardens at Kew.

[3]The sweet plantain (*Musa x paradisiaca*) is a plantain banana, for which the vernacular name is still 'Meia'; The 'apple', or 'Aoee' is the Malay apple (*Syzygium malaccense*). The Malay apple was a Polynesian introduction to the Pacific from the Malay Archipelago; 'Peah' is still called 'pia', or Tahiti arrowroot (*Tacca leotopetaloides*); The 'very large Yam' (*Dioscorea alata*) has tubers which weigh as much as fifty kilograms, and which may reach two metres in length.

[4]'Lansa' is the lanseh or langsat (*Lansium domesticum*). It is a popular fruit with edible arils; 'Mangostan' is the mangosteen (*Garcinia mangostana*); The durian (*Durio zibethinus*) is a famous foul-smelling yet delicious fruit; 'Jamboo' is the Jambu (*Syzygium* spp.); 'Nanca' is unclear; 'Tchamapadha' is the chempedak (*Artocarpus integer*). It is a relative of the bread-fruit and jack; 'Blimbing' (*Averrhoa bilimbi*) is a bitter fruit allied to the starfruit (*Averrrhoa carambola*); 'Jambolan' (*Syzygium cumini*). This is related to the Malay apple and jambu; 'Boabidarra' is unclear; 'Salac' (probably *Salacca* spp.). These are palms with edible fruits, especially *Salacca zalacca*; Black and long pepper (*Piper nigrum*, and *Piper longum*); 'Mountain rice' is almost certainly rice not grown under inundated conditions. 'Mountain rice' is now the name used for a species of North American *Oryzopsis*.

Banks seems to have drawn directly on his experience on the *Endeavour* expedition for these plants. For instance, he recorded many of them in his journal: *Endeavour Journal*, vol. II, pp. 206–217. The 'Nanca' and 'Boabidarra' appear in this section of the published journal with some editorial explanation.

[5]Dr. Alexander Anderson.

[6]Wiles decided to stay as Superintendent of the Government Botanic Garden at Bath, Jamaica: Wiles to Banks, 5/3/1793, M.L./D.L. Banks Papers Series 52.13. Christopher Smith, his assistant on the voyage, returned home.

[7]Smith to Banks, 1/8/1793, M.L./D.L. Banks Papers Series 52.15. For more of the correspondence arising from this voyage, and an account of its aims, organization and collections: D. Powell, *Bulletin of the Institute of Jamaica: Science Series*, 'The Voyage of the Plant Nursery, H.M.S. *Providence*, 1791–1793', Kingston (1973), no. 15, pt. 2. The point is made that the figures for plants collected and transported tend to vary, as is the fact that in Jamiaca the bread-fruit tree has flourished.

[8]William Aiton (1731–1793): the first superintendant at the Royal Botanic Gardens, Kew.

Letter 45

1792

To Sir George Macartney, 1st Earl Macartney

[Soho Square]
22 January 1792

My Lord,[1]

I send with this for your Lordships Perusal a Chinese book, which I have been in the habit of Calling their Encyclopedia.[2] It is said to consist of some hundred Volumes, & to contain an exact explanation with Figures of all the mechanical Contrivances & Handicraft tools usd in the Empire.

The 70 volumes sent with this Contain some matters that I hope will prove usefull, especialy in mechanics; &, as Cuts offerd an universal language, your Lordship will be as well able to understand the Point at which their science is now Fixd nearly as well as if you was acquaintd with the Chinese Language.

I send also the first volume of Hermans beytrage,[3] which Contains the Short Statistical account of the Chinese Empire I mentioned when last I had the honor of seeing your Lordship.

I confess I Feel much interest in the Success of an undertaking from whence the usefull as well as the ornamental sciences are likely to derive infinite benefit. The Chinese appear to me to possess the Ruin of a state of Civilization in which, when in Perfection, the human mind had carried all kinds of knowledge to a much higher Pitch than the Europeans have hitherto done.

What is there of the great inventions as we call them that are not known to these people? Gunpowder, Printing, Arabic notation of figures as we call it, Paper making, with infinite others upon which the very state of Science & Civilization absolutely depends, were only reinvented, if not perhaps stolen from that Country. Their Porcelane is a chef d'oeuvre of Chemistry, which Europeans have not yet been able to attain. Their very Tea depends upon a Chemical Process we are unable to imitate, which takes from the Leaves of the Plant not only an unpleasant taste, but, as Kaempfer[4] tells us, a deleterious Quality, & substitutes in its Room a flavor agreable to all Palates, leaving just enough of the Poison to afford possibly the most exhilarating medecine we possess.

To Learn these arts alone would be to give to Europe an invaluable blessing, but how many more of these wrecks of ancient wisdom China possesses we know not. Probably most of her manufactures stand upon the Same basis, and a few Learnd men admitted among their workmen might in a Few weeks acquire

Knowledge for which the whole Revenue of the immense Empire would not be thought a sufficient equivalent.

I sincerely hope that no stint whatever in Point of Expence will be sufferd to interfere with the Proper preparations for Commanding Respect & Procuring information. When a nation whose Annual revenue Amounts only to a Few Millions beyond what is necessary to pay the interest of a debt, & Consequently cannot be applied to magnificence, sends an ambassador to one whose net unmortgaged revenue is said to be 100,000,000, they ought not to be economical, but profuse in the extreme to Save themselves from becoming ridiculous in the eyes of a people before whom they must appear respectable if they would procure any kind of advantage.[5]

[S.L. Banks MS. C1:95; U.Y.S.L. Banks MS.]

[1]In 1791 Lord Macartney was appointed to lead a British Embassy to China, 1792–1794. Banks took a keen interest in Chinese natural history and culture, and was asked to assist with preparations for the mission. This was largely because of his association with Sir George Leonard Staunton FRS (1737–1801), who went as Secretary to the Embassy, and Henry Dundas, a minister and member of the Board of Control of the Honourable East India Company. Banks sought well-qualified personnel, and provided useful information where he could, but found tight financial constraints and secrecy difficult to cope with. The Embassy was rebuffed by the Chinese.
[2]There was more than one Chinese encyclopaedia available at this time. This one might have been the *Sancai tuhui*, which may be translated as 'The Illustrated compendium of heaven, earth and man'. This was a Ming (1368–1644) compilation in 106 juan (sections), with many illustrations. The British Library copy is bound western style, but would have consisted of some 60–70 volumes originally.
[3]Benedict Franz Johann Herman (b.1755): superintendant of mines, Siberia. He published *Beyträge zur Physik, Oekonomie, Mineralogie, Chemie, Technologie und zur Statistik besonders der Russischen und angränzenden Länder*, Berlin and Stettin, in 3 volumes, 1786, 1787 and 1788.
[4]Engelbert Kaempfer, the German traveller and artist, who was especially interested in Far Eastern culture and natural history.
[5]See Letters 46 and 48.

Letter 46

1792

To Josiah Wedgwood F.R.S.

[Soho Square]
6 February 1792

Ld. Macartney, who has accepted the post of Ambassador to China, has suggested the Propriety of takeing under the appearance of a servant a Person well Skilld in all the mysteries of Pottery,[1] who may, if opportunity offers, acquaint himself with any mode of manufacture usd by the Chinese which the artists of this Country are ignorant of.[2]

To you, sir, who have turnd the art of Pottery into a Science, I naturaly look for advice, & wish to know whether you are acquainted with such a person as we want; & whether [he] would be willing to engage for two years absence, which is as little as we can count upon.

Unfortunately, the Company[3] no sooner conceivd the Idea than they were in haste to Carry it into Execution, so that we shall not have so much time as in my opinion is necessary to Prepare, for which reason I wish for a speedy answer.[4] As the whole is kept secret, I must also request that you will not tell of it. All who realy know it, speak of it with doubt.[5] The Rumour, therefore, which always attends matters of this nature will dye away, and never gain a permanent Credit.

[S.L. Banks MS. C2:4.]

[1]Macartney to Banks, 17/2/1792, S.L. Banks MS. C2:43.
[2]Banks wrote to Thomas Percival FRS (1740–1804) in much the same terms to enquire after a dyer: Banks to Percival, 7/2/1792, U.Y. Beinecke Banks MS., S.L. Banks MS. C1:64.
[3]The Honourable East India Company.
[4]Wedgwood had referred to the problems caused by secrecy: Wedgwood to Banks, 13/2/1792, S.L. Banks MS. C2:44.
[5]Banks complained about the difficulties he laboured under, and felt that natural history was being neglected in the plans for the Embassy: Banks to Staunton, 17/4/1792, S.L. Banks MS. C1:87.

Letter 47

1792

To Jean Florimond Boudon de Saint-Amans (alias Amand)

Soho Square
27 February 1792[1]

Sir,

Accept my best thanks for the eloquent Eulogium[2] you have written in praise of that God of my adoration, Linné.[3] I congratulate the Linnean Society on possessing a pen so capable of doing justice to the merits of the man whose name they have set before them as an excitement to emulate his industry, & follow as nearly as possible the paths of his Genius.

I thank you, Sir, also particularly for having honoured me with copies of this & your other useful publications. The place they occupy in my Library will always fill my mind with gratitude to the Donor. I wish I could be good for any thing to you where I am, & am with perfect consideration,

J. Banks.

[N.H.M. B.L. D.T.C. VIII. 13; M.L./D.L. Banks Papers Series 73.037.]

[1]The Mitchell Library version of this letter in Banks's hand has been dated 28 February.
[2]Saint-Amans to Banks, 12/1/1792, B.L. Add. MS. 8098, f. 1. Saint-Amans had sent Banks a copy of his 'Éloge de Charles von Linné...', Agen (1791), *Extrait...du Journal des Sciences utiles ci-devant d'Hist. naturelle année 1790.*
[3]Carl von Linné (Carolus Linnaeus).

xi. **Carl von Linné (Carolus Linnaeus) F.R.S.**
Portrait by L. Pasch after A. Roslin, 1775. Copied for Banks, and presented by Robert Brown in 1853 to the Linnean Society, London.

Letter 48

1792

To Sir George Leonard Staunton F.R.S.

Soho Square
18 August 1792

My dear Sir George,

Herewith you will receive a Copy of the *Icones Kaempferianae*,[1] which I beg you to except, & enclosed in it such loose hints relative to Horticulture & Botany as I could conveniently throw together.[2] I beg you will Present them with my hertiest Good wishes to Ld. Macartney,[3] & request that his Lordship will in the Course of the Voyage[4] put them in such hands as he may find the most willing, & the most able to make use of them.

The Ceremony of taking Leave I have always considered a Painfull and unnecessary Struggle, & I therefore left our Friends last night before the time of retirement arrived. I must, consequently, beg you to explain to all of them that every Good wish I am capable of Forming will attend them during the whole course of their Peregrinations, and that they will find, if it Pleases God that I should continue as resident in this world at the time of their Return, one who will honor in the extreme the endeavours they make for the Promotion of Science, the increase of Commerce between distant nations, & the Real prosperity not only of their own Country, but also of that they intend to visit.

Relative to the Ipecacuanha,[5] which you were so good as to say you would search for in the Brasils, all I know on the Subject of it is that, as a valuable part of the Materia Medica, it is in some degree a disgrace to the national character of our Medical men that we are still wholly ignorant to what Family of Plants it belongs. Govr. Philip,[6] at my desire, procured for me when he was there specimens of the two sorts, the black & the white, which seemed to differ widely from each other; but, unfortunately, there were only leaves & roots, & you know that without Flowers & Fruits the distinguishing characters cannot be investigated. Either flowers or Fruits, however, would be a valuable addition to our Medico-botanical knowledge.

Adieu, my dear Sir George. Be assured of my heartiest good wishes for your success in your Present undertaking, & your happy return. If either Lord Macartney or yourself should favor me with such notices of your progress from time to time as you can spare time to make, you will lay me under infinite obligation. I beg also of you to hint to Dr. Gillan[7] that a few words from him relative to the progress of his department will be infinitely agreable, and that I shall feel gratefull to Dr. Scott[8] or any of the other gentleman who can spare time to Favor me with a Line when opportunity offers.

From this time till that of your departure my ad[d]ress will be Revesby Abbey, near Horncastle, Lincolnshire,[9] &, if there is anything I can do for you respecting either Public or Private Business, I shall have great Pleasure in hearing from you there. If not, give me leave to add adieu to the best wishes of,

Your Faithfull,
Humble Servant,

Jos. Banks.

[N.H.M. B.L. D.T.C. VIII. 68–70; M.L. Banks MSS. 743/3*.]

[1]A volume of engraved illustrations of Japanese plants copied from drawings by Engelbert Kaempfer in the British Museum. These were published by Banks in 1791 as *Icones selectae Plantarum quas in Japonica collegit et delineavit Engelbertus Kaempfer*. See Letter 35, Note 2.
[2]Banks sent an important document with the letter above. It contained over 3500 words of careful advice: 'Hints on the Subject of Gardening suggested to the Gentlemen who attended the Embassy to China', August 1792, L.S. MS. 115.
[3]Sir George Macartney.
[4]The Embassy to China, 1792–1794, led by Lord Macartney. Staunton was Embassy Secretary. On the voyage to China the Embassy ships, HMS *Lion* and the Indiaman *Hindostan*, stopped at Rio de Janeiro from 30 November 1792 to 17 December. They then sailed across the Southern Atlantic Ocean, rounding the Cape of Good Hope on 7 January 1793.
[5]'Ipecacuanha', a Tupi name for the roots of the medicinal plant *Psychotria ipecacuanha*. Research has shown that this Brazilian plant contains a powerful emetic and amoebicide. The bioactive compound, emetine, is now extracted from the plant roots and marketed as a drug. *Psychotria ipecacuanha* belongs to the family *Rubiaceae*. Staunton sent Banks parts of the plant when the Embassy was in Rio de Janeiro, but there were no flowers among them: Staunton to Banks, 9/12/1792, S.L. Banks MS. C 1:79; Banks to Staunton, 24/2/1793, R.S. M.M. 19 120. Some years later the plant was described with its flower by the Portuguese botanist Felix d'Avellar Brotero (1744–1828): *Transactions of the Linnean Society*, vol. 6 (1802), pp. 187–141, 'Descriptions of Callicocca Ipecacuanha.' Read 3/2/1801.
[6]Arthur Phillip, who was at Rio de Janeiro with the 'First Fleet' in August 1787.
[7]Dr. Hugh Gillan FRS (d.1798): physician; naturalist. Gillan was recommended by Banks to go on the Macartney Embassy to China following Dr. James Lind's withdrawal.
[8]Possibly Dr. Helenus Scott (1760–1821): physician, then on the medical staff of the Honourable East India Company at Bombay.
[9]Revesby Abbey, Banks's Lincolnshire estate home.

1792

To an Unknown Correspondent[1]

Soho Square
12 November 1792

My dear Sir,

It is with Deepest Concern I take up my pen to inform you that My Poor Uncle, Mr. Hodgkinson,[2] is no more. He died yesterday morn after a tedious & most painfull Complaint, in the Latter part of which his Physicians gave him no kind of hope, Dr. Austin, the person who Last attended him, having declard it his opinion that he had a stone[3] in his Bladder & water in his Chest. The opinion, however, respecting the Stone was never divulgd to him, it being the decided opinion of the Doctor that, as the water on his Breast must soon kill him, the painfull & horrible operation of Cutting for the stone ought not to be Proposd to him.

He has never held up his head since Mrs. Hodgkinsons death. Indeed, he always declard it his Solemn wish that he might as soon as possible Follow her.

It is with the truest sincerity that I condole with you & the General on the Loss of a Friend who you both have for so Many years respected in no ordinary mode. To me his Loss is severe. He was the Guardian of my nonage, & the friend of my youth, & I have ever lookd up to the manly sense of his opinions with veneration, as I have to his Kindness to me with gratitude.

I beg my best Compts to the General, who I understand is with you, & that you will beleive me,

Your Faithfull & Affectionate
Hble Servant,

Jos: Banks.

[U.Y. Beinecke Banks MS.]

[1]Neither the recipient nor 'the General' in the letter have been identified.
[2]Robert Banks-Hodgkinson FRS (1722–1792) succumbed to pleurisy and kidney failure on 11 November. His wife, Bridget, died shortly before, on 14 July 1792. Robert had been close to his nephew, and acted as Sir Joseph's guardian after the death of his father, William Banks-Hodgkinson (1719–1761).

[3]A stone was removed in the autopsy, which Banks studied carefully. It weighed '2 ounces 2 pennyweights 7 grains Troy'. By 26 January 1793 the stone weighed less, being '2 ounces all but 2 grains', and it was then delivered to Dr. Everard Home.

Letter 50

1792

To Count Alessandro Volta F.R.S.

[Soho Square]
[20 December 1792]

Sir,[1]

I have receivd through the hands of Mr. Cavallo[2] two Letters on the Subject of the nerves & the Muscles, which do infinite honor to your Sagacity as a Philosopher, & your reasoning as a Logician. I beg, Sir, to offer you my thanks for them as a Prelude to those of the Royal Society, which will in due time be forwarded to you officialy from the Secretary whose duty it is to return thanks for Papers read at the meetings of the Society.

We are at present employd in clothing them in an English dress, which is necessary in this Country to their being generaly understood. As soon as that Can be Compleated, they will be presented to your Confreres of the Society, who will, I am sure, receive them with welcome & applause.[3]

I thank you, Sir, for the Communication more particularly as it is upon a Subject intirely new, which has lately begun to occupy our people considerably. The utility of your Communication, therefore, is Particularly Evident as it prevents them from wasting time in determining facts already Known in your Country, which has had the honor of the discovery, & where enlightened minds like yours have been already Employd a Considerable time in pursuing it.[4]

[B.L. Add. MS. 8098, f. 102v.]

[1]This letter was drafted on the verso of: Volta to Banks, 30/5/1792, B.L. Add MS. 8098, f. 102.
[2]Tiberius Cavallo FRS (1749–1809): natural philosopher.
[3]*Philosophical Transactions*, vol. 83 (1793), pp. 10–44, 'Account of Some Discoveries made by Mr. Galvani, of Bologna; with Experiments and Observations on them. In two Letters from Mr. Alexander Volta, F.R.S. Professor of Natural Philosophy in the University of Pavia, to Mr. Tiberius Cavallo, F.R.S.' Read 31/1/1793. Volta's greatest contribution to science commenced this year.

In 1791 Luigi Galvani (1737–1798) had discovered that the muscles in dead frogs contracted when brass and iron were placed in contact with each other and the muscle. This prompted Volta to repeat the experiments using various metals, whole frogs and other animals. Volta also found that tin on the tip of his tongue caused an unpleasant taste when touching a silver spoon resting further back (a bi-metallic arc). This was all due to the effects of electricity. Volta concluded from his experiments that the source of the electricity was in the junction of the metals, and not a 'nerveo-electrical fluid' in the animals as Galvani had suggested. Volta went on to produce a list of metals in order of their electricity production, thus deriving the electromotive series. Afterwards he invented the 'voltaic pile', thus making it possible to produce high electric currents for the first time.

[4]See Letter 81.

Letter 51

1793

To William Windham M.P.

Soho Square
20 April 1793

My dear Windham,

The Questions now at issue between the Board of Longitude & the House of Commons appear to be as follows:[1] whether the House shall bestow the honorable distinction of a public reward on an inferior artist, manifestly to the injury & discouragement of those who are superior to him in the Same Line; & whether, in order to effect this purpose, they will supersede the jurisdiction long ago Solemnly given by them to the Board of Longitude.

On both these points I have now done what I conceive to be the duty of my station in such cases. I have stated in a Printed Paper,[2] signd by myself, those Leading Facts which in my opinion ought to influence the Conduct of the House; & I have taken measures that the names of the Persons who Constitute the Board of Longitude may be Known to the members. Having now done these things, I shall wish for no Farther delay on my own account.

In Answer to yours I must, however, Observe that if the Conduct of the Board of Longitude is on this instance subjected by the House of Commons to the examination of a Comm[itt]ee Above Stairs, I shall Consider the Board as sorely humiliated; & in Case it is so determind, & the members who consititute this Select Commee are not men as well informd on the subject matter of the business in Question as the members of the Board of Longitude are, I shall for Ever after Consider its decisions as liable to be superseded by the jurisdiction of an incompetent Tribunal, & its Functions as being thereby renderd useless to the Public, if not virtualy repeald.[3]

Beleive me, my dear sir,
unalterably Yours,

Jos: Banks.

[B.L. Add. MS. 37854, ff. 50–51; R.S. M.M. 7. 128; A.P.S. Miscellaneous Manuscript Collection.]

[1]The Board of Longitude was created in 1713 to assess and reward methods and inventions for calculating longitude accurately at sea. Banks was an active *ex officio* commissioner from his election as President of the Royal Society in 1778. Windham's 'Devonshire Friend' was Thomas Mudge (1717–1794), who produced his first marine timekeeper in 1774 to compete for the longitude prize. This amounted to £10,000. By 1793 prizes often went to those craftsman who made a chronometer accurate enough to be used to determine longitude reliably. Banks felt that John Arnold (c.1736–1799) produced better instruments than Mudge could, and certainly more of them. Arnold ought therefore to receive the major awards.

However, from 1792 a parliamentary Select Committee had also deliberated over such matters. It included Windham, and other powerful political figures such as William Pitt the younger. They favoured Mudge, who became ill in the early 1790s. His son, Thomas Mudge Jnr. (1760–1843), a lawyer and horologist, then took up the dispute over who deserved more credit. Banks feared that the Board of Longitude's authority had been undermined, and that the serious task of finding the best way to measure longitude was being disrupted.

[2]'OBSERVATIONS on Mr. MUDGE'S Application to Parliament for a Reward for his TIME-KEEPERS, which, agreeably to the Act of Parliament of 14th of George III had been unanimously denied him by the BOARD of LONGITUDE', 27/3/1793, R.S. MS. 7 100, with handwritten draft.

[3]See Letter 53.

Letter 52

1793

To Sir John Sinclair F.R.S.

[Soho Square]
23 April 1793

My dear Sir John,

I thank you for your plan of a Board of Agriculture received this day,[1] which I have perused with attention. You know I was always averse to the establishment of any board of that kind if it was to be armed with the powers of making regulations relative to the direction of agricultural industry. Yours, I observe, has no such view. It is calculated merely to examine the progress and encourage the improvement of Agricultural efforts, and, as such, is in my opinion far better suited for a private than for a public establishment.

Relative to the Royal Society, I beg leave to set you right if you suppose, as your readers will do, that Lands granted by the Crown are enjoyed by that body as any part of their present Fund. They had, it is true, at an early period of their institution a grant of Chelsea College,[2] but that was very soon resumed by the Crown, and the sum of £1500, and a gilded mace (as I remember),[3] given them as an equivalent. But true it is that since that time the Society has not enjoyed either Lands granted to them by the Crown, or any annual payment whatever derived from the public purse.

That the president of the R. Soc. ought to be an official member of your intended Board, I very much doubt. The present incumbent, indeed, as he has paid much attention to rural economy, might not hold a place there unprofitably, but, on a review of his predecessors, much doubt will arise whether any one of them would have made an useful member of an Agricultural Board.

Above all, it appears to me that the employment of such a board for the collection of a Statistical Account of England would prove itself an heterogenious arrangement.[4] If Government wishes for such a Collection, there are surely many modes of procuring it obviously far more likely to succeed in practice than that of employing 24 idle men, who are to have nothing given them for their pains, and who have abundance of other work set out for them to which their institution will direct them to give a preference. I fear indeed the expence of your Board, independant of all Idea of Emolument, will far exceed your calculation, for those who give their labour gratis are seldom frugal, &, I fear, you will never persuade 24 persons without salary to agree in dividing the Labour of their Functions equally, or indeed in uniformly promoting the interest of a public Trust.

Beleive me, My dear Sir,
Your Faithful Hble Servt,

Jos. Banks.

[I.A.C.R. Rothamsted Archive S192. 5 f. 127.]

[1]Sinclair to Banks, 23/4/1793, I.A.C.R. Rothamsted Archive S192. 5 f. 125–126. The plan appeared in the *Annals of Agriculture*, vol. 20 (1793), pp. 204–213, 'Plan for Establishing a Board of Agriculture and Internal Improvement.' Sinclair largely ignored Banks's advice.
[2]Chelsea College was granted to the Royal Society by a Royal Patent of April 1669 together with thirty acres of land. However, the college buildings were in poor condition, and the Society could not afford to repair them. When Charles II (1649–1685) expressed an interest in the site, which was wanted to build a military hospital on, the Royal Society decided to sell it back for £1300. This was done in 1682.
[3]Charles II presented the mace in 1663 as a sign of royal favour.
[4]Sinclair gained parliamentary approval for his proposals, and, as founder and first president of the new Board of Agriculture, he pressed ahead immediately with ambitious plans for a series of county surveys in England. The Board lacked sufficient staff and funds for this type of enterprise, and in practice had little influence with government. Nevertheless, Sinclair continued with expensive publications, which drained finances.

Banks was an official member of the Board, and along with others saw that such spending would lead to financial disaster. In June 1796 a number of members voted to stop all payments for the county surveys, and in March 1798 Sinclair was voted out of the chair. He was replaced by John Southey, 15th Baron Somerville (1765–1819). The Board was then managed more prudently; so much so that Sinclair could safely be re-elected to the chair in 1806.

Letter 53

1793

To William Pitt the younger, Prime Minister

Soho Square
16 June 1793

Sir,

Situated as I am in the chair of the R[oyal] S[ociety], and concluding that those who placed me there have a right to enquire in all matters where the progress of useful knowledge is likely to be impeded whether I have exerted myself in order to prevent an evil of so great magnitude, I trust you will not think me improperly troublesome in addressing this letter to you.[1]

When you gave your consent to the expenditure of public money as a reward to Mr. Mudge for his improvement of a Time-keeper, I conclude your consent was on condition that the allegations contained in his petition were proved, and that, in case of his not being able to prove the material ones, or of their being fairly disproved by the testimony of credible persons, you would no longer be bound by your consent; but, on the contrary, you would be compelled by your duty to the Nation, and by that fostering love of Science which Ministers are always believed to possess, to resist a claim upon the public Treasure wholly unfounded in fact, and to rescue Science from the disheartening precedent of the least meritorious being rewarded by the Nation, while those whose talents have been more successfully applied for the benefit of the country are neglected, and possibly despised.

Now, Sir, as the Evidence of the Select and Sub Committees is not yet printed, and probably may not be delivered to the Members till after the Motion intended to be brought forward by Mr. Windham[2] to morrow has been disposed of, I think it fair to conclude that you have not read it, and therefore to state to you that, as far as my Idea of Evidence goes, it has been fairly proved to the Committtee by men of acknowledged abilities and veracity that watches made by other artists have gone much better than those for which Mr. Mudge claims a preference; that if the method of calculation proposed by him as more proper than that adopted by the Commissioners of Longitude had been used instead of it, his Timekeeper would not even then have gone within the limits prescribed by the act as sufficient to entitle him to a farther trial; and that the Timekeepers of other makers have been brought into more general use than those of the petitioner.

These, among a multitude of additional facts tending to shew that Mr. Mudges claim to superior excellence is in no degree founded in fact, have been proved by witnesses brought by those who, from a love of Science, were desirous to ward off the evil likely to arise if the Council of the Nation granted a Reward to one who did not deserve it. But on Mr. Mudge's part, he has *not* proved that the Commissioners of Longitude ever requested him to prosecute his invention; that the principles of his watches are more permanent than those of others; that they require less attention to keep them from varying, or that their accuracy of going for a considerable length of time may be depended upon. These allegations must therefore be considered as not being founded in fact.

Under this view of the matter in question, I have no difficulty in asserting that Mr. Mudge has faild in the proof of every one of those allegations of his Petition on which you can have grounded your Consent to the appropriation of public money as a reward to his merit, & consequently no scruple in calling upon you to take in the House the part of Science against the misrepresentations of Mr. Mudge, & the zeal of his friends to whom the advancement of useful knowledge is avowedly of less importance than victory in the present contest. You, Sir, may save the Board of Longitude from the censure in which they must be implicated if the House reward a man whom they have declared less meritorious than other claimants upon the Funds entrusted by Parliament to their expenditure. You, Sir, may save the House from a decision which cannot but be disgusting to the understanding of their constituents, and may rescue Science from the discredit she must fall into if Public Rewards are given to those who have the greater interest in preference of those [who] have the most merit.

I have the honor to be, Sir,
with the most unfeignd respect & regard,
Your Obedient & most Hble Servant,

Jos: Banks.

I have usd the hand of a Clerk,[3] who lives with me and does the duty of a Secretary, to Copy this in order to save you the Trouble of decyphering my almost illegible hand writing.

[P.R.O. 30/8/100; R.S. M.M. 7. 138.]

[1]Parliament had already awarded Thomas Mudge £500 for a chronometer Banks believed was inferior to other designs. There was a vote in the House of Commons the day following this letter by which Mudge received a further £2500. Banks argued that politicians did not have the expertise or information available on which to base a decision, and weakened the authority of the Board of Longitude considerably with such awards.

[2]As a Member of Parliament William Windham had influence in a Select Committee which considered these issues.

[3]John Swan: clerk and secretary at Soho Square, c.1780–1795. Banks was suffering from gout, which affected his handwriting. This can be seen from the final comments and post script which Banks wrote in the P.R.O. document.

Letter 54

1793

To Professor Johann Friedrich Blumenbach F.R.S.

Soho Square
31 December 1793

Dear Sir,

What is the matter I do not exactly Know, but I have heard rumours that the King, displeasd at the great use which was made of his Quarterly messengers, has abridgd very much the Space allotted to their Packages; & I fear that the monthly messengers have Shard the Same fate, which I shall greive to Know as our Communication through their means usd to be so very Convenient.[1]

Immediately on my return to London from the Countrey, where I had been Confind by the Gout Considerably beyond the usual time of my return, I deliverd to Mr. Best a Perfect Otaheiti Cranium in hopes it might very shortly have set out for your Collection, & this was about the middle of November. He told me that he Feard he could not Send it till After Christmas, & you will best Know whether you have receivd it, as you must have done before this Comes to your hands if it goes by the monthly messenger. Best looks so secret on the subject that I did not think it Either Friendly or Polite to ask him any Questions.

I should have been able to send you my best Sentiments respecting the Countenances of the New Holland Savages more Correctly than I can do had Governor Philips[2] been in London. I should then have had an opportunity of viewing them at Leisure, probably amusd by some of their natural exercises, but I have waited for his return from Bath till my patience is nearly Exhausted. All I can say is that they did not in my opinion at all resemble the Portraits No. 6 & 7 of Mr. Webber.[3] That gentleman indeed was by Profession a Lan[d]scape painter, & what he has done in the portrait Line I have given little Credit to, for he Drew a Picture of Capt. Cook which did not in any degree resemble him, tho his hard & markd features were So Strong as to make it almost impossible for any one to miss his Likeness. At all Events, I am quite Sure that Lebruns figure of his New Guinea savage is much more like our New South Wales visitors than Mr. Webbers.

I shall rejoice to receive your Annotations on the subject of mummys, & to hear them read to the R[oyal] S[ociety]. I can have no doubt that [they will] also be Printed, but that, you Know, is decided by [a] Comm[itt]ee of Publication as it is Calld, in which [I] have only one voice.[4]

The Season of Printing lists of the R. Society is at hand. In a few days the new lists will be publishd. As soon as that is the Case I will forward a Copy, with one also of the Statutes, to Mr. Best for you.

Pray give my best respects to Mrs. Blumenbach. Tho unknown, the Puddings she was so good as to send were the very best I have tasted. A merry Christmas to you both says,

Your Obedient & Very
Faithfull Hble Servant,

Jos: Banks.

[U.G. — University of Gottingen, Col. Blumenbach III. Bl. 40–41.]

[1]George III was Elector of Hanover, and Banks's influence at court meant he had access to the diplomatic post between London and Hanover.
[2]When Arthur Phillip left Sydney in 1792 he took with him two aborigines, Bennelong and Yemmerrawannee. The former was introduced to George III, but neither liked London very much. They lived at Eltham in Kent for a while, and Yemmerrawannee died there in May 1794. Bennelong returned to New South Wales in 1795.
[3]John Webber RA (1752–1793): artist. Webber was a supernumary on Cook's third voyage in HMS *Resolution*, 1776–1780. He made numerous drawings and paintings, including some of native heads and figures. Webber was particularly involved in supervising the engravers for the official account of this voyage, which was completed under Banks's overall direction as *Voyage to the Pacific Ocean. Undertaken for making discoveries in the northern hemisphere...*, London (1784).

From 1784–1791 Webber exhibited finished works at the Royal Academy, which were mainly concerned with exotic South Sea subjects. He also produced more than one portrait of Cook. An example in which the explorer appeared quite severe was the oil painting of Cook's head and shoulders completed at the Cape of Good Hope, 1776, National Portrait Gallery, London. Webber based later portraits on it, such as an unsigned three-quarter length version, Museum of New Zealand, Wellington, New Zealand, Te Papa Tongareuva's Collection. For a fine sample of Webber's work on Cook, and the South Pacific: *The Art of Captain Cook's Voyages: The Voyage of the Resolution and Discovery 1776–1780*, vol. 3, R. Joppien and B. Smith, Oxford (1987).

The catalogue numbers given by Banks correspond to those he neatly laid out in an untitled list of engravings for *Voyage to the Pacific Ocean...*: '6 Man of Van Diemens Land' and '7 Woman of Van Diemens Land', N.L.A. Banks MS. 9/29. These illustrations were shown together under the title 'A Man and Woman of Van Diemen's Land' in that publication.
[4]*Philosophical Transactions*, vol. 84 (1794), pp. 177–195, 'Observations on some Egyptian Mummies opened in London. By John Frederick Blumenbach M.D. F.R.S. Addressed to Sir Joseph Banks, Bart. P.R.S.' Read 10/4/1794. Blumenbach spent more than three months of the winter of 1791–1792 in London, which was when he studied the Egyptian mummies and first met Banks in person.

Letter 55

1794

To an Unknown Correspondent

[Soho Square]
2 March 1794 Sunday Morn.

My dear Sir,[1]

I have, since I had the pleasure of seeing you in Frith Street, & of receiving your very friendly Letter, tried to arrange myself in such a manner as to be able to attend the club on Tuesday, but have not been able to affect it. I therefore trouble you with this line merely to say that I feel an inexpressible obligation to my friends, who have on my account arranged their intentions towards the */question/* bishop[2] in the way you express.

My feelings on that subject have, however, been dead for several years past. A man who does an injury, it is said, never forgives the party he has injured. If therefore the good Bishop had injured me, he probably would never have forgiven me; but as his 'telum' was in my case 'imbelle' & 'sine ictu',[3] he probably has by this time admitted to his Breast the return of Christian Charity.

For me, even had I been injured, custom would have allowed me to forgive, but, in truth, I was not. I was seated more firmly in my chair by the Doctor's attempt to dispossess me; &, in truth, our controversy ended as I told him it would on the very first day that war was regularly commenced. 'You have raised a storm, my good Doctor', said I, '&, trust me, I will ride upon it.'

It is now, I think, 8 or nine years since he has left me unmolested to Crow upon the dunghill he so valiantly disputed with me, & has not in that time made his appearance in the Society's rooms three times. To bear malice against him therefore would be insufferable. In truth, was it in my power to be present, I would certainly for conscience sake, & to prove to myself that no hatred remained lurking behind, ballot in his favour. How far he may be a Club-able man, every individual will for himself consider when he determines on turning his Elbow to the right or to the left; & whether I should have done the one or the other if he was a man whose former conduct had been wholly indifferent to me is a matter which I am not able to solve to my own satisfaction.

Believe me, My dear Sir,
with infinite gratitude for
your friendly conduct on this
occasion,

Most faithfully Yours,

Jos. Banks.

[N.H.M. B.L. D.T.C. IX. 28–28A; S.I.L. Banks MSS. 44A.]

[1]Note by Dawson Turner reads: 'no address — the original in the possession of Mr W. S. Fitch. (1833)'.

[2]William Windham proposed Bishop Samuel Horsley for membership of the Literary Club in early 1794. Horsley was Banks's chief opponent in the Royal Society dissensions of 1783–1784. Apparently, Banks was not alone in thinking Windham's proposal 'a matter of impropriety': Banks to Malone, 3/3/1794, U.Y. Beinecke Banks MS.

[3]Banks often has a particular phrase from Horace or Virgil in mind when he uses Latin phrases in his letters: Virgil, *Aeneid*, 2, 544, 'sic fatus senior, telumque imbelle sine ictu'. Banks appears to mean something like: 'his blows were without force.'

Letter 56

1794

To Evan Nepean F.R.S.

Soho Square
16 March 1794

My Dear Sir,

I have receivd Mr. Dundas's letter,[1] & sent it, with resolutions & a summons for a Meeting of Lincolnshire Gentleman resident in London on Thursday next, to the Newspapers.[2] My Grand Jury will, I am sure, feel inexpressible pleasure at finding themselves honord with the King's approbation, & I have every reason to expect that the subscription on Thursday will be proportioned to that which has met a reward invaluable to the hearts of those who feel true Loyalty.

I very much wish to be intrusted by his Majesties Ministers with their Ideas of the proper Measures to be adopted for the Defence of Lincolnshire before the Meeting on Thursday, &, if Cavalry are approved, I shall thank you for an account of the customary Levee Money for a Troop, expence of Horses &c.; in short such Documents as, with the account of pay now in my hands, will compleat the information I shall in that case have occasion for.

I cannot help remarking here that I sincerely hope the resolution of arming any part of the Force to be raisd with Pikes will be retracted. It seems a solecism in politics that those who hold the Public Purse, & are able to Furnish Musquets, which from their price are unattainable by the lower Classes, should from motives of Economy teach them how they may for two pence a piece arm themselves with a Weapon of which one Blacksmith will manufacture several hundreds in a day.

If you have any occasion for me at the Office, I shall come immediately on being sent for. The business now in hand has with me the Preference over all others, but I shall not, unless sent for, attend any more as my occupations will not any longer allow me to wait in the antichambers for whole mornings as I have formerly done.

I am, Sir,
Your faithfull Servant,

Jos: Banks.

[L.C.L. Banks MS. 11/1/14A.]

[1]Henry Dundas, who was Home Secretary at this time: Dundas to Banks, 14/3/1794, L.C.L. Banks MS. 11/1/12A.

[2]Banks was appointed High Sheriff of Lincolnshire this year. He thought carefully about the defence of his county homeland. Meetings, correspondence and resolutions show him at work, as does his detailed pamphlet: 'Outlines of a Plan of Defence against a French Invasion; Intended for the County of Lincoln; but applicable to all other Counties whose inhabitants are sensible of the Danger of the Present Crisis, and willing to subscribe a small Part of their Incomes for the Security of the Remainder', (1794).

1794

To William Pitt the younger, Prime Minister

Soho Square
3 December 1794

Sir,

Tho' I am well aware that in the present situation of affairs there is no occasion for the use of a Telegraph, it is very certain that a change might take place which will render that instrument essentially useful, & that such a change might take place rather suddenly. If these axioms are admitted, it is clearly desirable that a well digested plan for a Telegraph, easily & expeditiously to be erected, with proper arrangement of signals, & a collection of appropriate sentences to be conveyed by single signals, is a useful, if not a necessary precaution.

As I have had numerous applications from persons who wished to be employed in this line, I do not doubt but that you, Sir, have had many more. This day,[1] however, I received a letter from one of the most ingenious & diligent men of my acquaintance, who seems desirous to enter upon the business. He is a man of competent property, & not likely therefore to be expensive. His contrivance, for he has already thought much on the subject, is, he says, simple & very cheap. His motive for applying seems to be merely ambition. He would like, he says, to be made Telegrapher to the King. Do not smile, for a man motivated by that species of ambition may easily be made to exert his utmost vigour of mind at a very small expence in money.

For his ingenuity, his industry, & for his having already succeeded in some most difficult undertakings, I will readily vouch. If you, Sir, incline to make use of him, I will write to him to come to town immediately, from whence his residence is not distant.[2]

J. Banks.

[N.H.M. B.L. D.T.C. IX. 132; R.B.G. Kew Archives B.C. 2. 119B; P.R.O. 30/8/102.]

[1]Sir Charles Wilkins FRS (1749–1836): orientalist. Banks received and acted on Wilkins's letter rapidly: Wilkins to Banks, 2/12/1794, N.H.M. B.L. D.T.C. IX 130–131.
[2]Pitt's reply has not been identified.

Letter 58

1795

To Sir Henry Hawley

Soho Square
6 July 1795

My dear Sir,

Had I receivd any thing like the Customary notice of the time when the honor I have newly receivd was to have been conferrd upon me, I should not have Sufferd you to Read the first account of it in the Papers. I am too well aware of the Kind interest you take in Every thing that is agreable to us not too have taken great pleasure in communicating the news to you, but from circumstances, the Particulars of which I have not thought it necessary to Enquire into, I Learnd only on Tuesday that I was to have my Ribband[1] on Wednesday, & Gen. Abercrombie, who was to have been my Companion in honor, actualy did not receive his [in] Time enough to take advantage of it.

While I was kneeling on the Cushion before the King, & the Sword which had [dubbed] me a Knight was Still hanging over my Shoulders, the King Said to me in a Low voice, 'Sir Jos: I have many years wishd to do this'. A mark of distinction so flattering has made me pleasd with an honor which, as it came without Sollicitation, you may Easily beleive not to have been any Object of my wishes. It had in the First instance been made palatable by coming in a direct Course from the pure Fountain of honor without any portion of Ministerial Contamination,[2] but this Latter instance of Gracious Condescension had made it inexpressibly valuable to my Feelings.

Lady Banks, My Mother[3] & Sister desire to be rememberd in the Kindest manner to you & to all yours. I beg, my dear Sir, you will beleive that I am sincerely in all good wishes, & that I am with

Sincere Affection & real Regard,
Most Faithfully yours,

Jos: Banks.

We Shall be happy if you can give us Good news respecting our Cousin, whose life was so long despaird of, & whose cure, if it is Effected, will not be owing to the Skill of London Surgery.[4]

[L.A.O. Hawley MS. 6/3/4.]

[1]Banks received the Order of the Knighthood of the Bath on 1 July 1795 at the Court of St. James, but was given one day's notice of the event. Sir Ralph Abercromby (1734–1801) found this too short to appear.
[2]Banks had first been offered this honour in 1794, while he had been working to raise funds for volunteer foot and cavalry forces in Lincolnshire. His aim was to supplement the existing militia at a time when the French threatened Britain with invasion. As High Sheriff of Lincolnshire he felt unable to accept the knighthood. He feared the suggestion 'that my Political & not my Literary Character is the Object of the distinction granted to me': Banks to Dundas, 5/4/1794, U.Y. Beinecke Banks MS., de Beer Collection 1.
[3]Sarah Banks (*née* Bate), (1719–1804).
[4]One of Hawley's daughters had been unwell earlier in the year: Hawley to Banks, 26/1/1795, B.L. Add. MS. 33980, f. 2. She must have been from his second wife, Anne (*née* Humphreys), (d.1829), whom he married in 1785. Their second daughter was called Louisa.

xii. **'The great South Sea Caterpillar, transform'd into a Bath Butterfly'.**
From a caricature, 1795, by J. Gillray. British Museum, Department of Prints and Drawings, *Catalogue of Political and Personal Satires*, 1935, vol. 7, 8718.

Note: Banks was depicted as a butterfly emerging from a chrysalis by the satirical cartoonist James Gillray shortly after receiving the red ribbon of the Order of the Bath in 1795. Banks was shown gazing up at the Sun, which symbolized the royal favour by which he gained the honour. The star and sash denote the prestigious order. Banks was also decorated with plants, crustacea and insects. This imagery was not unusual, and it drew on Banks's reputation as a naturalist. So did the reference to his earlier travels in the South Seas. The comments underneath imitated the style of a collector, interested in curiosities only for their rarity and market value. They contrast with the direct and technical language of a truly scientific account. The humour of the image comes mainly from the comic inversion of collector and collected.

Letter 59

1795

To Sir James Bland Burges

Soho Square
29 December 1795

My dear Sir,

Noe,[1] the Gardener who was sent to Russia,[2] is Just now arrivd after a passage of 11 weeks. By his account it appears that the success of his management was very good. He Lost very few Plants indeed, & when he Carried his Cargo to Paulesski[3] he found that only 3 of all those he brought had before been seen in Russia.

The Grand Duchess[4] Receivd them with all possible honor. 15 Coaches were sent to Carry them from the water to the Palace, &, as they arrivd there in the night, the Garden was Lighted up with Lamps for the facility of unpacking them.

The Grand Duchess was in the Garden by Six the next morn, & gave directions respecting them. Before morn Noe was sent for into the Palace, & had the honor to exhibit the Plans of the hothouses at Kew, & the Drawings of Plants he had been chargd with, to the Grand Duke and Duchess. At 2 they Came to the Garden with 70 attendants.

The G. Duchess orderd the King of Englands Mark, G. R., to be inscribd by Noe on Every Pot he had brought that they might not be confounded with her own, & She Every day Spent an hour in Learning the names of the Plants.

When her imperial highness removd from Paulessky to a Palace About 30 miles from it, she orderd Noe to attend with Every Plant that Should Flower, & she with her own hand made a drawing of Each.

The most Hansome overtures were made to Noe if he would Consent to Enter into her Sevice, but he declined on account of his having had a recal[l] sent to him from Wurtemburgh,[5] where he says he is sure he never shall receive so much pay as was offerd him by the Grand Duchess.

He Receivd, when he went away, a hansome gold watch, & 100 ducats as a Present.[6]

Whenever you Come to Town, my dear Sir, if you will let me Know I will meet you at the office that we may finish this affair.

Beleive me, my dear Sir,
with infinite Regard & Esteem,
very Faithfully Yours,

Jos: Banks.

He has brought a large Collection of Seeds & some curious Plants for Kew. Many more, I am Sure, will follow.

[R.B.G. Kew Archives B.C. 2. 130.]

[1]George Noe (fl. 1790s): from Stuttgart, a foreman-gardener, Royal Botanic Garden, Kew.
[2]In 1793 Catherine II (1729–1796), Empress of Russia, requested seeds and plants from the King's Botanic Garden at Kew for the gardens she was making at the palaces of Tsarskoe Selo and Pavlovsk, both south of St. Petersburg. It seems that Sir Charles Whitworth (1752–1825), who became ambassador at St. Petersburg this year, encouraged the idea initially. By 1795, with a defensive treaty between the Coalition Powers and Russia in prospect, arranging such a present was an appropriate diplomatic gesture. At a meeting in Kew in April 1795 Banks was therefore instructed by George III to select and prepare a collection of exotic plants to be sent to Russia. A gardener from Kew was to go and instruct the Russians in the best way of cultivating the plants, and plans and elevations of the hothouses at Kew were prepared as part of the gift. Banks started work at Kew immediately. A month later he visited James Bland Burges at the Foreign Office to initiate the official business of choosing a ship. A small vessel, the *Venus*, was eventually provided for the trip. Banks designed a two-tier structure for the hold, where more than three tons of material from Kew would be stored. This was situated amidships.

The cargo was assembled and loaded ready for a final inspection by Banks early in July. The *Venus* could then sail, with George Noe in charge of the plants inside. Its convoy was guarded by HMS *Daedalus*, and so arrived safely at Kronshtadt, near St. Petersburg. The plants were unloaded and placed in hot houses at the Pavlovsk gardens by 11 August. For the ensuing month Noe explained how best to cultivate them there. He left St. Petersburg on 11 September, and was back in London by 27 December. The letter above was written on the day he reported to Banks in Soho Square. The mission had been a success, but Banks received small thanks from the Foreign Office for his efforts. Indeed, his personal costs, amounting to some £162, were not fully repaid for three years.

Correspondence and other documents relating to this episode were published in: H.B. Carter, 'Sir Joseph Banks and the Plant Collection from Kew sent to the Empress Catherine II of Russia 1795', *Bulletin of the British Museum (Natural History)*, London (1974), vol. 4, no. 5, pp. 283–385.
[3]'Paulesski', by which Banks meant the palace at Pavlovsk.
[4]Maria Feodorovna, Grand Duchess, and wife of Paul (1754–1801), Grand Duke and later Emperor of Russia. The palace and gardens at Pavlovsk had been under construction since 1781 for the Grand Duke.
[5]Baden-Württemberg, province in south-west Germany, of which Stuttgart is the capital.
[6]Burges received an expensive snuff box as well, but Banks was not sent a gift.

1796

To Sir Henry Hawley

Soho Square
8 January 1796

My dear Sir,

I confess I feel not a little disturbed at the idea of your son's[1] 'Compliance to give up his favorite inclination', as from the words of your letter[2] it would seem as if he complied, not through a conviction of the truth of the argument used against the measure, but from the idea of submitting his reason to the opinion of another; a measure I confess I do not think a proper one in any case, or by any means likely to produce future comfort or quiet.

I must therefore, my dear Sir, intreat you by no means to suffer your Son to alter his intentions on my account unless he is himself convinced that his intentions were likely, if executed, to produce unpleasant consequences. What I said I meant as advice to the Father, not to the Son, as we pray to God to keep us out of temptation, well knowing that however we may by reason defend ourselves against the effects of temptation, we are not able to conduct ourselves in such a manner as to steer clear of innumerable trials.

His reasons against entering into a Marching Regiment are good, and your reasons for his going into the Guards are also good. I had formed an idea that the inducement was from some of his co[n]temporaries being in the Corps.

Why then should my fears prevail? I told you upon what foundation they rested, but as Gold is more purified by being tried in the fire, why should not your son stand the trial of a purification in St. James's Street?[3] The more trials he successfully combats when he is young, the fewer dangers he will have to encounter when he comes to a more mature age.

I have only one thing to entreat, and that I do entreat most tenaciously, which is that he will not on any account follow my advice, or by any means act in conformity to it, unless he is satisfied that the consequences of it will make his life more pleasant. To him, then, it would be otherwise, for I dare not Expose myself to the mental Burthen which would be laid upon me if, by following my advice, my young Relation Should abandon an occupation in which he was willing to Employ himself, & become afterwards an idle useless member of the Community; a mere 'fruges Consumere natus',[4] a Species of Vermin I have been always usd to hold in more detestation than Foxes or Carrion Crows. I had rather he

Should Run all other risques than the danger of falling into the Yawning Gulph of Idleness.

[U.Y.S.L. Banks MS., Sir Joseph Banks Papers.]

[1]Sir Henry Hawley, 2nd Baronet (1776–1831). Banks was related to the Hawleys through his grandparents, Joseph Banks II (1695–1741) and Anne Banks (*née* Hodgkinson), (d.1730). Their second daughter, Elizabeth (1720–1766), married Dr. James Hawley (1705–1777) in 1744. Banks's correspondent was their son, a cousin, and 1st Baronet of Leybourne Grange, Kent.
[2]Banks had not expected that his opinions would prove decisive in the choice of military career taken by another man's son: Hawley to Banks, 6/1/1796, U.Y.S.L. Banks MS; Hawley to Banks, 10/1/1796, U.Y.S.L. Banks MS.
[3]The young Hawley wanted to join a 'Guards' regiment based in London, with officers in it who knew him and would help him to make progress. He wished to avoid being stationed somewhere remote from his family and friends, unless during war time.
[4]Horace, *Epistles*, 1, 2, 27: 'fruges consumere nati'. A devourer of provisions; a squanderer; a parasite. This manuscript is a draft copied by an amanuensis until the last two sentences, which are in Banks's hand.

Letter 61

1796

To Jaques Julien Houttou de La Billardière[1]

Soho Square
9 June 1796

Sir,

I have spoken to different members of our administration on the subject of Restoring the Collections[2] made by you & your Companions in your late voyage, & have had the pleasure of Finding them much inclind to acquiesce in my opinion of the Propriety of doing so. I was Promisd to be made acquainted with the Determination of the Cabinet on the subject last Sunday, but owing to the pressure of other business it has not yet been taken into Consideration.

I am Confident I Should before this time have been empowerd to Restore them had not an application been made for them in the name of the Brother of the Late King of France.[2] I have combated this claim with all diligence, and I hope not without success.[3]

Whatever the event may be, & I entertain Good hopes that it will be Ultimately favorable, rest assurd of my unwearied Diligence. That the science of two Nations may be at Peace while their Politics are at war is an axiom we have Learned from your Protection to Capt. Cook, & surely nothing is so likely to Abate the unjustifiable Rancour that Politicians frequently entertain against Each other as to See Harmony and good will Prevail Among their Brethren who Cultivate Science.

Beleive me, Sir,
with Real Esteem and Regard,
Your Faithful Hble Servant,

Jos: Banks.

Remember me affectionately to Broussonet[4] if he is in Paris.

The bearer, M. Le Chevalier,[5] has been indefatigable in his attentions to every thing that Could promote the object we both wish for. You may be sure that he has gained my esteem by the Confidential manner in which I have always Conversd with him on the subject which I Consider as a matter of no Less importance than that of making a Peace between the Science of the Two Nations.[6]

[N.M.M. GAB/2; A.T.L. Banks Collection MS-Papers-0155-19, which is a photostat copy of the original at the National Maritime Museum.]

[1]Banks wrote this letter during a period of prolonged European conflict and revolution. Wars between Britain and France lasted from 1793 to 1802, and from 1803 to 1814.
[2]In 1785 the navigator and explorer Jean-François de Galaup de la Pérouse (1741–1788) set out on a French expedition to the South Seas, which was partly inspired by Captain James Cook's example. It was doomed to failure. La Pérouse was wrecked at Vanikoro Island in the South Seas. In 1791 Joseph-Antoine Bruni d'Entrecasteaux (1739–1793) led a mission in the ships *La Recherche* and *L'Espérance* to discover his fate. The mission included a number of botanists and naturalists, whose job was to collect and observe on the way. La Billardière went as a botanist.

Disaster struck when d'Entrecasteaux died during the voyage, and command passed to d'Auribeau, a royalist. He led officers loyal to the French monarchy, and in 1794 turned pro-revolutionary La Billardière over to the Dutch authorities in Java, along with his colleagues and collections. La Billardière was released early in 1795. However, his collections had been removed separately by one of the loyal French officers, Elisabeth-Paul-Edouard de Rossel (1765–1829). He hoped to offer them to exiled King Louis XVIII (1755–1824). Rossel's plan was to travel to Holland in a Dutch ship.
[3]The British Navy caught Rossel off the Shetland Isles in 1795. The collections were seized because Holland had been annexed to France, and so was now at war with Britain. Rossel protested that he had no idea war had been declared. The British government therefore agreed to let King Louis decide what should be done with the collections, and he offered them to Queen Charlotte. After an initial inspection by Banks in March, she accepted much of the botanical material.

By now La Billardière had found his way home via Mauritius. He and the French Directory applied for the collections to be returned. They argued that the collections had not legitimately been in the gift of the French crown. Banks accepted this view, and set about the awkward task of persuading the British government and Queen that everything should be given back. Many of the type specimens collected by La Billardière's on this mission are at the Herbrarium Universitatis Florentinal, Firenze. However, the type specimens of lichens and algae are at the Muséum National d'Histoire Naturelle, Paris.
[4]Pierre Marie Auguste Broussonet.
[5]Possibly Jean-Baptiste le Chevalier (1752–1836): traveller.
[6]Banks wrote numerous letters in pursuit of such a 'Peace'. Letters 63, 64 and 71 were among them.

Letter 62

1796

To William Wyndham Grenville, 1st Baron Grenville, and Secretary of State for Foreign Affairs

Soho Square
20 July 1796

My Lord,

When I had the honour near three weeks ago of waiting upon your Lordship, by your appointment, on the business of M. de Billardiere,[1] I was in hopes I had convinced your Lordship that the measure of returning to that gentleman the collections of natural history he had made during his employment as a naturalist on the voyage of discovery sent from France for the purpose of inquiring into the fate of the late M. de Peyrouse,[2] was a measure likely to do honour to the national character of the English as a people loving science and abounding with generosity, as well as with justice, and liable to no reasonable objection whatever.

I was in hopes also that your Lordship would consider it as creditable to His Majesty's Ministers to grant in this instance a truce to the unfortunate animosities at present subsisting between England and France by following the precedents of their predecessors in the case of M. de Condamine[3] of the French nation, under their late form of government in that of Captain Cook,[4] and under their present one in the mistaken instance of M. Spillard.

I hope I have not been mistaken, though your Lordship will allow that I have reason to fear the contrary, because you promised me a speedy answer, and I have not heard from your Lordship since. Respecting the opinion of M. Billardiere having received any special commission, or enjoyed any salary from the late King of France,[5] I have made every inquiry in my power without learning anything to make me believe that to have been the case. The late King did certainly draw upon private instructions for M. de Peyrouse, and this has probably been the origin of the mistake.

Allow me then, my Lord, to request a speedy answer to this interesting subject, and to deprecate a refusal. M. de Billardiere is, as I am informed by printed documents, at this time Director of the Botanical Garden at Paris, at the head of his department of science, and in a country where, however humanity may have been outraged by popular leaders, science is held in immeasurable esteem. He will have it in his power to appeal to Europe if in this case the justice is refused which was formerly granted by us to De Condamine, and by his countrymen to Cook; and I fear Europe, if such an appeal is made, is more likely to take part with the complainant, than with a nation which for the first time refuses a reasonable indulgence to science in alleviation of the necessary horrors attendant on a state of warfare.

As I possibly may have occasion to correspond with yout Lordship on another subject similar in principle to that now under consideration, I take the liberty to state as follows. The French either have, or will soon solicit from His Majesty's Ministers, a passport for a ship intended to be sent to Trinidad for the purpose of bringing away a collection of living plants deposited there for fear of capture. I hope, my Lord, that this request will be readily granted. The credit Europe has given to the English for having brought useful plants from the South Seas to their colonies in the west, has fully shown that all good men respect the extensive benevolence of increasing the food of mankind by removing useful plants to countries where nature has not provided them; and our amiable Monarch has set the example of sending useful plants from his Botanic Gardens to the East, to the West Indies and to Africa.

Besides, my Lord, the very application virtually offers, during the horrors of a war unprecedented in the mutual implacability of the parties engaged, an unconditional armistice to science. Surely, my Lord, such an offer should not be neglected. The ready acceptance of it may be the signal of the return of the dawnings of good will towards men, and produce consequences, in the present position of Europe, valuable beyond appreciation to all the nations who inhabit it.

I have the honour to be, my Lord, with due respect and unfeigned esteem,

Your Lordship's Obedient,
humble servant,

Jos. Banks.

[Lord Brougham (Ed.), *Lives of Men of Letters and Science who flourished in the Time of George III*, London (1845–1846), vol. II, pp. 387–388.]

[1]Jacques Julien de La Billardière. Banks spoke to Grenville at the end of June on the subject of La Billardière's collections, portions of which had been wrongly given to Queen Charlotte. On 4 August Banks was officially told that he could make arrangements for La Billardère's property to be returned to France: Banks's note of the meeting, R.B.G. Kew Archives B.C. 2 146.

[2]Jean-François de Galaup de la Pérouse.

[3]Charles Marie de la Condamine FRS (1701–1774): French scientist; explorer. La Condamine took part in an expedition to verify Sir Isaac Newton's hypothesis that the Earth is flattened in the Polar regions. This left France for Peru in 1735. La Condamine eventually journeyed across Peru to explore the Amazon.

Banks appears to be mistaken in suggesting La Condamine travelled freely under a truce. The Frenchman claimed that a British privateer fired on his ship on 25 September 1745 as he returned to Europe.

[4]Captain James Cook, whose ships of discovery had been exempted from attack by French vessels.

[5]Louis XVI, King of France.

Letter 63

1796

To Major William Price

[Soho Square]
4 August 1796

My dear Sir,

Since I had the honour of writing[1] to you on the subject of the Collection of Curiosities offered by the Duc d'Harcourt[2] as a present to the Queen, the whole of the business relating to those things has taken a very different turn. I sincerely hope it will not be productive of any disappointment to Her Majesty, & I feel it my duty to do all in my power to obviate as much as I am able all possibility of that being the case.

When the Collection was offered to the Queen, it was supposed by all who were concerned in making the offer that it belonged to the present King of France. It was believed that the late King of France[3] interested himself personally in directing the outfit of the Voyage, & that His Majesty actually employed in his own service the persons engaged to make Collections.

An application having since been made by the directory of France requesting that the Collection might be returned to the Collector, M. de Billardière, in the same manner as M. Ulloa's[4] Papers were returned to him when captured by an English Vessel, in order that he might be enabled to publish his Observations for the advancement of knowledge as M. Ulloa had done, farther inquiry was made, in the course of which it appeared that the late King of France took no particular interest in the outfit of the expedition in which M de Billardiere sailed. The mistake of his Majestie's having so done probably originated in the fact of his actually having done so on the Voyage of the unfortunate M de Peyrouse,[5] who, it is confidently said, was honor'd with private instructions in the King's own handwriting.

Under this view of the business, his Majesties Ministers have thought it necessary for the honor of the British Nation, & for the advancement of Science, that the right of the Captors to the Collection should on this occasion be wav'd, & that the whole should be returned to M. de Billardiere in order that he may be enabled to publish his Observations on Natural History in a complete manner.

You will easily see, my dear Sir, that by this determination the right of the Duc d'Harcourt to make a present of the Collection to the Queen is entirely done away, & that it would bring a discredit on this Country if the whole Collection unimpaired in the smallest degree is not actually returned.

By this Her Majesty will lose an acquisition to her herbarium which I very much wish'd to see deposited there, but the National character of Great Britain will certainly gain much credit for holding a conduct towards Science & Scientific men [which is] liberal in the highest degree.

Have the goodness, My dear Sir, to state these matters to the Queen on the first convenient opportunity, & do me the farther favour of acquainting me as soon as you can with Her Majesty's pleasure on the subject. Add, if you please, that it is in my power, I verily believe, to make an addition to Her Majesties Collection as valuable at least as the one in question, [and] that I shall feel myself honor'd in the extreme if I have permission to do so. In case [I] am so fortunate, I will, the moment I return from my annual journey to Lincolnshire, undertake the business of preparing & arranging the plants in such manner as I think the most likely to render them worthy [of] the honor I solicit for them, and for me.

J. Banks.

[N.H.M. B.L. D.T.C. X (1). 60–62; B.L. Add. MS. 33980, f. 72.]

[1]Banks to Price, 2/4/1796, B.L. Add. MS. 33980, f. 62v, N.H.M. B.L. D.T.C. X(1) 35.
[2]François-Henri, Duc d' Harcourt: Ambassador of Louis XVIII at the Court of St. James's. Louis XVIII instructed Duc d'Harcourt to present La Billardière's collections to Queen Charlotte. However, the British government had agreed that everything should be returned to France because, as Banks had argued, none of it ever belonged to the French King.
[3]Louis XVI, King of France, who was executed in 1793.
[4]Antonio de Ulloa FRS (1716–1795): Spanish statesman; soldier; scientist. He left Lima on *La Notre Dame de Bonne Déliverance* in 1744. In August 1745 he landed at Louisburg in Nova Scotia, a port the British had taken a short while before. Ulloa was captured along with his papers and taken to England in December, but then released. His papers were returned to him by Martin Folkes, PRS, (1690–1754).
[5]Jean-François de Galaup de la Pérouse. La Pérouse was wrecked at Vanikoro in the South Seas.

Letter 64

1796

To John King, Under-Secretary of State at the Home Office

[Revesby Abbey]
7 November 1796

My dear Sir,

Little did I think when lately I had the honor to write to you on the subject of the Militia Returns, & when I assurd you of the loyalty of the neighbourhood in which I dwell, that a Riot[1] had actualy taken Place at Castor a few days before by which the Deputy Lieutenant[2] had been prevented from filling up the vacancies in the Northern battallion of our Militia, & that another would take place at Horncastle[3] within a few days after.

So little had our magistrates here any Idea of any material dissatisfaction among the People that they actualy appointed the 5th. of Novr., being market day, for the meeting, & it was not till a day or two before that they had reason to apprehend what followd.

On the 5th, when they attended About 11 O'clock, they found the Town occupied by bodies of young men, & that Parties of 20 or 30 were Coming in from all Quarters. They Spoke to several of these parties, whom they overtook on the way, and askd what they wanted. The answer was, 'We are told that it will Cost us 2 or 3 guineas a peice, & we are come to Say that we will neither go nor pay.' On being Reasond with, however, & told that it was very unlikely it Could Cost them a guinea a peice, these parties appeard Satisfied.

When, however, the Parties who had been spoken to by the magistrates & those who had not met in the Town, it seemd to be Spedily Resolvd that their original Plan should be Carried into Execution, which was to Seize all the Lists from the Constables, & by that means deprive the D[eputy] Lieut[enant]s from the Possibility of proceeding to any ballot.

This was immediately done. Some Posted themselves in the avenues to the Town, & stopd all Constables who came in, while others searchd the alehouses; & if any Constables refusd to deliver their Lists, the mob proceeded to those punishments which our laborers here inflict upon one another, lifting them up by Ears &c. so that in a Short time they posessd themselves of all the Lists, except 4 or 5 which had already been deliverd to the Lieuts. These they demanded with clamor, & the Lieuts were Obligd to comply to prevent farther mischeif, after which they left the Town. They were Mobd a little in their retreat, which was made through the Public street in no haste, but nothing outrageous was done to them. We suppose the numbers of the mob from 4 to 500.

I spent yesterday in examining the young men of my Parish who had been at Horncastle on Saturday — about eleven in number. Three of them, I found, had joind the mob. The others Seemd to have kept out of it. They all, however, agreed that they had been told it would Cost Them Two or Three guineas, but none would tell me who it was that told them. Everyone Said it was the Common talk.

The District concernd in this unpleasant business is Lindsey, which Raises more than $\frac{1}{2}$ the militia of the County, pays one half of the County rate, & actualy is the Larger half of the County of Lincoln. It is bounded by the Humber, the Trent & the Witham. If you Carry your Eye from the Humber's mouth up the Humber & Trent to Torksey, from thence to Lincoln, & so along the Witham to Boston, you will see that it is as large as an ordinary County, possibly not much less than 2000 Square miles. Part [is] Rich & Populous, another part [is] poor & ill peopled, but upon the whole answering Probably about to the medium population of England.

The persons who Composd the mob are uniformly servants of the Larger Farmers, & sons of the smaller ones. All the middle agd people, as I am informd, are Loyal in the extreme, except those few Jacobines who are to be found in every market Town I beleive in England. The mob itself, indeed, was loyal, for tho their Cry was 'No requisition', they repeatedly said, 'We will Fight the French whenever they Come. Why do not the gentlemen come out? We will go with them any where.' In the Evening the more daring bought Ribbons, & yesterday some of these were worn in the Churches as I hear.

The origin of this dissatisfaction I myself attribute to the very unusual mode of summoning which has been usd, possibly in order to Obtain the very Returns you were so much in need of. Tho a few persons, we are told not more than 15, are wanting in our district to Fill up the vacancies by death &c., Every Parish was summond, and an alarm created by that means which none of us Know how to allay, for so little Communication have we with our battalion that none of us Know Either the Reason of the summons being so exceded, or the Real business of the meeting.

Once [and] for all, give me leave to say here that I never have had a deputation from the Ld. Lieut,[4] & that I therefore have never at all interferd till the present moment in the miltia business. It is possible these very meetings ought to have been held sooner, but of that I Know nothing.

We are, as you may suppose, under Considerable alarm. The whole power of executive Justice has been seizd out of the hands of our magistrates by these young men, & no one can tell in what manner young men will use Power while in their hands. We expect on any Slight Cause of dissatisfaction, Real or imaginary, to see the young gentlemen with their Ribbons in Force again on a very short notice.

We think that our Lord Lieut ought to have been among us before now as he had notice of the Castor business on the 1st. inst, & we have seen in Norfolk the Marquis of Townshend[5] & all the gentlemen of the County assembled together on a very short notice in a business that we do not think was of so extensive a consequence as ours.

I am the only non-resident gentleman of this County now here, &, as the magistrates do me honor to think that I may be of use, I certainly shall not Leave them till all is settled; except that I must run to London for a few days to the Anniversary of the Royal Society on the 30th of this month.

In the Present state of affairs, when the supplementary militia is so soon to be Raisd, I think there Cannot be a doubt of the Propriety of using all possible dispatch in subdueing the spirits of these young men. My own opinion in respect of the district in which I reside [is] that Cavalry should be Sent to Quarters in The Towns of Horncastle, Spilsby, Louth & Alford, with orders to assist the Peace officers when calld upon for that Purpose by the magistrates.

In the mean time, we are getting information against the most turbulent of those who were guilty at Horncastle of assaulting the Constables in the execution of their duty.

These, as soon as the Soldiers are arrivd, we will send for separately, & if they resist the Civil, we will by means of the Military assistance bring them to our justice meeting, & Commit them to Lincoln Gaol.

In the mean time, Precepts may Issue for a new meeting, & I have no doubt that before they can be held, we shall have so far intimidated Individuals as to Prevent any serious attempt at mobbing. However, if it is our misfortune not to be Able to Prevent it by these means, it will be our duty to suppress it, which we shall have the Power to do, &, I trust, not want the inclination to do it with every possible attention to humanity.

If this Plan is followd, the other district of Lindsey must likewise be furnishd with Troops. Castor, Raisin & Brig would Probably be a sufficient number of Stations, but I say little of that as it must be arrangd with magistrates who will engage to act. Mr. Ellison,[6] newly come into Parliament for Lincoln, is the best Person you Can consult on the subject. I will be answerable both for his Courage & for his discretion.

[L.C.L. Banks MS. 3/1/5.]

[1]Home defence against invasion was arranged through 'voluntary' militia at a time when war with France stretched most of the professional forces Britain could muster. Each parish supplied men, who were armed, clothed and given a shilling a day for exercises. Those who could afford it, and wanted to, paid others to take their place, or simply bought out of service altogether. Poorer labourers and farm workers had no such option. Once chosen, they lost free time in training, and faced the prospect of combat if the French came. Unrest at the way militia lists were being collected threatened Lincolnshire with county-wide turmoil in late 1796. Matters were particularly bad in parts of northern Linconshire, like Lindsey.

A few vacancies had appeared through death and other normal circumstances, but concern grew when it seemed that the lists would be used to draft men for more than the usual militia duties. This became alarm at the thought of being enlisted to fight overseas. Banks and Coltman were extremely active, both in combating this impression, and in bringing the worst rioters to justice. Offenders were tried publicly by them, and in front of crowds which certainly included many people who took part in the disturbances. Although the Somerset Fencible Cavalry arrived to maintain order, Banks and Coltman agreed to enter the towns unarmed and unguarded to show there was no fear of more rioting. However, Banks blamed much of the panic on poor leadership and organization, rather than a few lively malcontents. With uncharacteristic harshness, he complained 'I feel more than I Chuse to express in seeing all the Spirit which I Left in full vigor when I came from home ouzing out by imperceptible degrees through a Leak I cannot stop because our L[ord] L[ieutenant] is a Duke & a Fool': Banks to King, 1/12/1796, L.C.L. Banks MS. 3/2/1. He also kept a detailed diary of these events, 'Diary of Riots in Lindsey 1796', which is listed at the County Library, Lincoln, as Banks MSS. 3/1/27.

[2]Thomas Coltman (1745–1826): Lincolnshire landowner; magistrate.

[3]Banks referred to a number of places in this letter, many of which were frequently mentioned in his correspondence to do with Lincolnshire and its surrounding area. Revesby, the village and parish, is some 6 miles to the south-east of Horncastle town. Both are in Lindsey, as is Caistor to the north. Further north still is the Humber estuary, into which both the Ouse and Trent drain. Central to Lincolnshire is the Witham. It rises south of Grantham, and flows north-east to Lincoln, before turning south-east towards Boston, and then to the Wash.

As an important Lincolnshire landowner, with a seat at Revesby Abbey, all these places were well known to Banks. Most of the towns he suggested as possible quarters for troops had previously been the scenes of disturbances at public meetings. Brigg, Caistor and Market Rasen appear to have been misspelled by Banks.

[4]Brownlow Bertie, 5th Duke of Ancaster and Lord Lieutenant of Lincolnshire.

[5]George Townshend (1724–1807), 4th Viscount and 1st Marquis Townshend.

[6]Richard Ellison (1754–1827): Lincolnshire landowner; MP for Lincoln; banker of Ellison and Co.

Letter 65

1797

To Steno Ingemar Lung

[Soho Square]
[8 January 1797]

Sir,

Having been taught with the earliest rudiments of Science that the proportion between the sexes of mankind is nearly the same over the whole earth, & being in the habit of thinking that if the number of males was materially to exceed in any Country that of females infinite evil would ensue, not only to the morals of that nation, where the most detestable of all vices must naturally /prevail/, but to the neighbouring Countries also, who would be overpower'd & conquer'd, probably for the sake of their women, by the superior number of men that would assail them, I am little inclined to believe that nature has intrusted the key of so dangerous a secret as your friend[1] supposes himself to have discovered to any of the sons of men.

Readily admitting, therefore, that the Children you have begotten have always proved of the sex you expected, I do not feel any kind of conviction. Accident will occasionally cause imaginary experiments to succeed to a much greater extent than your friends' have done, as all who have play'd at Cards or at Dice must know.

Respecting the cure for Barrenness, we have some practioners here who pretend that they can irritate the uterus by injections, & render it more fit for conception than nature had done, but we do not consider this doctrine as in any shape experimentally proved.

If your friend has an intention to make profit of his invention, he must practise the arts he has discovered, & disclose to individuals, upon certain terms & under restrictions in point of secrecy, the mode of obtaining what they may want. Many are the persons in all Countries who wish ardently for children,[2] & many who would pay a large sum for a Male in preference to a female. To such individuals he must apply, & if he has but a moderate share of success he will not fail to persuade people to try his experiments, & to reward him liberally if they succeed.

J. Banks.

[N.H.M. B.L. D.T.C. X (1). 3–4; R.B.G. Kew Archives B.C. 3. 10.]

[1]Eric Norén to Banks, 29/9/1796, B.L. Add. MS. 8098, ff. 372–373; Lung to Banks, 26/10/1796, B.L. Add. MS. 8098, ff. 369–370. Lung had forwarded to Banks a dissertation by Norén about the determination of the sex of children. Banks could not accept Norén's theories.

[2]Banks died without legitimate issue.

Letter 66

1797

To Charles Jenkinson, 1st Baron Hawkesbury and 1st Earl of Liverpool

Soho Square
14 March 1797

My dear Lord,

I have been this evening employd in looking a little into the former Coinages of halfpence,[1] & I find that Your Lordships principle of making the real value of the Copper, with the Cost of the workmanship added to it, coincide as nearly as possible with the nominal value of the Coin has been recognisd by Parliament as early as 1699. I find also that the first Coinage of halfpence of the present Tower Standard, which was made by Queen Caroline[2] in 1729, yielded a profit of £24.13.4 a Ton only; not much more than what must have been necessarily allowd for putting them into Circulation.

I have procurd a Copy of Dances[3] Drawing, but he has since left with me, for your Lordship, two others varying but little from the first one, which he prefers, & indeed so do I. The 3 accompany this.

One remark only occurs on the subject of the Reverse. Britannia has hitherto been constantly represented on our Coins with her Lance resting on her left Shoulder, & bearing in her right hand a branch of Olive; that is, with War kept back, & Peace held forward.[4]

In the present case, the Trident in her right hand threatens successful war at sea, but still she ought, I think, to hold the Olive Branch in her left to shew that she has not abandond the idea of peace.

I have the honor to be,
with sincere regard, respect
& esteem, your Lordships
Obedient hble Servt,

Jos: Banks.

P.S. I doubt we have run the matter a little too near in point of size in the Pence as, if I calculate right, the Ounce of Copper which Boulton[5] proposd for a Penny will, at the Rates he has today stated, cause a loss of two Guineas a Ton at the present price of Copper. The error originates, I believe, in its having been originaly thought right that Government should pay for circulating them, which was changd this Morning, & the addition of Commission, which was not before thought of.

[S.L. Banks MS. Co. 2:41.]

[1]Banks was sworn in as a Privy Councillor on 29 March 1797, and he was immediately appointed to the Coin Committee. By the end of March this committee had produced a report for the King and government on measures to be taken for procuring a supply of copper coinage to meet a severe shortage. This was caused by a run on the Bank of England's gold reserves following a failed invasion by the French.
[2]Caroline of Anspach (1683–1737). She married George II (1683–1760) in 1705.
[3]George Dance the younger (1741–1825): architect; city surveyor, 1768–1815; professor of architecture, Royal Academy, 1798–1805. Dance drew Banks in 1803, and his pencil sketch was engraved and published in 1811, *A Collection of Portraits sketched from the Life, since the year 1793, by Geo. Dance, esq., and engraved in imitation of the original drawings by Will. Daniell, A.R.A.*, London.
[4]The image of Britannia offering peace by extending the olive branch, while holding back war in the form of a trident, was retained on Banks's advice.
[5]Matthew Boulton. A contract for 500 tons of twopences and pence was formally placed with the Soho Mint, Birmingham, by June 1797. In August Banks visited Boulton to see the presses at Soho. He wrote an excellent report on Boulton's operation, and commented on its capacity for minting silver coinage: Banks to Jenkinson, August 1797, N.H.M. B.L. D.T.C. X(2) 122–125, B.L. Add. MS. 38423, ff. 20–22.

Letter 67

1797

To William Pitt the younger, Prime Minister

Soho Square
17 March 1797

Sir Joseph Banks presents his respectful Compts to Mr. Pit, & requests the honor of a few minutes conversation with him at any time that may best suit Mr. Pit's convenience on the subject of opening a communication with Paris for the reception of the Literary Productions of the Members of the *Institut National* & other scientific persons,[1] some of which are highly interesting to the Royal Society, & for sending in return the Philosophical Transactions, the Greenwich Observations &c., which have not been sent for some years past. Sir Jos. has an opportunity of doing this by the means of M. Charretié,[2] Commissary here for the Exchange of Prisoners, if Mr. Pit has no objection to the measure, but wishes for an opportunity of explaining the particulars.[3]

[N.H.M. B.L. D.T.C. X (1). 115; P.R.O. 30/8/104.]

[1]Napoleon Bonaparte (1769–1821) first emerged as a brilliant military leader in the Italian campaign of 1796–1797 with a series of crushing victories over Sardinian and Austrian forces. He gained territory and concessions as he went, capturing a string of cities. Among them were Milan, Modena, Bologna and Ferrara. Mantua fell in early 1797 after offering stubborn resistance. The First Coalition against France had been broken.

Britain now faced the possibility of invasion, and so exchanging scientific information with the French became a delicate matter. Indeed, strict British laws forbade correspondence with France. Banks's influence at high political levels was crucial in overcoming such difficulties.

[2]When the opportunity to restore scientific contact with the French came, Banks lost no time in seizing it. Letter 68 contains his thanks to Charretié for trying to renew communications between the Institut National and the Royal Society. It followed Banks's request to see the Prime Minister by just one day.

[3]For more correspondence on this subject: Banks to Jean [or Josef] Charretié, 4/2/1797, S.L. Banks MS. A 4:56 (for Banks's initial suggestion of exchanging publications); Charretié to Banks, 16/3/1797, R.B.G. Kew Archives B.C. 2 157; Banks to Charretié, 18/3/1797, N.H.M. B.L. D.T.C. X(1) 116–117.

xiii. **The Library at Soho Square.**
The interior of the main library room overlooking Dean Street. Sepia wash drawing, c.1828, by Francis Boott. The General Library, Natural History Museum, London.

xiv. **The Study at Soho Square.**
Sepia wash drawing, c.1828, by Francis Boott. The General Library, Natural History Museum, London.

Letter 68

1797

To Jean [or Josef] Charretié, French Commissary in London

Soho Square
18 March 1797

Sir,

I lose not a moment in returning to you my best thanks for the zealous & effectual steps you have taken to open an intercourse between the *Institut National* & the Royal Society. Such communication cannot but be of material use to the progress of Science, & may also lay the foundation of a better understanding between the two Countries in future than, unfortunately for both, has of late years taken place.

I hope our Ministers, to whom I have applied for an audience on the subject, will not throw difficulties in the way of its continuance; & I have reason to expect that they will not.

The moment I have my instructions from them I will write fully to you in answer to M. Allamant's[1] obliging paragraph, which you were pleased to communicate to me in your last.[2] In the mean time, I forward to you by the bearer those volumes of the Philosophical Transactions which M. Lalande[3] mentions as not having been received in the usual course of presents, and also copies of the Greenwich observations in the same predicament for the *Institut*, the National Library & the Observatory.

I return, at the same time, with many thanks, the Books you were so good as to lend to me some days ago, from the perusal of which both myself & several of my friends have received great satisfaction.

I am, Sir,
with real esteem & regard,
Your faithful & obedient,

Jos. Banks.

[N.H.M. B.L. D.T.C. X (1). 116–117.]

[1]Nicolas de Conteray de Lallemant (1739–1807).
[2]Charretié to Banks, 16/3/1797, R.B.G. Kew Archives B.C. 2 157.
[3]Joseph Jérome le François de Lalande.

Letter 69

1797

To John Hunter, Governor of New South Wales

Soho Square
30 March 1797

My dear Sir,

I congratulate you on the state in which you found your Colony,[1] & I felt infinite satisfaction in reading your account of it.[2] I know it will prove an incentive to you to press forward the improvements. The Climate & Soil are in my own opinion superior to most which have yet been settled by Europeans. I have always maintaind that assertion grounded on my own experience, but have been uniformly contradicted, except by Govr. Philips,[3] till your last favors have taken away all doubts from the minds of those who have been permitted to peruse them.

You have a prospect before you of no small interest to the feeling mind: a Colony just emerging from the miseries to which new Colonists are uniformly subjected. To your abilities it is left to model the rising state into a happy Nation, & I have no doubt you will effect your purpose. Here matters are different. We have of late seen too many symptoms of declining prosperity not to feel an anxious wish for better times. I keep up my spirits & those of my Family as well as I am able, but in truth, my dear Sir, could it be done by Fortunatus's wishing cap,[4] I have no doubt that I should this day remove myself & Family to your quarters, & ask for a grant of Lands on the banks of the Hawksbury.[5] My next [letter] I hope will state better hopes.

The recovery of your Cattle, for I consider them to be recoverd tho they are not caught, is a matter of no inconsiderable importance to you.[6] I see the future prospect of Empires & Dominions which now cannot be disappointed. Who knows but that England may revive in New South Wales when it has Sunk in Europe.

Whenever prosperity returns, I shall sollicit the King to establish a Botanist with you. The Plants we have receivd, which are now tolerably numerous, make a most elegant addition to the Gardens. I trust, good Sir, that when you make your excursions, or when you send parties into new districts, you will not forget that Kew Garden is the first in Europe, & that its Royal Master & Mistress never fail to receive personal satisfaction from every Plant introducd there from foreign parts when it comes to perfection.

Respecting the Political State of things here, it is nearly the same as when you left it. Pit rules.[7] Fox grumbles.[8] The French beat all whom they attack, &

the King of Prussia threatens all who assist the Emperor. The chief change is made by the death of the Empress of Russia. The new Emperor seems honest & inclind to peace.[9] The best hope we have, however, is that he will keep the King of Prussia in some check.[10] From Prussia, however, peace is to come at last, for he is the only power situate near the seat of war who is not exhausted. As soon as it comes you shall hear from me. I am a bird of peace. My business as an encourager of the transport of Plants from one Country to another is suspended during war, and then, as I am no politician, I am the least employd when all other people are in hurry & bustle.

Accept my best thanks for the favor of your correspondence. Be assurd I put a due value upon the continuance of your friendship, & beleive me, as in truth [I] am, with real regard & esteem,

Your faithfull hble
Servant,

Jos: Banks.

[M.L./D.L. Banks Papers Series 38.04; N.H.M. B.L. D.T.C. X (2). 93–95.]

[1]Hunter was governor of New South Wales, 1795–1801.
[2]The last letter Banks received from Hunter on these matters was: Hunter to Banks, 20/8/1796, N.H.M. B.L. D.T.C. X(2) 12–19, M.L./D.L. Banks Papers Series 38.03.
[3]Arthur Phillip.
[4]Of medieval legend, Fortunatus possessed an inexhaustible purse along with his wishing cap. He is associated with good luck and wishful thinking.
[5]The River Hawkesbury drains into Broken Bay, north of Sydney. Its rich floodplains were being more widely settled and cultivated in the middle to late 1790s.
[6]Hunter was one of a party that found the cattle at Cowpastures in November 1795. These were the descendants of two bulls and four cows, which strayed from Port Jackson in July 1788. The herd, consisting of some forty animals, was approached, and one bull proved so ferocious it was shot. This allowed Hunter to verify that the cattle were a Cape breed he had brought out in HMS *Sirius*. Hunter made it an offence to interfere with the herd or its pastures.
[7]William Pitt the younger, Prime Minister.
[8]Charles James Fox, leader of the Whig opposition in the House of Commons.
[9]Catherine II, Empress of Russia, died in 1796. She was succeeded by her son, Paul I, Emperor of Russia.
[10]Frederick II (1744–1797), King of Prussia. Frederick William III (1770–1840) followed him.

Letter 70

1797

To Jean [or Josef] Charretié, French Commissary in London

Soho Square
1 May 1797

Sir,

Encouragd by the Friendly civilities I have repeatedly Receivd from you, & the Uniform protection given by your Nation to all attempts for the increase of human Knowledge, I have no hesitation in making the following Request.

The African Association,[1] a List of which I Send with this, was instituted in the year 1788. It has no President, but its affairs are managd by a Treasurer & a Secretary, both which offices happen at Present to Rest in me since the death of Mr. Beaufoy,[2] but the duties of the Latter are now Executd by Mr. Edwards,[3] who will be Electd into the office at the next anniversary meeting.

The business of the association is to Send Travellers to explore the interior regions of Africa, in which Service they have already Lost 3 well Qualified men[4] without having much Elucidated the internal Geography of the Country, but they Still persevere. They have one[5] at Present whom they hope is on his Return from Tambuctoo[6] towards the Gambia, & they have another Ready to Set out for Alexandria, from whence, by the road of Cairo, his [objective] is to attempt a passage to Cashna.[7]

In order to arrive at Marseilles, or Leghorn,[8] or any other place where he can get Shipping for Alexandria, it is necessary he should pass through the dominions of the French Republic. My Petition is therefore that he may be supplied with a Passport, & Such [a] Rout assignd to him as the French Government Shall think fitting.

His name is Frederic Hornemann,[9] native of Hildesheim[10] in L[ower] Saxony. He has had regular Education, & has Studied expressly at Gottingen for the Purpose of Preparing him for his voyage at the expence of the Association.

With this, Sir, you will also receive a Copy of the Publications of the association, which I beg you to accept of as a Present from them. If you feel the least inclination to send it to anyone of your Literary friends in France, I Shall immediately on being acquainted with that circumstance have the honor of supplying you with another for your own use,

I am, Sir,
with Real Regard & Esteem,
Your Hble Servt.

[S.L. Banks MS. A. 4:58.]

[1]The African Association was formed in 1788 at a time when little was known about the interior of the continent. It emerged from the Saturday's Club, of which Sir Joseph Banks was an influential member. The Association actively promoted exploration in Africa until 1831, when it was merged with the Royal Geographical Society of London. This society had been established in the previous year, on 16 July 1830. As the letter above suggests, Banks was an important figure in the African Association. He was a founder member, Treasurer, 1788–1805, and Secretary twice, 1795–1797 and 1799. Furthermore, Banks remained a committee member for thirty-two years until he died in 1820.
[2]Henry Beaufoy FRS (d.1795): Whig politician; member of the Saturday's Club; founder member and first Secretary of the Association, 1788–1795; Secretary of the Board of Control for India, 1791.
[3]Bryan Edwards FRS (1743–1800): merchant; banker; member of the Saturday's Club; Secretary of the Association, 1797–1800.
[4]John Ledyard (1751–1788): an American traveller. He sailed as a British marine on HMS *Resolution* under Captain James Cook, 1776–1780. Later he undertook a mission for the African Association, but died at Cairo of a 'bilious complaint' in November 1788 before attempting to cross Africa.

Simon Lucas (fl.1760–1800): oriental interpreter to the British Court. Prior to this he had been Vice-Consul and Chargé d'affaires in Morocco. He offered to travel from Tripoli to Fezzan in the African Association's first mission, but turned back, arriving in London on 26 July 1789.

Daniel Houghton (c.1740–1791): army major; former captain in the 69th regiment. Houghton was a Consul in Morocco in 1772. He was chief engineer and fort-major at Goree, after the French island fell to the British. Houghton became an explorer for the African Association with a mission to investigate the River Niger, and travel to Hausa and Timbuktu. In 1791 he bravely reached Bambouk, but vanished on the way to Timbuktu. Houghton may have been taken as a spy by Moors, who murdered him, or he may simply have died of illness.
[5]Mungo Park (1771–1806): surgeon; explorer. Park went on the first successful mission organized by the African Association to explore the interior of Africa. He was instructed to reach the River Niger, to ascertain its course, rise and termination, and to visit the principal towns in its neighbourhood. Park sailed from Portsmouth on 22 May 1795, and returned to England again, arriving in the *Chesterfield* at Falmouth on 22 December 1797.
[6]Timbuktu, Tombouctou or Timbuctoo, in Mali, West Africa. It is situated to the north of swamps and channels connected to the Niger: see Note 7 below.
[7]Friedrich Conrad Hornemann referred to Kashna, or 'Cashna', in the Hausa territory. A map is provided with an explantion of the area on pp. 111–112 of his journal (see Note 9 below). Maps of the north and west of Africa in this volume show Kashna as a 'Fine Country', its main town also bearing the name Kashna. Kashna was also known as Katsina.
[8]Leghorn, seaport and capital of Livorno province, north-west Italy.

[9]Friedrich Conrad Hornemann (1772–1800): of Hildesheim, a talented student of theology at Göttingen University; explorer of Africa. He was recommended to Banks by Johann Friedrich Blumenbach, an anthropologist and comparative anatomist at Göttingen. Hornemann proposed exploring from Cairo to Fezzan, and on to the River Niger for the African Association. The passport necessary for this was duly received, and Hornemann could depart. He was caught up in Napoleon Bonaparte's invasion of Egypt in July 1798, but managed to obtain Bonaparte's blessing to continue to Africa. Hornemann travelled into Fezzan, and then to Tripoli, from where, on 29 November 1799, he forwarded a version of his journal in English: Hornemann to Banks, 29/11/1799, N.H.M. B.L. D.T.C. II 324–328. His next destination was to be Hausa. However, the last letters from him were written in Fezzan. Hornemann's journal was published as *The Journal of Frederick Hornemann's Travels from Cairo to Mourzouk, the Capital of the Kingdom of Fezzan......in the years 1797–8*, Sir W. Young (Ed.), London (1802).
[10]Hildesheim, in Hanover district, south-west of Hanover. Göttingen is a university town in the same district, further south of Hanover.

Letter 71

1797

To Jaques Julien Houttou de La Billardière

London
15 July 1797

My dear Sir,

The bearer of this, Abbe de la Rue,[1] has Conciliated so large a Portion of my Friendship that I can not but feel deeply interested in his Fortunes. He has livd almost wholly in my Library during his residence here, & never interferd in politics even in Conversation. If you are so good as to be of use to him, you will Confer a favor upon me which I Shall not Easily forget.

I hope you have by this time made some progress in arranging the Glorious Collections which I had the Good Fortune to Send over to you.[2] I Envy you the posession of Some of them, which I Saw accidentaly, & hope when Peace returns to procure from you Some duplicates, for I can assure you as a man of honor that not a single Specimen was on any account retaind here, Either by my self or, I firmly beleive, by any other person here. The Whole intire Quantity were transmitted under my Seal to the hands of Mr. Charretié.[3] I hope we shall soon hear of your Publications.

We labor here as usual. We had a large Collection of Plants from China brought by Lord Macartney,[4] & Some very valuable ones from Masson,[5] who Travelld on the Kings account for many years at the Cape, & who now is going to upper Canada.

We have here also receivd a most valuable Collection of Plants from Sierra Leone Collected by Afzelius,[6] [who was] Educated under Linnaeus.[7] Heaven send that Peace may at last arrive, & allow us to Compare our Collections.

Beleive me, dear Sir,
with Real Esteem & Regard,
Very Faithfully yours,

Jos: Banks.

[Ub. Uppsala Ur.]

[1]Abbé Gervais de la Rue (1751–1835): scholar; antiquary.
[2]Banks had negotiated the return of La Billardière's captured natural history collections, part of which had been offered to Queen Charlotte: Banks to Charretié, 10/8/1796, R.B.G. Kew Archives B.C. 3 5; Charretié to Banks, 10/8/1796, B.L. Add. MS. 33980, f. 73.
[3]Jean [or Josef] Charretié.
[4]The Embassy to China, which Lord Macartney led, had returned in September 1794. However, Banks was disappointed by the botanical collections made on this mission.
[5]Francis Masson (1741–1805): gardener and botanist. Masson was the first collector funded by George III to be sent from the Royal Botanic Gardens at Kew. He collected in South Africa, North America, Spain, Madeira, the Canaries, the Azores and the West Indies.
[6]Adam Afzelius FRS (1750–1837): professor of medicine, Uppsala. Afzelius was botanist for the Sierra Leone Company. He kept Banks supplied with West African seeds and specimens while stationed at Freetown.
[7]Carl von Linné (Carolus Linnaeus).

Letter 72

1798

To Reverend Richard Shepherd F.R.S.

[Soho Square]
6 March [1798]

Sir J. B. thanks Dr. S.,[1] but freely declares that he does not think a Layman need spend much of his time in studying St. John.

He feels no doubt that his Faith is sufficient, if his actions prove aceptable, to conduct him to his Lord hereafter; believing that the Strait Path pointed out by the other Evangelists will lead him as safely to his object as the intricate one of St John, which Dr. S. has so laudably attempted to Elucidate.

[N.H.M. B.L. D.T.C. X (2). 232.]

[1]Shepherd wrote to Banks offering 'an unpublished Book', which he explained was 'to separate from Christianity the unscriptural Doctrine of three Divine Subsistences': Shepherd to Banks, 19/2/1798, N.H.M. B.L. D.T.C. X (2) 222–224. Shepherd published *Notes, critical and dissertatory, on the Gospel and Epistles of St. John...*, London (1796)

1798

To Dr. Everard Home F.R.S.

Spring Grove
30 September 1798

My dear Sir,

The Strongest & most Beautiful instance of the Reasoning of a bird I have seen was exhibited a few days ago in the garden under my bedchamber window, from whence I was Looking at a Pond Covered with Plants of Zizania palustris,[1] the corn of which was Ripe, as I was Recovering from a Severe Fit of Gout.

A number of Sparrows & Linnets were feeding on the grain. The Latter rested on the Straws, which in some Cases were Strong enough to bear Sparrows. An old Sparrow, however, observed a Ripe Ear of Corn supported by a Slender Straw, which grew near the bank, but hung over the Pond.

I saw him Contemplate the Object, Standing on the bank near it. He flew up on [a] sudden, & seized the Ear, but, on its bending, instantly quitted his hold, & Returned to the bank.

He Sprang up again, &, seizing the Ear with his bill, closed his wings, & threw himself towards the bank. The Ear, however, escaped from his hold, & he fell on his side on the bank.

This he repeated many times with better or with worse success, till, at last, he effected his Purpose. He fell to the ground with the Ear in his mouth, & instantly, starting up, fixed his Feet upon it, & began to eat the Corn.

Here is a variety of Reasoning which a human being need not be ashamed of:

He deduced first, & confirmed by experiment, that the Straw was too weak to bear him.

He next concluded that, if he bent it down by his weight in the direction of its natural Bend, he would fall with it into the water.

He next drew all the Consequences of his swerving it from its natural Bend, & drawing it by his weight down upon the bank.

He finished by fixing his weight upon it, & prevented it from Rising till he had eat the Corn.

[N.H.M. B.L. D.T.C. XI. 102–103; de Beer Collection 15.]

[1]*Zizania aquatica*. This is a wild rice, which was introduced to England by Banks in 1790.

1798

To William Philip Perrin F.R.S.

Soho Square
19 November 1798

My dear Perrin,

I have offerd your beautifull Specimen of Selago[1] to Lady Banks as a Chelensk, but she is such a Fool that She Likes diamonds better, & Cannot be persuaded to wear it as a botanists wife Certainly ought to do.

Dickson,[2] whose sagacity in all matters that Respect moss is scarce credible, said the moment he saw it that it must have come from the neighborhood of Leith Hill.[3] He tells me that Several northern plants are found there, particularly Vaccinium Vitis Ideae[4] in such Abundance that the Fruit is Annualy brought to London, & Cried in the Streets in Paniers.

I have been Employd this Summer in making at Spring Grove an artificial Spring & a ducking, or at least a wet bog, which in Gardening I do not Know has before been attempted. I mean to grow in it Droseras, Eriophorums, Schoenus, Parnassias, Osmundes,[5] & many more which we cannot have in our Gardens, or scarce see without wading. If you will provide me 2 or 3 Selagos with the Earth adhering to their Roots, & send them by the Coach in a box, I Shall be Thankfull, & I will try to Cultivate Them as I have now both a wet & a dry bog in my Garden.

My Peyrouse[6] is at the book binders so I Cannot say any thing Correct on the subject of his remarks. It is natural for a French man at the Present moment to say any thing likely to Reflect disgrace on an Englishman; and, after the Gross Falsehoods circulated in France respecting our Treatment of Prisoners, I can beleive any thing of that mendaceous nation. In Truth, however, as Europeans Continue to visit the S. Sea Islands, the Venereal disease must be Communicated to them, so it is scarce worth disputing who first left it. The probability is that some Islands had it from one, & others from the other nation.

Beleive me, my dear Perrin,
with infinite regard & Esteem,
Very Faithfully Yours,

Jos: Banks.

[D.R.O. FitzHerbert MS. D239M/F15889.]

[1]'Selago' (*Huperzia selago*), which is not a moss (*Bryopsida* [*Musci*]), but a clubmoss (*Lycopsida* [*Pteridophyta*]).
[2]James Dickson (c.1737–1822): botanist; nurseryman and seedsman of Covent Garden, London; founder member of the Horticultural Society, 1804. He is chiefly noted for his studies on mosses, fungi and grasses.
[3]Leith Hill, Surrey.
[4]Cowberry (*Vaccinium vitis-idaea*). A relation of cranberries, which Banks was later to plant on a market-garden scale. Banks's paper on this: *Transactions of the Horticultural Society*, vol. I (1808), pt. II, pp. 75–78, 'An Account of the Method of cultivating the American Cranberry *Vaccinium macrocarpum*, at Spring Grove. By the Right Hon. Sir Joseph Banks, Bart. K.B. P.R.S. &c.' Read 1/3/1808.
[5]'Droseras', Sundews (*Drosera* spp.); 'Eriophorums, Schoenus' are sedges of different kinds (*Eriophorum* and *Scirpus* spp.); 'Parnassias', Grass of Parnassus (*Parnassia palustris*); 'Osmundes', Royal Fern (*Osmunda regalis*).
[6]Possibly Jean-François de Galaup de la Pérouse: Milet-Mureau, *Voyage de La Pérouse autor du monde* ..., Paris (1798).

1799

To John Hunter, Governor of New South Wales

Soho Square
1 February 1799

My dear Sir,

I have to thank you for innumerable marks of your attention, for which I beg you to be assured that I am very grateful. I have received Seeds & specimens by Capt. Neven, & since by another ship, the former of which were sent to the Royal Gardens as a present from you, the latter I keep in my own collection.[1]

You have been very friendly in writing to me the very particular account I have received from you of the state of things in your Colony. I grieve to observe that matters go on so ill, & I am mortified that so little has been done towards putting you more at your ease; but be assur'd that the situation of Europe is at present so critical, & his Majesty's Ministers so fully employed in business of the deepest importance, that it is scarce possible to gain a moment's audience on any subject but those which stand foremost in their minds; & colonies of all kinds, you may be assured, are now put into the back ground.

Perservere, however, my good Sir, in the manly, honest & open conduct you have hitherto held, & you must in time prevail. Your Colony is already a most valuable appendage to Great Britain, & I flatter myself we shall, before it is long, see her Ministers made sensible of its real value. Rest assured in the mean time that no opportunity will be lost by me of impressing them with just ideas of the probable importance to which it is likely before long to attain, & to urge them to pay it that degree of attention which it clearly deserves at their hands.

I have recommended to you by a separate letter a young man[2] whom I have sent out to assist in investigating the produce of your Country. He is hardy & bold. His skill in Botany is much beyond what his appearance promises, & he has some smattering in other sciences. The Duke[3] has been pleased also to recommend him to your good offices, & I hope & trust he will not prove unworthy of the patronage I trust you will afford him, or disgrace the recommendation I have ventured to give him.

With Capt. King[4] you will receive, I hope, a considerable supply of plants likely to be usefull to your growing Colony if the bitter frost, which has now vexed us for near a month while the Ship has been out of my reach, has not destroyed them. You will have Grapes of the sorts from whence the valuable wines of Europe are made. These, I hope, will encourage you to plant Vineyards, & some of them surely will produce marketable wine.

Allow me to recommend the person[5] who has been employed to attend to these Plants[6] & carry them out. He has had no pay from Government, but has undertaken this task merely on condition of being allowed to go out as a Settler. He is an honest, hard-working, industrious & ingenious lad. He wishes to settle as a Market Gardiner, & carries out with him a good assortment of seeds. If he pleases you, & you can be so good to him as to settle him in some low spot of good land convenient for water carriage to transparent his goods to your town, I think I may venture to promise that he will thrive. You will find him honest & trusty, so that I hope he may in other matters make himself useful to you.

I have one more favour to beg of you. There are among the Convicts who proceeded in the Hillsborough two Brothers of the name of Wilkinson, who come from the town of Consby[7] in my neighbourhood in Lincolnshire. They were convicted of maiming a Cow, & transported for that offence. I do not mean to vindicate the act, but I really do think it a venial crime when compar'd to Robbery, Burglary & most of those which your subjects have been guilty of. These young men must make their Country amends by working with you for the Public. All I have to request is that, if you find them as I hope they are, not ill disposed, & if they conduct themselves as lads whose Morals have not been quite destroyed ought to do, that you will now & then give them a cheering word, & tell them that by continuing a good conduct they may sometime revisit their Country & their Friends.

[N.H.M. B.L. D.T.C. XI. 187–189.]

[1]Banks's correspondence with Hunter for 1798 contained letters in which the state of the colony at New South Wales was described, animals preserved in spirits were organized to be sent to England, and Banks made arrangements for gardeners and settlers to go to Australia: Hunter to Banks, 12/3/1798, N.H.M. B.L. D.T.C X(2) 240–248; Hunter to Banks, 5/8/1798, M. L. Banks Papers Series 38.10; Banks to Hunter, [13]/12/1798, M.L. Banks Papers Series 38.13.

[2]George Caley (1770–1829): gardener and botanical collector. Caley was employed by Banks as a collector, 1799–1810. He was based at Parramatta, New South Wales during this period. Afterwards Caley became superintendent of the gardens at St. Vincent, 1816–1822. He was a temperamental man whom Banks treated with kindness and patience.

[3]William Henry Cavendish Bentinck FRS (1738–1809), 3rd Duke of Portland: statesman; Home Secretary 1794–1801; Prime Minister, 1783 and 1807–1809.

[4]Philip Gidley King (1758–1808): captain, RN; Commandant of Norfolk Island, 1788–1790, and Lieutenant-Governor of Norfolk Island, 1791–1796; Governor of New South Wales, 1800–1806. He was leaving England to replace Hunter as Governor.

[5]George Suttor (1774–1859): market gardener; settler, New South Wales and the Baulkham Hills.

[6]Banks originally wanted Suttor to carry out plants for the colony on board HM Armed Vessel *Porpoise*. Misgivings about this ship were expressed by Banks in mid-1798, but preparations for the voyage went ahead anyway, with a plant cabin being situated on the

quarter deck. When it was brought from Deptford to Portsmouth in late January 1799, the *Porpoise* was obviously crank. So there was a considerable delay for alterations before it sailed as part of a large convoy, only to sustain a damaged rudder in the Bay of Biscay. The Navy Board condemned the ship once it had returned to England.

A captured Spanish prize, the *Infanta Amelia*, was re-fitted and re-named HMS *Porpoise* instead. It was in this vessel on 17 May 1800 that Suttor sailed with his charge for a second time. However, many of the plants died during the voyage. King and Caley went in a different ship altogether. They left on 12 November 1799 in a whaler called *Speedy*, and arrived at Port Jackson on 15 April 1800.

For an indication of the plants and seed Banks sent, there are a number of documents to consult, for example: M.L./D.L. Banks Papers Series 19.11-44, from George Suttor, including a list of grapes.

[7]Coningsby, Lincolnshire.

Letter 76

1799

To Adam Wolley

Soho Square
4 February 1799

My dear Sir,

I am Much Obligd to you for Sending[1] up the Curious horns & bones which have been lately discoverd at Matlock, [2]& have much pleasure in stating to you what I happen to Know on the subject of them.

Mr. Barker,[3] as you mention, found a pair of Horns in the Tuft at Alport,[4] & sent an account of them to the Royal Society in the year 1785. By this description it appears that those horns intirely resembled yours in their Structure so that there is no doubt of both having belong'd to the Same species of animal.

Mr. Barker, who seems to have Consulted the Park Keepers in his neighborhood, Entertains no doubt of their being the horns of the Same Species of Stag as now is found wild in Scotland, & is Kept for the purpose of hunting in Several Parks in England. Their dimensions were nearly Twice as Great in thickness as those of an ordinary stag, yet Mr. Barker found extraordinarily large Stags horns, which were nearly $\frac{7}{10}$ of the diameter of his fossil one

I find also that a horn about $\frac{1}{10}$ larger in diameter than yours, & apparently of the Same species of animal, was found several years ago near Holker in Lancashire.[5] I have also seen an account of a Stags horn found in digging a well in France, which seems to have been Quite as large as yours, & of the Same animal.

I Shall not trouble you with more Quotations, having already given you Enough to Show That Horns, apparently of Stags, have been found in several places as large or larger than yours, but only remark that as your Horn is mutilated it is impossible to Know what The Real length of it has been. In Mr. Barkers Case, however, who was able to measure the Length of his horn, it appears not to have exceeded the dimensions of the actual Stags horn in Length nearly in So great a proportion as in breadth, the Stags horn being 2 feet $8\frac{3}{4}$ inches, & the Fossil horn 3 feet $3\frac{1}{2}$.

This difference has been generaly Observd between fossil & Recent bones &c., & has given occasion for some persons to suspect what may Possibly be the Cause: that Bones & Horns increase in thickness by laying for a number of Centuries in damp Earth, a conjecture which is renderd in some degree more probable by the fossil being more spongy than the Recent Object.

For my own part, I think the great size of these fossil Horns must be accounted for Two Ways, for as the excess in Leng[t]h is Considerable, & this Could not have been owing to increase after the death of the animal, it seems Quite Clear that the Ancient animal was larger Considerably than the modern one; & it is Curious to Observe that Mr. Barker finds his Old Horns, which he supposes to have hung where he found them for 2 or 3 or perhaps more Centuries, were materialy larger than those of the 7 years old Stag which he measurd.

There is no doubt that the animal in Question had been an inhabitant of the forest of the Peak, and probably he livd before the forests of his Tribe had been much molested by the intrusions of mankind; whether before the Romans workd mines there, or since, I conclude it Quite impossible to determine, for I beleive the Peak has not been deaforested more than 5 or 600 years.

In the deer Kind, as in some others, nature seems to have provided for a regular improvement of the breed in size & Strength where ever they are numerous, & man does not interfere with them. At the season of the Rutt, the Strongest & Largest males always take the lead, and possess themselves of the does. Nor Can any weak or young deer get access to a doe till the strong male, who first Cohabited with her, is exhausted. The largest Quantity of Fawns, then, will naturaly be begotten by the largest & Strongest males so that, if there are abundance of males, whenever a Giant appears among them, he is very likely to propogate his Species extensively.

In Parks again, where 10 or 12 Brace of Stags is generaly Considered as a large Stock, the Chance of a Giant being born & improving the size of the herd is diminishd in an immense proportion; and as nature, if She has been by accident Enlargd, is always inclind to resume her natural Size, the Probability is that in Small herds the Stags will diminish in size below The Standard to which they are raisd by Living together in Large ones.

The moose deer of America differs from The Stag in having horns that End above in a broad Palm Somewhat Similar to that of a fallow deer,[6] but Broader. It is in my opinion The Same animal as the Elk now found in Hungary, & some of the northern parts of Europe.[7]

That your animal has not been a moose is very Evident, & in my opinion it is as clear that it has been a Stag. The moose, however, has been an inhabitant of this Isle, for a distinct pair of moose horns were about 50 years ago dug up in a mass in Yorkshire.

I have Troubled you with a Terrible long letter, my motive for which was to Shew you that I am very sensible of your Attention to me in Sending up the horns, for which I beg To thank you. They came in most Excellent Condition, the Package being frozen into one hard lump.

Beleive me, Sir,
with due Consideration & Regard,
Your Obedient Hble Servt,

Jos: Banks.

[B.L. Add. MS. 6669, ff. 220–222; B.L. Add. MS. 56301(7), ff. 40–42.]

[1]Wolley to Banks, 26/1/1799, I.C.C.L., Annan Collection, item 52, 'Letters to Sir Joseph Banks, 1792–1800', pp. 499–502: 'About a fortnight ago some labourers were getting tuft (Tophus - wch in this neighbourhood is improperly called marl) near the river Derwent, at a small distance from Matlock New Bath. They discovered, at the depth of nine or ten feet from the surface, in stratum of that substance, the bones of a deer of a very uncommon size.'

[2]Matlock, Derbyshire.

[3]Reverend Robert Barker FRS, who was vicar of Youlgreave, 1770–1797. *Philosophical Transactions*, vol. 75 (1785), pp. 353–355, 'An Account of a Stag's Head and Horns, found at Alport, in the Parish of Youlgreave, in the County of Derby, in a Letter from the Rev. Robert Barker, B.D. to John Jebb, M.D. F.R.S.' Read 14/4/1785.

Banks's terms require some explanation. He appears to use 'stag' to refer to the European red deer (*Cervus elaphus*). By 'horns' he means 'antlers'. The late Pleistocene fauna of Britain included very large, heavily antlered red deer (*Cervus elaphus*) and elk (*Alces alces*). In earlier, colder times there were also giant deer, *Megalocerus giganteus* (sometimes known as the Irish elk). More details of elk and moose are given in Note 7 below.

[4]Alport, Derbyshire.

[5]*Philosophical Transactions*, vol. 37 (1731–1732), pp. 257–258, 'An Extract of a Letter from Mr. Hopkins to Mr. John Senex, F.R.S. concerning an extraordinary large Horn of the Stag Kind, taken out of the Sea on the Coast of Lancashire.'

[6]Fallow Deer, *Dama dama*.

[7]'Elk' is the English term and 'moose' is the American term for deer of the species *Alces alces*. This species is widely distributed in areas of cool, temperate woodland throughout North America and Eurasia, including Scandinavia, Eastern Europe, Russia, Siberia and Mongolia.

Letter 77

1799

To George John Spencer F.R.S., 2nd Earl Spencer, and 1st Lord of the Admiralty

Soho Square
6 May 1799

My dear Lord,

We have now Finaly Arrangd, I beleive, our Institution[1] matters. Ld. Bessborough,[2] who is one of the Visitors, is to Change Places with your Lordship, & Ld. Winchelsea[3] is to be chosen a manager in order that he may be our President, which his Lordship has undertaken to be, I am told, with some symptoms of Satisfaction.

In the meantime, I am directed by the managers to Request Permission to Continue the use of your Lordships name as a manager till the Charter is Seald,[4] which is proceeding as expeditiously as Possible, & I Realy do not see any inconvenience Likely to happen to your Lordship by granting us this indulgence. Every thing yet done by The managers has been approvd warmly by the Proprietors at their Last meeting, & another meeting must soon be Calld to Elect Ld. Winchelsea an additional manager. Till the Charter is Seald no material business will be Transacted. Now the house is Purchasd,[5] & so well do we prosper in Point of Finance as well as of increasing members, that we hope in a very short time to have enough to pay the Purchase of the house, which your Lordship Remembers is £4500.

I beg that my sincerest good wishes may be acceptable to Lady Spencer,[6] & hopes, I hope not unreasonable ones or unlikely to be fulfilld, that the French Fleet will soon find its way into an English port under English Colors.

Count Rumford has of late kept himself intirely in the back ground; nor do I Think he will ever venture forward again if it has cost him a Fit of Sickness to find his way from the Ideal preeminence of his Character to his actual situation in the opinion of this Country.[7] But I beleive he has now satisfied himself, & will be, as he ought to be, an extremely usefull inventor of machines, & governor of machine makers.[8]

I have the honor to be,
with inifinte Regards & esteem,
your Lordships obedient,
Hble Serv,

Jos: Banks.

[B.L. Althorp G32, 'Miscellaneous Letters 32, 1799.']

[1]The Royal Institution of Great Britain, which was established in 1799. It was largely Count Rumford's idea. In his view the purpose of the Royal Institution was to promote the practical benefits of science and technology in society. Banks was a subscriber, and one of the managers. The first meeting of the managers took place on 7 March at Banks's home in Soho Square. See Letters 92 and 96.

[2]Frederick Ponsonby (1758–1844), 3rd Earl of Bessborough.

[3]George Finch FRS (d.1826), 5th Earl of Winchelsea. As First Lord of the Admiralty, Earl Spencer was fully occupied with fighting the French at sea. On 1 October Winchelsea was therefore elected a manager in his place. This prepared the way for Winchelsea to become President of the Institution.

[4]The charter received the Royal Seal on 13 January 1800.

[5]A house in Albemarle Street was purchased, and the Royal Institution is still there today. The building was comprehensively equipped with a lecture theatre, repository, library and laboratories.

[6]Lady Lavinia Spencer (*née* Bingham), (d.1831).

[7]Rumford did not attend any meetings of the managers from the middle of September 1799 until February 1800. He was feeling unwell, and went on an extensive tour of the British Isles.

[8]Spencer approved of this letter, and noted on the verso in a corner: '7 May Ans ? That I was very glad to have so good a bargain as was made & that I had no objection to My Name remaining as desird. P.S. inclosing a Draft of £50 as My Subscription to the Institution.'

Letter 78

1799

To Charles Jenkinson, 1st Baron Hawkesbury, and 1st Earl of Liverpool

Soho Square
8 June 1799

My Lord,

I have waited some days in hopes of having had the pleasure of meeting your Lordship, and delivering with my own hands the enclos'd paper, which I have been directed by the Commee of the African Association to deposit in your Lordship's [hands] in hopes that your Lordship will give it due consideration; &, if you think the Project feasible, & likely if successful to prove advantageous to our Commerce, & honorable to our national character, that your Lordship will give it your Recommendation to the other Members of the Cabinet.[1]

As the trade with the Negroes for manufactured Goods is already firmly established, and as Gold, the home investment, is found in abundance in all the torrents which fall into the Joliba,[2] I confess I feel strongly impressed with the Hope of success should the project be fairly tried. An expenditure, apparently considerable, must however be encountered in the outset; but as Science has never yet been applied to the search of Gold carried down by Torrents, & as the Theory of this operation seems manifest, and has been materially elucidated by the Discovery of Gold in the Wicklow Mountains lately,[3] I feel sanguine hopes that the produce of that valuable Metal may, under proper regulation, be increased in Africa to almost any given extent.

Should the undertaking be fully resolved upon, the first step of Government must be to secure to the British Throne, either by Conquest or by Treaty, the whole of the Coast of Africa from Arguin to Sierra Leone; or at least to procure the Cession of the River Senegal,[4] as that River will always afford an easy Passage to any rival Nation[s] who mean to molest the Countries on the Banks of the Joliba.

Should the Experiment be made, I have little doubt that in a very few years a trading Company might be established under the immediate controul of Government, who would take upon themselves the whole expence of the measure; would govern the Negroes far more mildly, and make them far more happy than they now are under the Tyranny of their arbitrary Princes; would become popular at home by converting them to the Christian Religion, by inculcating in their rough minds the mild morality which is engrafted on the Tenets of our faith, and by effecting the greatest practicable diminuition of the Slavery of mankind upon the Principles of natural Justice & commercial benefit.

That these effects are likely to take place may be seen by the whole Tenor of Mr. Park's Book,[5] in which it uniformly appears the small superiority in useful acquirements which the Moors possess is sufficient to induce the Negroes ardently to embrace the Tenets of the Alcoran; not because these Tenets appear in themselves wise, but because those People who teach them are supposed to have gained from them their superiority in the use of the Art they teach, which is the knowledge of writing. How much more, then, would the more intelligible Doctrines of the Scripture, if made part of an Education contrived to teach the more useful branches of European Mechanics, be followed, and how much more completely would those who are taught be civilized? But I am led away too far by this Idea. Allow me only to add that if your Lordship wishes any further information, I shall be happy to furnish all in my power. Having the honor to be, with sincere regard and real Esteem,

Your Lordship's Obedient
and faithful Humble Servant,

J. Banks.

[N.H.M. B.L. D.T.C. XI. 233–235; B.L. Add. MS. 38233, ff. 94–95.]

[1]At the General Meeting of the African Association, 25 May 1799, Banks argued that there were "advantages to be derived from Mr. Park's discoveries, and more immediately in a commercial point of view. He [Banks] said, 'a gate was now opened into the interior of Africa, which it was easy for every nation to enter, and extend its trade and adventure from the west to the eastern side of the great continent." *Proceedings*, Association for Promoting the Discovery of the Interior Parts of Africa, London (1802), vol. II for 1799, p. 4. Banks proposed a British establishment on the west coast of Africa, which would have to take the form of a "commercial and military station". He perhaps recalled earlier and unsuccessful attempts to settle there peacefully. For instance, in 1794 Freetown in Sierra Leone was attacked and wrecked by privateers, and in 1796 a mission to send a consul to Senegambia was indefinitely postponed without a ship of the assembled convoy having sailed.

It was therefore resolved at the General Meeting "that it be recommended to the Committee, to take again into consideration, the plan of appointing a Consul for the district of Senegambia; and the sending there a sufficient force to take possession of a station on the banks of the Joliba, and from thence exploring the interior of Africa; and that the Committee be empowered to lay before Government, for their information, such particulars respecting the importance of that object, as they may think expedient." *Proceedings*, p. 4. This is from Sir Everard Home's copy, in the General Library, The Natural History Museum, London.

On 29 May Banks helped to draft a memorandum based on his ideas, which was delivered with the letter above to Jenkinson for the consideration of the Privy Council Committee for Trade and Plantations.

[2]The River Niger.

[3]Wicklow, county in Leinster province, Ireland. News reached Banks of gold in the Wicklow mountains in 1795. Details of this appeared in *Philosophical Transactions*, vol. 86 (1796), pp. 34–37, 'An Account of the late Discovery of Native gold in Ireland. In a Letter from John Lloyd, Esq. F.R.S. to Sir Joseph Banks, Bart. K.B. P.R.S.' Read 19/11/1795; pp. 38–45, 'A mineralogical Account of the native Gold lately discovered in Ireland. In a Letter from Abraham Mills, Esq. to Sir Joseph Banks, Bart. K.B. P.R.S.' Read 17/12/1795.

[4]Both Sierra Leone and Arguin Bay are on the west coast of Africa. The River Senegal rises in the Fouta Djallon massif, not too far from the origins of the River Niger in the Loma mountains. The Senegal drains north-west before turning to the west. It reaches the ocean near St. Louis on the west coast.

[5]Mungo Park undertook an epic journey of exploration to the banks of the River Niger for the African Association from 1795 to 1797. His published account appeared in 1799 as *Travels in the interior districts of Africa: performed under the direction...of the African Association in 1795, 1796 and 1797; With an Appendix...by Major Rennel*, London. It was subsequently referred to by many travel writers in works on Africa. In 1805 he returned to Africa and the Niger to explore, but died there.

Letter 79

1799

To Sir William Hamilton F.R.S., Ambassador at the Court of Naples

[Soho Square]
8 November 1799

My dear Sir William,

It is now long since I have written to you, or had the pleasure of receiving a Letter from you. I have therefore the greater satisfaction in saluting you under better circumstances & in pleasanter times than those which have passed away, & in hoping sincerely that the intercourse which is now opend between us may never again be interrupted by the success of our Enemies.

The more immediate object of this Letter is the present situation of Dolomieu. He is confin'd, as I am told,[1] by the Government of Naples under a charge which we here find it difficult to comprehend, and is in some danger of being tried as a Criminal, tho his situation appears to a common observer clearly to be that of a Prisoner of War.[2]

You have no idea how much sensation his confinement has made in the Literary world here, and how anxious men of Science feel in all parts of Europe for his Liberation. In my little way of correspondence I have already had more than 30 signatures of men of Letters anxiously enquiring whether it is possible anything can be done to save him.

Whatever Dolomieu has done while under the command of Bonaparte by that General's order surely cannot be considered as Criminal. Who is there who can, or indeed who ought to resist the command of the General of the army to which he is attach'd? If he does resist the commands of a man whose vengeance has hitherto uniformly preceded his justice, and in whose breast mercy (as appears by his conduct in Syria) has no place, what he can he expect? Our favorite English Laws, my dear Sir William, The Laws of reason, & law of nations agree to throw a veil over crimes committed under the influence of irresistible power; & surely the power of Bonaparte at the head of his Egyptian army, stained with the innocent blood of half Italy, hungering & thirsting for the treasures of the Order of Malta, must be admitted as irresistible.

That Dolomieu has long ago devoted himself to Science is well known. In the pursuit of it he has brav'd the danger of Calabrian[3] Banditti, & encounter'd a thousand other difficulties which very few men would have voluntarily subjected themselves to. That his reason for going to Egypt was scientific & not political,

I have no doubt. That he is at this moment animated by a sincere wish to advance human knowledge, & that he is able from his observations in Egypt to do it much service is clear. Where then can be the objection to release him? Should he be inclined to Political interference, what can his influence do? He is not among those who have grasp'd at power, [n]or is he ever likely to obtain it. No mischief can then accrue from his release; but if the gratitude of thousands of learned men, whose eyes are now steadily fix'd on his situation & its circumstance, is an object to the King of Naples,[4] or to you, it may be immediately secured by this measure. Solicit his release then, my dear Sir William, & try to obtain his liberty. You have obliged me on a thousand occasions, as I am willing to testify, & I hope I have been ever ready to obey your commands with zeal, but there is nothing I have ever ask'd of you in which I have taken a tenth part of the interest I now feel in the hopes of Dolomieu's being by your means restored to the service of science, and his Egyptian Observations being rescued from the Oblivion of an untimely grave.[5]

[N.H.M. B.L. D.T.C. XI. 313–315; M.L./D.L. Banks Papers Series 73.061.]

[1]Nion to Banks, 18/9/1799, Biblio du Mus Nat d'Hist.
[2]Déodat de Gratet de Dolomieu was among the scientists Napoleon Bonaparte took on his expedition to Egypt in 1798. By 1799 he was unwell, and decided to return to France. On the way home a storm forced his ship into Taranto, and he was made a prisoner by the Neopolitans. Dolomieu was carried to Messina, where he was accused of assisting the French in taking Malta. As he was a Knight of Malta, he received little mercy from his captors, among whom were brother Knights.
[3]Calabria, province of southern Italy.
[4]Ferdinand I, King of Naples and Sicily.
[5]Letters 80 and 85 refer to this incident.

Letter 80

1799

To Lady Hamilton

[Soho Square]
8 November 1799

My dear Lady Hamilton,

I have ever been convinc'd of your friendly dispositions towards me, but the readyness of my friend, Sir W[illia]m,[1] to anticipate on all occasions my wishes has been so constant hitherto that I have never before been under the necessity of putting your Ladyship's friendship to the tryal.

Now, however, a business has occurred which cannot, I am convinced, be effected unless both your influences are united in soliciting it. If it is possible to engage the Gallant Admiral[2] to join the Trio, which is in the power of no one but yourself, what may I not hope from the warmth of your friendship, if the debilitating climate which you have lately inhabited has not damp'd its ardor?

What I wish for is the liberation of Dolomieu, who, for the honor of the kingdom of Naples, ought to be Liberated because he is, & must [be] considered as a Prisoner of War, & cannot be made answerable by the Law of nations for actions done by him under the order of his Commander in Chief. I need not trouble you with details on the subject as Sir William will, no doubt, shew you my Letter to him by which you will find how lively an interest the scientific persons in Europe take in his fate.

Allow me, my dear Lady, to be indebted to you for a favor which no one but yourself can effect. Your influence is, as it ought to be, unbounded. Gloriously as you are now circumstanc'd, having at least a third part of the merit of replacing the Crown on the heads of their Sicilian Majesties,[3] undertake the business with your usual spirit. It will be for the honor of the Court of Naples, for the advancement of Science, for the benefit of humanity, & will fix an indelible obligation on one who has the honor of signing himself,

with sincere regard & real esteem,
your faithful Hble Servant.

[N.H.M. B.L. D.T.C. XI. 316–317; M.L./D.L. Banks Papers Series 73.062.]

[1]Sir William Hamilton, Emma's husband, and British Ambassador at the Court of Naples. Banks wrote to him about Déodat de Gratet de Dolomieu in the previous letter: Letter 79.
[2]Admiral Horatio Nelson, 1st Viscount Nelson, and Duke of Bronté.
[3]Ferdinand I, King of Naples and Sicily, and Maria Carolina, Queen of Naples and Sicily. Emma Hamilton was close to the Queen, who was an influential woman at the Court of Naples.

Letter 81

1800

To Martjin van Marum F.R.S.[1]

Soho Square
14 June 1800

Sir,

I have of Late so Seldom had an opportunity of writing to you that I take the present opportunity of hazarding this Letter with much pleasure, hoping it may Come into your hands, & that it may give you some Pleasure in return for the many valuable Communications with which you have favord me.

Dr. Herschel[2] has made a discovery of Some Considerable importance. He has found that not only Luminous or visible rays Come from the body of the Sun, but invisible Rays also, which have more heat than the Luminous ones, & are more refrangible. This he demonstrates by passing a sun beam through a Prism, & he finds a place about an inch beyond the Red Rays, where no Color is Perceptible, but where a Thermometer will rise higher than it will in any part of the Spectrum of Colors; & he finds also that the heat of all the Colord rays is different, decreasing from the Red, which is the hottest. Hence it seems probable that Light & heat are Caused by Distinct Rays, which, however, Emanate together from all Luminous bodies that afford heat, but are not equaly refrangible.

Mr. Charles Hatchet[3] has made Some very Curious experiments in the analysis of Animal matter in which he finds Phosphate of Lime, or Carbonate of Lime, to be the universal cause of hardness in bones & Shells of fishes, & Discovers an abundance of membranous tissue even in the hardest Shells. He has usd this Test with great advantage in tracing new affinities between Some Sea productions which were not suspected to have any Relation to Each other, & he seems also to prove that all soft animal matter, [such] as flesh, skin, horn, hoof, nails, hair /*scales*/ &c. &c., may be, after the Calcereous matter is taken from them, reducd into three Component parts: albumen, Fibre & Gelatine, which he almost considers as the Elementary Principles of animal Structure.

Volta[4] of Como has discoverd a Very Curious property of Galvanism, for by Placing a Certain number of Plates of Silver & Zink alternate upon each other, with a peice of Pasteboard of a Similar Shape between Each pair, he procures a Stream of the Electrical Fluid always Issueing between the Top & the bottoms /of the Pillar/, which will give repeated Shocks like a Small Leyden Phial if the hands of the operator are made wet & applied to the Top & the bottom plate by a metallic Conductor. Plates of the size of the Larger Silver money answer very well, & 20 pair produce a very Sensible Effect.[5]

I am, Sir,
With real esteem & Regard,
Your Very Hble Servant,

Jos: Banks.

[A copy of the original letter is at the Natural History Museum, London, in the Banks Archive. The original is in the archive collections of the Hollandsche Maatschappij der Wetenschappen, at the Rijksarchief in North-Holland. See also: B.L. Add. MS. 56301(1), ff. 28–29.]

[1]Some scholars have suggested this letter was for Freiherre von Nicolaus Joseph Jacquin FRS (1727–1817): botanist and chemist.
[2]Dr. (later Sir) William Herschel, the astronomer, published his observations on the solar spectrum: *Philosophical Transactions*, vol. 90 (1800), pp. 255–283, 'Investigation of the Powers of the prismatic Colours to heat and illuminate Objects; with Remarks that prove the different Refrangibility of radiant Heat. To which is added, an Inquiry into the Method of viewing the Sun advantageously, with Telescopes of large Apertures and high magnifying Powers. By William Herschel, LL.D. F.R.S.' Read 27/3/1800; pp. 284–292, 'Experiments on the Refrangibility of the invisible Rays of the Sun. By William Herschel, LL.D. F.R.S.' Read 24/4/1800; pp. 293–326, 'Experiments on the solar, and on the terrestrial Rays that occasion Heat; with a comparative View of the Laws to which Light and Heat, or rather the Rays which occasion them, are subject, in order to determine whether they are the same, or different. By William Herschel, LL.D. F.R.S.'. Read 15/5/1800, and 'Part II' of this paper, pp. 437–538. Read 6/11/1800.

Herschel found that there were temperature differences between various regions of the spectrum, and established that the hottest radiation was not within the visible range, but in the region now known as the infrared. He also showed that heat obeys laws of reflection and refraction analogous to those of light. This was an important early contribution to the science of stellar photometry.
[3]Charle Hatchett FRS (c.1765–1847): chemist. Once again, these details relate specifically to a Royal Society paper: *Philosophical Transactions*, vol. 90 (1800), pp. 327–402, 'Chemical Experiments on Zoophytes; with some Observations on the component Parts of Membrane. By Charles Hatchett, Esq F.R.S.' Read 12/6/1800. Hatchett based this paper on work commenced in 1799. He analysed calcified tissues of shells and bone, finding an organic matrix, which was left when the material had been dissolved in acid. In bone and mollusc shells this matrix is proteinaceous. The outer organic layer of mollusc shells is keratin-like. It resembles finger nails, hair and so on.
[4]Alessandro Volta, the physicist, announced his invention of the 'Voltaic pile' or 'compound Galvanic circle' in letters sent to Banks: *Philosphical Transactions*, vol. 90 (1800), pp. 403–431, 'On the Electricity excited by the mere Contact of conducting Substances of different kinds. In a Letter from Mr. Alexander Volta, F.R.S. Professor of Natural Philosophy in the University of Pavia, to the Rt. Hon. Sir Joseph Banks, Bart. K.B. P.R.S.' Read 26/6/1800.

[5]The sequence of Banks's comments followed that of the papers referred to in Part II of the *Philosophical Transactions* for 1800. The last paper was read after the letter above was sent. Advanced news of Volta's work would certainly have been of interest to Van Marum.

Letter 82

[1800]

To George John Spencer F.R.S., 2nd Earl Spencer, and 1st Lord of the Admiralty

[Soho Square]
[December 1800]

Hints respecting the Rout that may be pursued with advantages by the Investigator discovery Ship.

Mr. Baudin,[1] who Saild from Havre about the middle of October last, Carried with him an astronomer, a Botanist, a Zoologist, an anthropologist & a mineralogist. Hence it is clear that he intended to proceed upon some business of Discovery. It seems, however, probable that he had other */business/* service to perform before his scientific plans were to be Carried into execution, October being the worst possible time of Sailing for a Ship bound to discover in New Holland, as she must naturaly arrive there when the Commencement of winter renders an investigation of the Southern Parts very inexpedient, & when the S[outh] E[ast] Monsoon renders the Northern parts almost inaccessible.

The object of this voyage was stated, when a Pas[s]port[2] /from the british Government/ was askd for, to be the discovery of the N[orth] W[est] Portion of New Holland, a part of the Coast better known to navigators than any other. Hence it is probable that Mr. Baudin is orderd to touch at the Isles of France & Bourbon,[3] & that the alledgd destination is intended as an excuse in case the Ships are met with by our Cruizers in the neighbourhood of these Islands.

If this Political manoeuvre takes place, he will probably stay two months at least in these Isles, & if he, who has Two vessels with him, is a Fortnight longer in his Passage than the Investigator that goes alone, our vessel will arrive on the Coast of New Holland as soon as Mr. Baudin, if she Sails in January.[4] But it is probable that Mr. Baudin will not attempt the Coast till October at the Soonest, when the monsoon will be favorable for discovery on the northern, & when it will be summer on the Southern Parts of this vast mass of Land.

Having premisd thus much for the Probability of the Investigator reaching her Destination before the French */are upon the Coast/* reach theirs, I shall proceed to Point out what appears to me the Best Rout for her to take in her outward bound passage.

In a voyage which will deprive the persons engagd in it of the usual Comforts of a European life So long, it is expedient to allow them to touch outward bound at some port where they may lay in a stock of comfortable necessaries at a Cheap Rate. For this Purpose Rio [de] Janeiro seems well fitted. Port, wine, spirit, Tobacco, sugar, Lemon juice as an antiscorbutic, & many other such matters may be procurd there very cheap & very Quick. A week will be long Enough to do all the Crew can want to do at that Place.

As I judge the voyage from England to Rio [de] Janeiro will take up about 8 weeks, & that 14 or 15 more will Carry them to the Coast of New Holland, they of Course will arrive there in a very stormy & Turbulent season. That, however, they must Execute; & in order to secure themselves from being anticipated by the French, they must in my opinion Run down the Coast from the Place where the French Saw it Last in 130 E[ast] Longitude to Basses Straits.[5]

Before they enter upon this undertaking, they may put into Geo[rge] 3rds Harbor,[6] & Refresh themselves if it is Found necessary, but, whether or not that is done, I have no doubt that by standing off shore in hard gales, & Coming in towards it in moderate weather, a survey of this Coast may be made even at that Season Sufficient to anticipate the French, & possibly to discover some harbors that may be of infinite utility when the Coast is afterwards */to be/* minutely & accurately examind.[7]

In case of any opening being Observd in this Shore likely to Lead to an inland Sea or a Streight, the Commander Should have descretionary orders to examine it or not as he may find Most expedient. A Rough Sketch of an object of Such importance taken without his Tender[8] would ensure to him, & to the English nation, the honor of the discovery, & he may at his Leisure examine & Lay down the Particulars minutely after his Return from Sydney Cove.

Howsoever important his discoveries may be in this boisterous Season, it is likely that he will arrive at Sydney before the winter is at an End, & Consequently find the Lady Nelson (for which he must Carry out a Comm[issio]n as she is new & [a] Transport) still in harbor. Here, then, he must Remain, Refreshing his people & Refitting his vessels, till The Proper Season for Entering upon the more detaild Survey arrives, which can only be executed in low latitudes in the summer months.[9]

By the middle of October, or at latest by the first of November, the Ship must proceed on the business of Surveying, & must in [the] first place accurately investigate the Coast */She/* visited before in the winter season. Every creek & opening must now be lookd into in hopes of finding a Red Sea, a Persian Gulf or a Strait leading through the Land; & in Case a River is Found, The Tender must enter it, & proceed as far up it as shall be found Practicable, & this must be done on all parts of The Coast that are examind.

After having Carefully Surveyd the whole of the Coast from Basses Straits to King Geo[rge] 3ds Harbor, the next business will certainly be to explore the Gulph of Carpentaria,[10] & those parts to the westward of it between the 130th & 140th degrees of East Longitude, where the Coast of New Holland is not Laid down as Continuous. The time of doing this must, as I conceive, be left to the discretion of the Commander, provided only that he is directd to seize the Earliest possible opportunity when the Season & prevalent winds are favorable for visiting these Seas.

Next to this, a Carefull investigation & an accurate survey of Torre's Strait,[11] & [of] the Coast of New South Wales as low down as the Isles of Direction,[12] a part of which was not seen by Capt. Cooke, appears the most interesting. After that, the whole Remainder of the N[orth] & N[orth] W[est] Coasts of New Holland, from Cape Vandieu[13] in Longitude 131 East /to Cape Chatham/,[14] and Especialy those parts of the Coast most likely to be Fallen in with by Indiamen in their outward Bound Tracks.

When these things are Finishd, nothing will Remain but a more Carefull Examination of the East Coast seen by Cooke from Cape Flattery to the Bay of Inlets,[15] where, as I remember, Lieut. Flinders's Survey Ends.

In the whole of this business, much discretion must be left to the Commander. I trust that Lieut. Flinders will be Found worthy of that indulgence, & he may be allowd, in order to Refresh his People & give the advantage of some society to the Painters, to touch at the Fegees,[16] or some other of the South Sea Islands.

In order to Favor science as much as may be in this undertaking, & at the same time to Render the survey more than usualy accurate, it may be adviseable to instruct the Commander to use the Tender as much as possible in the survey, Leaving the greater vessel in Harbor, & moving her onwards when another Harbor is discoverd. This will give the naturalists time to Range about & Collect the Produce of the earth, and also allow the Painters Quiet & Repose, even for finishing a certain Quantity of their works on the Spot where they have been began.

Thus, my Lord, have I Scetchd out hastily such Ideas as occur to me relative to the Conduct of the intended Expedition. Much, however, of Detail remains to be done, especialy respecting the Engagements of the Persons to be employd as artists, & their instructions. This, however, & whatever Else I am Capable of executing, I shall most readily undertake, if your Lordship shall honor me with your Commands, as soon as I return from Lincolnshire, which will be before the End of the Christmas holidays.[17]

I have the honor to be,
with infinite Esteem & high
Respect & Regard,
Your Lordships Obedient
& Faithfull Hble Servt,

Jos: Banks.

[P.R.O. Adm. 1/4377.]

[1]Nicolas Thomas Baudin (1754–1803) left Havre on 19 October 1800 in command of *Le Géographe and Le Naturaliste.* His mission was to sail for the South Pacific, and explore the coast of Australia. The names chosen for his ships seemed to represent the cartographic and scientific aims of his voyage. However, British suspicions were aroused at what appeared

to be a French attempt to occupy portions of Australia, and thereby assert wider claims in the southern oceans.

Such fears did not impel speedy action though. HMS *Investigator* departed from Spithead on 18 July 1801 under Lieutenant Matthew Flinders RN, FRS (1774–1814), some nine months after Baudin had left. The orders Flinders carried were dated 22 June 1801: P.R.O. Adm. 1/1800; N.M.M. FLI/3. They contained many of the suggestions Banks sent to Spencer.

[2]A passport was necessary for ships of discovery to travel unmolested by the warships of enemy powers. Banks had generously arranged a passport for Baudin's voyage: Otto to Banks, 9/6/1800, Biblio du Mus Nat d'Hist; Banks to Otto, 13/6/1800, Biblio du Mus Nat d'Hist.

[3]Bourbon Island, now Réunion, Archipel des Mascareignes, Indian Ocean.

[4]Baudin did not hurry to his task. His ships reached Mauritius (Ile de France) in March 1801, and remained there for forty days. In total it took Baudin some eight months to reach the Australian coast, during which time Flinders gained on his French rival.

[5]Bass Strait, separates Australia from Tasmania (Van Diemen's Land). It is named after George Bass (d.1812), who explored it in 1798.

Flinders sailed to Australia in 1795 in HMS *Reliance* under second captain Henry Waterhouse RN (1770–1812). Their task was to convey the new Governor, John Hunter, to Sydney. This done, Flinders was given various naval duties, including the circumnavigation of Tasmania with Bass late in 1798 and into 1799. Flinders returned to England in 1800 a worthy candidate to take charge of HMS *Investigator*.

[6]The ship's seams were re-caulked at the Cape of Good Hope. HMS *Investigator* set out from False Bay on 4 November 1801 to cross the Indian Ocean. On 6 December the south-west coast of Australia was sighted near Cape Leeuwin. In keeping with Banks's instructions, Flinders hurried to King George Sound where repairs were made, supplies replenished and exploration of the area mounted. On 5 January he quit the Sound to start a precise survey of the south coast of Australia.

[7]HMS *Investigator* sailed east from King George Sound along the Great Australian Bight, and rounded Cape Catastrophe on 20 February 1802. From here the ship proceeded to explore Spencer's Gulf, where Port Lincoln was examined. It was established that Spencer's Gulf did not lead to a strait separating Australia into parts. Afterwards, the first of two stays was made at Kangaroo Island, and on 24 March Investigator Strait was crossed. St. Vincent's Gulf followed, with the second, longer stay on Kangaroo Island being made from 2 to 7 April.

Then on 8 April a sail was sighted in the east. This was Baudin in *Le Géographe*, which had lost contact with *Le Naturaliste* in early March. Encounter Bay commemorates the meeting place of the British and French ships. After taking leave of each other on 9 April, Flinders and Baudin continued in opposite directions so that Flinders next reached Port Phillip and Westernport. On 3 May HMS *Investigator* left Port Phillip, and by 9 May the ship had arrived at Port Jackson, where Flinders reported to Governor Philip Gidley King.

[8]The 60 ton brig *Lady Nelson* was fitted with a sliding keel, or centreboard, according to the designs of John Schank (1740–1823). Such a vessel seemed well-suited for coping with heavy open seas and close inshore work. *Lady Nelson* was to be used to explore shallow waters, bays and even to penetrate the Australian continent by means of rivers. She was waiting to join HMS *Investigator* at Port Jackson.

[9]Preparations to continue the survey of Australia's coast were made in June and July. It was decided the ships should sail northwards towards Torres Strait, and explore the Gulf of Carpentaria as well as the north-east coast of Australia. On 21 July HMS *Investigator*

and *Lady Nelson* left Port Jackson. However, *Lady Nelson* proved so slow and unreliable that Flinders ordered her back on 18 October.

[10]Gulf of Carpentaria, on the north coast of Australia, between Cape York Peninsula and Arnhem Land. Flinders entered the Gulf of Carpentaria on 3 November 1802 only to discover that many timbers in HMS *Investigator* were rotten. He had little choice but to return to Port Jackson, but decided to do so after completing a survey of the Gulf. He travelled back via Timor, the west and south of Australia. He eventually reached Sydney on 9 June 1803 — Australia had been circumnavigated for the first time.

[11]Torres Strait, lies at the northern extremity of Australia, between Cape York Penninsula and Papua New Guinea. It is a maze of shoals, reefs and islands. HMS *Investigator* entered Torres Strait on 28 October.

[12]Islands of Direction, off Cape Flattery, on the Cape York Peninsula, now called South Direction, North Direction and Lizard Islands.

[13]Possibly Cape Van Diemen, which was not part of the mainland as Dutch charts indicated. It was an island, which was given the name of Mornington Island.

[14]Cape Chatham, on the south-west coast of Australia, east of Point D'Entrecasteaux, as shown on 'Chart of Terra Australis By M. Flinders, Commr. of His Majesty's Sloop Investigator. South Coast, Sheet 1, 1801. 2. 3.'

[15]Bay of Inlets, the coastal area including Broad Sound, Thirsty Sound and Shoal Water Bay. The name is no longer used.

[16]'Fegee', Fiji Islands, South Pacific.

[17]A note in the bottom right-hand corner of the final page reads: 'Investigator. Sir Jos: Banks Rd. Jan: 1801.' Taken with Banks's final remarks, it suggests these instructions were drafted in December, prior to his visit to Revesby Abbey.

Letter 83

1801

To Sylvester Douglas F.R.S., Baron Glenbervie

Soho Square
1 February 1801

My dear Lord,

As your Lordship has done me the the honor to ask of me in what manner the general interests of Science can be best promoted by you during your intended residence at the Cape, I shall take the liberty to state briefly such methods as appear to me likely to effect that purpose with the least interruption possible to those short periods of leisure which the multifarious duties of the Governor of a newly conquered Colony[1] will allow your Lordship to enjoy.

I shall begin with Botany, a Science in which I know your Lordship takes a warm interest, because a due encouragement of this branch will enable your Lordship not only to obey the commands, but also to administer to the private amusement of our beloved King, whose Talents and Virtues your Lordship & myself well know how to appreciate, &, of course, most Sincerely to admire.

As every part of the Colony of the Cape has been diligently searched by a variety of Botanical Travellers, & more particularly by Mr. Masson,[2] who resided there many years, & visited some of the most remote corners of it at the King's expence, I do not think it in any degree adviseable for your Lordship to be at any charge whatever in sending persons in Search of Plants as, except in the case of the Timber Trees, it is certain that much the greater part of the most beautiful & valuable kinds produced there are already to be found in Kew Gardens.

As, however, the renewal of these very kinds is often desirable, some of them already tending by age to decay, & others not being perhaps so numerous in Kew Gardens as their beauty or curiosity deserve, I by all means advise your Lordship to accept such collections of Seeds or Roots as may be offered to you, & to instruct persons employed in Government's Service at distant places, & to request the Missionaries more particularly, to collect Seeds & roots when they have opportunities. It is scarce possible that a collection from the Cape should not produce something very acceptable, & if out of a hundred sorts of seeds one or two Plants are added to the Royal Collection, the whole trouble of Sowing, Planting & attending them is amply & satisfactorily recompensed.

From Africa, Pliny[3] tells us, something new is continually brought home, & this inexhaustible Continent still retains the character it had in Pliny's time. It will therefore happen occasionally, if your Lordship & Lady Glenbervie[4] patronise the pursuit, that beautiful Plants will be brought from the interior to Cape Town such as have not before been seen by the people of the Cape, & are uncommon in the Countries from whence they are brought. Such Plants your Lordship will have opportunities of acquiring; & there is great reason to hope that a large proportion of them will be highly acceptable at Kew, for very beautiful Plants are seldom forgot by those who have once seen them, & some Plants must have escaped the diligence of the persons who have searched the Country by flowering at seasons when no Botanist happen'd to be upon the spot.

Bulbous & Tuberous-rooted Plants are particularly interesting because many of them are difficult to increase, & some are not of long duration. These, if taken up when their Leaves have decayed & before any part of the flowering stem appears, & dried in the shade, bear the Voyage to England as well as seeds. Collections of these may easily be procured by the good will of the Boors who live in distant districts, and, when planted in the Cape Garden, may be taken up, dried & prepared for their transport in a proper season.

Forest Trees I particularly wish to recommend to your Lordship's notice. The vast Forests beyond Muscle bay[5] have not hitherto been of any use to the Colony. These will certainly be visited by your Lordship's order, & made accessible to the Colonists. The persons employed in this service will therefore be easily able to collect Tree Seeds, & these are almost certain of being new as we are ignorant even of the sorts of Trees that grow at Grote Vader's bosch,[6] so careless have our Travellers been hitherto in that respect; or, rather, so few have visited the woods at the time when Tree seeds are ripe. But the Seeds of the Timber of Groot Vaders bosch may be obtained immediately on your Lordship's arrival, either from the Moravian Missionaries who live about half way between it & Cape Town, or from those people who visit the wood in order to cut down for use the few remaining Trees now to be found there.

I have been told that the beautifull botanic Garden at the Cape has been sacrilegiously converted into a Horse pasture, & I fear that the thick cliped Hedges I saw there in 1771,[7] the best possible shelter for Plants in a windy climate, have been eaten up by those Locusts they were expected to inclose. I hope, however, they are not wholly annihilated, & that your Lordship will find some proper place there for your own Kitchen Garden, & some sheltered corner of a few Perches that you will appropriate to the keeping of such Cape Plants as you receive from time to time. Their Culture to a Cape Gardiner will be as easy as that of a Cabbage to an English one; & from thence they may, when in full health, be planted in proper casks & boxes, & be ready for a Voyage whenever a good opportunity occurs of sending them home.

The easiest method I know of sending home living Plants is in Casks about half filled with Earth, & the wider the Cask is the better it will suit this purpose. Such a Cask, placed upon the Quarter Deck of a Ship with a covering of netting to keep out the paws of Monkies & of mischievous people, & another covering of Canvas, with orders to the Quarter master of the watch to put this on whenever the sea rises into a foam in any degree, & to take it off when the weather is mild & the wind gentle, is almost a certain conveyance. Plants intended for the Kings Garden should be shipped on board the Vessels most likely to be ordered to the [Thames] River on their return home, & if addressed to me on his Majesty's Service, will always be taken out the moment the Vessel arrives, for there is a sailing boat at the Custom House always ready to go on the Kings' business as far as Long Reach.[8]

In all cases where specimens can be obtained of Plants sent home, they will be of great use to those who direct the management of the Royal Gardens. It will shew them at once the nature of the Plants they receive, & enable them to plant them in the most proper Soil, & place them in the kind of Climate that will best suit their respective temperaments. If, however, difficulty arises in procuring Specimens, the plants ought nevertheless to be sent as the great experience of the Cape plants that have of late years been obtained renders it practicable to conjecture very nearly what will be the proper kind of management, even of an unknown vegetable, brought from thence.

I shall proceed next to Geography, & here I must take leave to remark that in Mr. Barrow's[9] Map of the Colony, just published with his Book, the Longitudes of several places differ materially from those given of the same places in the Letter press. This is no doubt owing to mistake as Mr. Barrow is extremely capable both of taking the observations & calculating them. I am sure, however, it would be a most agreeable present to the Geographers here, & would tend materially to increase the number of Mr. Barrow's well-wishers, if he will Send home by your Lordship's hands all the observations he has made in his Journeys to be calculated here. It will give a confidence in his Map to all persons, which at present none but those who know his abilities can feel for it, & will place his name at once among those whose observations abroad have fixed Geographical points for the advancement of General Science at home. If Mr. Barrow should decline this invitation, which I think will not be the case, I take the liberty to advise your Lordship on all occasions when you employ persons on distant service, to request those who are able to make observations as often as is convenient, & to Send such observations home, for at present the public want confidence in the Longitudes of the different parts of the Colony, & those who know Mr. Barrow's merit, & wish to rest upon his authority, can scarce defend their opinions against the objections that are stated.

In the Science of Zoology, Pliny's observation of the constant novelties produced by Africa is particularly exemplified, for, numerous as the List of Quadrupeds already known at the Cape is, Mr. Barrow observed several new ones, the skins of which would make a valuable addition to the collections at the British Museum. That gentleman also gives some good reasons why it is probable that the prototype of our Heraldic Unicorn, which is asserted by an early Traveller to have been seen alive at Mecca in the beginning of the 16th. Century, is actually in existence beyond the Bambos berg.[10] I need not, I am sure, remind your Lordship how highly interesting the discovery would be, or urge you to profit by any opportunity that may offer of encouraging the Elephant hunters, or any others inclined to pass the Mountains & examine the Country beyond them. Here let me petition on my own account that your Lordship will, if an opportunity occurs, be so good as to provide for me one or more skulls of the African Wild Hog (the Sus Ethiopicus of Linnaeus).[11] The older the animal has been, or the younger, the better it will suit my object, which is an examination into the mode of Dentition, which in this Animal somewhat resembles that of the Elephant.

For comparative Anatomy, I beg to call your Lordship's attention to a very ingenious medical man whom you will find at the Cape, a Mr. Somerville, who is able, & no doubt will be willing under your Lordships auspices, to undertake such enquiries in that branch as may present themselves. I do not at present recollect any enquiry that particularly deserves to be distinguished, but the Anatomy of most of the Cape Quadrupeds is very superficialy known. Many opportunities will therefore present themselves, of which Mr. Somerville will be enabled to take the advantage.

Ornithology, Ichthyology & Entomolgy should not be quite neglected, but, as I wish your Lordship to pay attention to the more interesting branches of Science, & consider these to be in some degree subsidiary, I shall say no more of them than Collections of Birds, fishes or insects may come to your Lordships or to Lady Glenbervie's hands, & will in that case certainly be very acceptable to the British Museum.

In Geology Mr. Barrow will be able to give very good instructions. If any gentleman who is to be employed in distant parts chuses to undertake observations in that branch, & in Mineralogy also, he will be an useful adviser. It is certainly very desirable that Specimens should be sent home of all the kinds of Rocks in the Colony, or wherever the country adjacent may be visited, & also all Sorts of crystals, & of substances contained in Mineral veins, & of Earths & Salts likewise. The real Prehmite[12] is a scarce mineral in Europe. A parcel of it would be very acceptable to our Collectors, & it is more than probable, if an opportunity offers of Sending home Specimens of Earths, Stones, Crystals &c., that other novelties of still more interest will be found among them. The valuable discovery of Coal[13] in the Colony will make Mineralogy a most interesting pursuit, & I have little doubt, as it is now known to exist there in small quantities, that the probability arising from this circumstance of Veins sufficiently thick for working being discovered will induce Government to send out at the public expence a person used to search for that Mineral, & versed in the art of working Collieries, if such a one can be obtained upon reasonable terms.

The probability of finding coal may also be materially increased by offering Premiums of some value to those who shall discover it. These may be proprotioned to the vicinity the discovery shall have to the Cape Town, or to some navigable river to which coasting Vessels can have access. If a printed statement of the account & the conditions of those premiums is given to each Boor who shall visit the Cape Town, with a specimen of Coal to make him acquainted with the appearance of the Substance he is to search for, an immediate discovery may possibly be the consequence, for as it is now clear that Coal does exist in the southern parts of Africa, Beds of it may be already known to some of the Inhabitants of the Interior,[14] who, never having seen this mineral applied to use, are utterly ignorant of its nature & its qualities.

[N.H.M. B.L. D.T.C. XII. 171–179; S.L. Banks MS. A. 5:33; M.L/D.L. Banks Papers Series 73.067.]

[1]Banks's long involvement with African exploration was well known in government circles. He had been a member of the Saturday's Club, and subsequently a directing force in the African Association. Both of these bodies promoted expeditions to the continent, particularly to West Africa, and in 1831 the latter merged with the Royal Geographical Society.

Thus, when Lord Glenbervie was appointed Governor of the Cape of Good Hope, he approached Banks for advice on projects which might usefully be undertaken in South Africa. The letter above is therefore one of two in which Banks surveyed the extent of existing knowledge about Africa, and comprehensively suggested how a new governor might add to it. The other is: Banks to Glenbervie, 22/12/1800, S.L. Banks MS. Africa 1800 A5:32. Glenbervie never took up his post, settling instead for a political career in England.

[2]Francis Masson collected for the Royal Botanic Gardens at Kew, working at the Cape of Good Hope on two occasions, 1772–1774 and 1786–1795.

[3]Caius Plinius Secundus, (Pliny the Elder), (23–79): a Roman scholar, who produced the encyclopaedic *Historia Naturalis* in thirty seven volumes.

[4]Lady Catherine Anne Douglas (*née* North), (d.1817).

[5]Mossel Bay or Mosselbaai, Western Cape (former Cape Province).

[6]'Groot Vaders bosch' appears on Barrow's map in the 'District of Zwellendam', east of Cape Town.

[7]Banks spent time at the Cape in March and April 1771 when HMS *Endeavour* was returning to England. He took the opportunity to observe the area, including the Dutch East India Company's garden: *Endeavour Journal*, vol. II, pp. 254–255.

[8]Long Reach, Reach of the Thames river, above Gravesend, Kent.

[9]John Barrow. In 1795 a British expeditionary force successfully attacked the Dutch colony at the Cape of Good Hope. Two years later, in January 1797, Lord Macartney embarked from Portsmouth in HMS *Trusty* as Governor of the Cape. Macartney knew Barrow, who travelled with his Embassy to China, and took him to Africa as a private secretary.

While at the Cape Barrow explored and collected, acquiring the detailed knowledge of the colony which he published in 1801 as *Travels into the interior of Southern Africa, in which are described the character and condition of the Dutch colonists of the Cape of Good Hope, and of the several tribes of natives beyond its limits; the natural history...the geography...and a topographical and statistical sketch of the Cape Colony...*, London. It included Barrow's 'General Chart' of the colony. In due course, Barrow assumed an important role in African (and Arctic) exploration. He was largely responsible for the amalgamation of the African Association with the newly formed Royal Geographical Society in 1831.

[10]'Bambos berg' appears on Barrow's map beyond the limits of the colony, to the north-east of the 'District of Graaff Reynet' or Graaf-Reinet, Eastern Cape (former Cape Province).

[11]Wart-hog (*Phacochoerus aethiopicus*).

[12]Given as 'Prelornite' or 'Prehnite': S.L. Banks MS. Africa 5:33. Prehnite is named after the Dutch explorer, Colonel H. von Prehn (1733–1785). It is found in Germany and France,

but greater quantities exist in South Africa. The chemical composition of this mineral is $Ca_2 Al_2 Si_3 O_{10} (OH)_2$, Calcium aluminum silicate hydroxide.

[13]Banks's belief that valuable mineral deposits lay in South Africa was, of course, substantially correct. Coal is now the primary fuel source produced and consumed in South Africa, and is also its second largest export in terms of foreign exchange, after gold. Banks hoped that coal would be found near Cape Town. Some had indeed been discovered as early as 1699 in the Franschhoek Valley, which is thirty miles from Cape Town, but its low quality and quantity precluded mining. So from 1814 the British actually imported coal from Wales, until substantial coalfields in Natal, Transvaal and the Orange Free State were identified. Some three quarters of South Africa's current coal reserves are therefore located in north-eastern areas which were largely unexplored by Europeans in Banks's day. The main ones are at Waterberg, in the Northern Province (former Northern Transvaal), and at Witbank and Highveld in Mpumalanga (former Eastern Transvaal). These contain mainly bitumous coal, with a relatively high ash content, and a low sulphur content.

[14]African tribesman were aware of these deposits, perhaps from the Middle Ages, when they first penetrated the region.

Document 84

[1801]

To Robert Brown F.R.S., Ferdinand Lucas Bauer, William Westall A.R.A., Peter Good and John Allen

[Soho Square]
29 April 1801

Memorandum[1] of agreement signed by all the persons employed as scientific assistants on board H.M. Ship the Investigator for the purpose of exploring the country of New Holland.

In order to prevent all misunderstanding between the Lords Commissioners for Executing the Office of Lord High Admiral of the United Kingdoms, and the Persons employed by their Lordships as Scientific Assistants on board His Majesty's Ship the Investigator for the purpose of exploring the Country of New Holland, Their Lordships have been pleased to Issue the following instructions and commands to be obeyed by all Persons so employed. It is expected that every Person so employed do sign his name to the same in testimony of his acquiescence in the terms on which their Lordships are pleased to employ him.

1st. Their Lordships require every Person employed as a Scientific Assistant on board the Investigator to render Voluntary Obedience to the Commander of the Ship in all Orders he shall from time to time issue for the direction of the conduct of his Crew, or any part thereof.

2dly. Their Lordships require that all Persons so employed do on all occasions conduct themselves peaceably, quietly and civily to each other, each readily Assisting the other in his respective department to the utmost of his ability, [and] in such manner as will best promote the success of the Public Service in which they are jointly engaged, and unite their individual endeavours into one general result.

3dly. Their Lordships require the Draughtsman employed for Natural History to pay due attention to the directions he shall receive from the Naturalist, and the Draughtsman employed for Landscape and Figures to pay regard to the Opinion of the Commander in choice of Objects most fitting to be delineated; and their Lordships moreover require the Gardener and the Miner to pay obedience to the Naturalist in all such Orders as he shall think fit to give them.

4thly. Their Lordships consider the Salary allotted to each Person employed as full compensation for the whole of his time. They expect therefore that all Journals, remarks, Memorandums, Drawings, Scetches, Collections of Natural History, and of Habits, Arms, Utensils, Ornaments &c. of every Kind be delivered up on the return of the Ship to such Persons as their Lordships shall direct to receive them.

5thly. In order, however, to encourage the Persons engaged in this undertaking to exert themselves to the utmost in accomplishing the Object of their Mission, Their Lordships hereby declare That, if the Information collected during the Voyage is deemed of sufficient importance, it is their Lordships intention to cause it to be published in the form of a Narrative drawn up by the Commander[2] on a Plan similar to that pursued in the Publication of Captain Cook's Voyage,[3] and to give such pecuniary Assistance as their Lordships shall see fitting for the Engraving of Charts, Plans, Views, Figures, &c. In such case the most interesting observations of Natural History, & the most remarkable Views of Land, & delineations of People &c. will be inserted therein.

6thly. Their Lordships moreover declare that, in case the Persons employed in this undertaking as Scientific Assistants are industrious in their several departments, civil and obliging to each other, & co-operate together on all occasions in making the general work in which they are jointly engaged compleat by Assisting each other and uniting their efforts for the advantage of the Public, it is intended that the Profit derived from the Sale of the said Publication shall be divided between the Commander and the Assistants in proportion to the good Conduct each shall have held during the Voyage, and the comparative Advantage the Publication shall in the Opinion of their Lordships derive from the Labours of each Individual.

7thly. Their Lordships moreover declare that after such descriptions, Drawings and Scetches as shall be found necessary for the Illustration and Embellishment of the intended Publication shall have been Selected by such Persons as their Lordships shall be pleased to appoint, and such Specimens of Natural History, Arms, Implements, Habits, Ornaments &c. as their Lordships think fitting shall have been applied to such purpose as their Lordships shall approve, the remainder of the descriptions of Plants and animals &c., and the Scetches of all Kinds, shall be at the disposal of the Persons who have made them for the purpose of being Published by them whenever it is thought proper, at their own risque, and for their own advantage, provided, however, that all such Drawings as shall be finished during the Voyage, and such Scetches as their Lordships shall Order to be finished after the return of the Ship, shall be considered as the Property of the Public, and lodged in the Depot of the Admiralty when required so to be, and that the remainder of the Collections of Natural History, Arms, Habits, Implements, Ornaments &c. shall be at the disposal of the Persons who have collected them.[4] All this, however, on condition that each Person shall during the Voyage have behaved himself with propriety to the rest, Their Lordships reserving to themselves the Power of punishing all deviations from good humor and perfect harmony among the parties by witholding from the Persons offending such parts of the benefits above described as they shall think proper.

We, the undersigned, Robt. Brown — Naturalist, William Westall — Landscape & Figure Draughtsman, Ferdinand Bauer - Botanic Draughtsman, Peter Good — Gardener, and John Allen — Miner, in testimony of our concurrence in the above terms, and as a Pledge for our Obedience to all such Instructions and Commands as their Lordships shall be pleased to issue to Us during the time we shall be in their Lordships employ, have signed our names to this Engagement on the Twenty Ninth Day of April in the Year of our Lord 1801:

Robert Brown
Ferdinand Bauer
William Westall
Peter Good
John Allen

Signed by all the parties in my presence

Jos. Banks.

[P.R.O. Adm. 1/4379; M.L./D.L. Banks Papers Series 63.08 and 63.09, for which see *Historical Records of New South Wales*, F.M. Bladen (Ed.), Sydney (1896), vol. IV, pp. 349–351.]

[1]Banks was deeply involved in the arrangements for the voyage of HMS *Investigator* under Matthew Flinders, 1801–1803, and paid scrupulous attention to every detail. Having determined who the scientific personnel should be, he drew up their instructions. These reveal Banks's awareness of the problems and opportunities which might arise for such a group on a Royal Navy vessel.

This document carries the note 'Mr Pearce', and was clearly circulated after being signed. The Mitchell Library drafts, one in Banks's hand, were made in January, and Banks recorded that 'I deliverd this undertaking signd by all The Parties & witnessd by myself to Mr Nepean Augt 21 1801 he told me he should send it to the navy office'.

[2]Banks was given charge of the engravers, draftsmen, sketches, charts, journals and other manuscripts for work on the account of the *Investigator* voyage: Barrow to Banks, 15/1/1811, M.L./D.L. Banks Papers Series 70.03. This work could not commence properly until Flinders had returned. He was a capable author, who could be relied on to produce the narrative. In addition, he had amassed a considerable quantity of cartographic, scientific and descriptive material during his long absence. This was all used along with the resources gathered at 32 Soho Square. By 1814 the combined efforts of those involved culminated in: *A Voyage to Terra Australis; undertaken for the purpose of completing the discovery of that vast country, and prosecuted in the years 1801, 1802, and 1803, in His Majesty's Ship the Investigator*..., London.

[3]Banks referred to James Cook's third Pacific voyage in HM Ships *Resolution* and *Discovery*, 1776–1780. Banks managed the publication of the account of this mission: *A Voyage to the Pacific Ocean. Undertaken for making discoveries in the northern hemisphere*..., London (1784).

[4]The natural history collections from the voyage had arrived in London by late 1805, and Banks was in negotiations with the Second Secretary at the Admiralty, William Marsden FRS (1754–1836), over their value and future: Banks to Marsden, 19/10/1805, N.H.M. B.L. D.T.C. XVI 149–150. Early in 1806 Robert Brown, Ferdinand Bauer and William Westall were assembled in London too, so that Banks was able to lay early plans for the years of work to come: Banks to Marsden, January 1806, B.L. Add. MS. 32439, ff. 237–241, *Historical Records of New South Wales*, F.M. Bladen (Ed.), Sydney (1898), vol. VI, pp. 16–19; and a further undated letter to Marsden this month, P.R.O. Adm. 1/4379.

For Brown and Bauer this involved some 3600 plant specimens, together with more than 2000 sketches, a number of animal skins, 3 boxes of minerals and 1 case of insects. The animal skins, minerals and insects were destined for the British Museum where they could be cared for by 'persons more conversant in these branches of natural history.' However, Banks, Brown and Bauer selected 200 plant species to be shown in coloured botanical drawings, and everything would also need to be arranged and classified. Banks thought this would all take three years.

In the event, the systematic problems of dealing with so many new species led Brown to adopt and develop Antoine Laurent de Jussieu's system. This, and the sheer quantity of plants to deal with, meant a delay until 1810, when a single volume appeared: *Prodromus Florae Novae Hollandiae et Insulae Van-Dieman, exhibens characteres planterum quas innis 1802–1805...collegit descripsit R. Brown...*, London. Meanwhile, Bauer prepared some of the finest botanical illustrations and engravings yet made, some of which he published in three small fascicles of his own: *Illustrationes Florae Novae Hollandiae: sive icones generum quae* in *Prodromo Florae Novae Hollandiae et Insulae Van-Diemen descripsit Robertus Brown/ Ferdinandi Bauer*, London (1813–1816).

Despite the near loss of his drawings on Wreck Reef in 1803, by 1806 Westall was set to the task of finishing the sketches, some of which could then be engraved for the Admiralty. About 200 survive at the National Library of Australia, and 9 were included in *A Voyage to Terra Australis....*

Letter 85

1801

To Déodat Guy Sylvain Tancrède Gratet de Dolomieu

Soho Square
16 July 1801

Sir,

I seize with pleasure the opportunity which the return of Mr. de Drais[1] affords me of congratulating you on your safe arrival in your own Country,[2] on the re-establishment of your health, and, more partlicularly, on your having escaped the persecution which the ill-judged conduct of the Court of Naples subjected you to.

I felt anxiously desirous to assist in bringing about your liberation, and, as I had the good fortune to enjoy an intimate acquaintance with Sir W[illia]m. and Lady Hamilton,[3] and I trust some share of Lord Nelson's[4] friendship, I took every measure in my power, and used every argument that I could suggest to interest these my Friends to sollicit your release; and they did, I am sure, their utmost endeavours to obtain it. Their applications to the Government of Naples, as well as to the King and Queen,[5] were numerous and pressing, and, altho' I was grieved to learn that nothing was obtained by their intercession, except that your confinement was rendered less strict and severe, I was consoled by finding that their endeavours did not cease, and that even after they returned to England your hard fate was the subject of Letters from Lady Hamilton to the Queen of Naples, then in Germany.

Tho', Sir, your unfortunate situation caused a visible regret in all men of Science here who were acquainted with the active part you have always taken in promoting the increase of human knowledge, and tho' all of them in their private capacities, I have no doubt, did all in their power to promote an interference of our Government in your favor, yet, Sir, it is necessary I should inform you that the Royal Society, as a body, took no measure whatever, or indeed ever publickly took cognizance of your situation. We in England, Sir, are as firmly attached to Regal Government as you can be to Republican, and whatever our private wishes might be, we did not think it proper or decent to speculate upon the conduct of a King whose motives we could not be acquainted with, or, as a public body, to take matters of a political nature in any shape under our consideration. I hope and trust, therefore, that you have not in any public manner noticed your belief of the Royal Society having, as a body, interfered in your favour as I am quite sure you have no well grounded reason for such an opinion.

I shall be glad, Sir, to hear that you have resumed your pen, and the more so if you employ it in illustrating the Geology of Egypt. The treasure of knowledge which you have brought from thence must be invaluable, for, as that Country is now likely to be again abandoned to its illiterate and uncivilized inhabitants, ages may pass away before Science again gains an opportunity of investigating the things you have seen and are able to describe. We English, tho' much attached to Science, have not, as your Chief Consul did,[6] sent learned men with our army. Our successes therefore, if Heaven should favour us with success, will be productive of political advantages only, while Science, unthought of by our Rulers, must look to France alone for having blended Learning with her Arms, and gathered knowledge beneficial to the whole race of men with those Laurels which to our Commanders will be the fruitless ornament of successfull valour.

I am, Sir,
with much consideration and regards,
Your faithful and humble servant,

Jos Banks.

[N.H.M. B.L. D.T.C. XII. 240–243.]

[1]Étienne Gilbert, Marquis de Drée (1760–1848): mineralogist. Drée was Dolomieu's brother-in-law. In 1801 he had pleaded with Banks to muster British support for Dolomieu's cause: Drée to Banks, 13/4/1800, B.L. Add. MS. 8099, f. 54.
[2]Dolomieu was released at the end of March. This was one of the first conditions Napoleon Bonaparte insisted on for an armistice. Negotiations subsequently led to the Treaty of Amiens, but Dolomieu never recovered from his harsh imprisonment, and died later in the year.
[3]Banks had written to Sir William and Lady Hamilton: Letters 79 and 80.
[4]Admiral Horatio Nelson, 1st Viscount Nelson, and Duke of Bronté.
[5]Ferdinand I, King of Naples and Sicily, and Maria Carolina, Queen of Naples and Sicily.
[6]Napoleon Bonaparte.

Letter 86

1801

To William Forsyth

Soho Square
31 July 1801

Sir,

I Return with this Mr. Wedgwoods Letter & his Plan for an horticultural Society.[1] I approve very much the Idea. I Know of no Trade that Conceals So many Valuable branches of Knowledge as that of a Gardiner, & Few subjects where the Public will be more benefited by the disclosures which such a Society will immediately occasion. I Shall be Flatterd if the Gentleman who are to Arrange the Plan do me the honor to Set me down as an Original Member.[2]

I am, Sir,
your Very Hble Servt,

Jos: Banks.

[Lindley Library, Royal Horticultural Society. Also: *Journal of the Royal Horticultural Society*, vol. 79 (1854), figure 123.]

[1]The inaugural meeting of the Horticultural Society was held on Wednesday, 7 March 1804, at Hatchard's, 187 Picadilly. Present were: John Wedgwood (1766–1844): Josiah Wedgwood's eldest son, and founder of the Horticultural Society; William Forsyth; Sir Joseph Banks; Charles Francis Greville; Richard Anthony Salisbury (formerly Markham) FRS (1761–1829): botanist and Secretary of the Horticultural Society, 1805–1816; William Townsend Aiton; James Dickson.

[2]In 1809 the Horticultural Society was established by Royal Charter as The Royal Horticultural Society of London. The matter of a charter was originally discussed at committee meetings chaired by Banks early in 1808. Along with other prominent Society figures, he made arrangements for one, which arrived under the date 17 April 1809.

Letter 87

1802

To The President and Secretaries of the Institut National des Sciences et des Arts

London, Soho Square
21 January 1802

Messieurs,

I beg of you to offer to the National Institut my most gratefull Thanks for the honor[1] they have Conferrd on me by Electing me an associate of that Learned & distinguishd body.[2]

I Request you also to assure my much Respected Associates that I consider this mark of their Esteem as the most Gratefull, as well as the most dignified Literary honor I possibly Can Receive. To be the first Elected, of the first Class, of the first Literary Society in the world is more than my utmost Ambition Ever permitted me to hope. The attainment of this honor is therefore a gratification for which I cannot be too thankfull to the Nation of which they are the Literary Representatives; a nation which, throughout the terrible convulsions of the Late Tremendous Revolution, I have never Ceasd to Esteem, well aware, even in the worst Periods, that Good men in Abundance were Still there, & that these would in time Resume their Superior Stations, & Replace Virtue, Justice & honor in the hearts of their Countrey men.[3]

I beg leave also, Gentlemen, to Return you thanks for the Obliging manner in which you were pleasd to Communicate to me the Good news,[4] & I am,

with Sincere Esteem for your distinguishd Talents,
Your Obedient
& Faithfull Hble Servt,

Jos: Banks.

[M.L./D.L. Banks Papers Series 73.075.]

[1]Banks was elected a Foreign Associate of the Institut National des Sciences et des Arts in the First Class of Science, Physics and Mathematics: President and Secretaries of the Institut National to Banks, 26/12/1801, B.P.L. Banks Collection; Tenon to Banks, 26/12/1801, B.L. Add. MS. 8099, f. 207. This award followed the release of Déodat de Dolomieu,

whom Banks had struggled to help: See Letter 85. The gradual improvement in Anglo-French relations continued, and on 25 March 1802 the Peace of Amiens was signed.
[2]Two earlier drafts of this letter were made by Banks following the arrival of his certificate in January 1802: Banks to the President and Secretaries of the Institut National, 21/1/1802, M.L./D.L. Banks Papers Series 73.075.
[3]A translation of Banks's letter above was printed in French in the *Moniteur*, 18 March. This was then translated back into English, and used as the basis for an attack on Banks in a letter in William Cobbett's (1762–1835) *Weekly Political Register*. The attack was written by an anonymous figure called 'Misogallus', and expressed 'disgust at this load of filthy adulation'. April saw the appearance of Misogallus's letter as a pamphlet entitled 'A Letter to the Right Honourable Sir Joseph Banks K.B.' As before, Cobbett and Morgan were the publishers. Another letter, dated 4 November, appeared in the *Weekly Political Register* when the annual elections for President of the Royal Society came. This time the annonymous writer thought Banks unworthy of his place as President given the nature of his acceptance letter to the Institut National.

Banks ignored Misogallus, who reacted to his re-election as President on 30 November with a further letter, 7 December. Banks had been unwell, and tended to avoid ugly public exchanges anyway. Misogallus never declared himself openly, but William Windham or John Alexander Woodford (d.1817) might well have known who he was: Smith to Banks, 16/12/1802, N.H.M. B.L. D.T.C. XIII 329–331; Banks to Aiton, 21/12/1802, M.L. Banks MSS A80/4.*
[4]Banks had already received the news of his imminent election in November 1801: Rumford to Banks, 22/11/1801, B.L. Add. MS. 8099, ff. 107–108, N.H.M. B.L. D.T.C. XII 281–282.

Letter 88

1802

To Mme. Marguerite Ursule Fortunée Briquet (née Brennier)

[Soho Square]
[April 1802]

Madam,

My Feeble endeavours to Break the Chains of the Meritorious Dolomieu[1] have not been wanting, nor will my mind, I hope, Ever Remain inactive when there are hopes of Releiving Genius, when oppressed, from the Tyranny of those who know not how to Estimate its value.

From you, madam, I have Receivd my Reward. The Portrait[2] you have sent me revives in my worn out Frame the Remembrance of what Beauty Effected upon it when it was endowd with Sensibility & youth. The Idea that the Form of a face is, in your Case, endowd with the mind of a Muse has given me some Idea of Perfection inhabiting this Lower world.

Accept then, madam, this Effusion of Gratitude for the Pleasure I have receivd in viewing your Portrait & reading your verses,[3] & beleive me,

with sincere Gratitude,
& perfect admiration,
Your Most Hble Servt,

Jos.

[N.H.M. B.L. D.T.C. XIII 10–11; U.Y. Beinecke Banks MS.; de Beer Collection 25.]

[1]Déodat Guy Sylvain Tancrède Gratet de Dolomieu.
[2]Briquet to Banks, 24/3/1802, B.L. Add. MS. 8099, f. 153.
[3]*Ode sur la Mort de Dolomieu.*

Letter 89

1803

To Dr. James Edward Smith F.R.S.

Soho Square
10 March 1803

My dear Dr.,

Having nearly arrivd at the Bottom of our box of Excellent Biscuits,[1] which Arrivd here a month ago without any notification of the Place from whence it Came, we yesterday Discoverd a Letter, & by so doing Learnd that we had inadvertently opend the bottom instead of the Top of the Box, & we also discoverd a Parcel for Dixon,[2] which was deliverd to him directly.

This, I hope, you will accept as a justifiable excuse for my Silence. Had I Known that I owed the Good Cakes to your friendly recollection of me, I should not have left a day in acknowledging your kindness. At present the Ladies like them So well that they wish for a Constant Supply. Will you therefore be so good as to desire the maker to write me a line saying at which Rate he will undertake to Furnish them, for as they Keep well I Should wish to have $\frac{1}{2}$ a years supply at Least at a Time from him.

We are in alarm here least Bonaparte[3] Should invade us, & in Truth it is but too probable that he will make the experiment. His ambition is so inordinate, & his means of Satisfying it so profligate, that every thing is to be Feard, & nothing to be hopd. If an attack is inevitable, the Sooner it Comes the better. We are united heart & hand to oppose it, & when it has faild, which it must do against a united people, we Shall then have peace & Quiet — but never till then.

Pray Return Lady Banks's & my best thanks to your mother[4] for her Kind Care of us in Procuring the Principal Delicacy of our Christmas Feasts. I think the Turkeys have been this year better than Ever.

I am sorry to hear you have been so indifferent. I sincerely hope you will Soon recover your health. The present Cold weather Pinches me much, but warm weather must now be very near.

Beleive me, my dear Dr.,
Very Sincerely Yours,

Jos: Banks.

[L.S. Smith MS. 301. 1. 123.]

[1]The Smiths sent gifts of turkeys, biscuits and buns to the Banks family for many years.
[2]James Dickson, the botanist.
[3]Napoleon Bonaparte could not invade. The Royal Navy prevented it. Instead, he tried to blockade Britain.
[4]Frances Smith (*née* Kinderley), (d.1820).

Letter 90

1803

To Samuel Tyssen Jnr.

[Soho Square]
21 March 1803

My dear Sam,[1]

I am not, as you well know, your Guardian,[2] & I therefore do not intend to take any charge of you, but I Shall be ready at all times to give you any advice when you chuse to ask for it; notwithstanding the Old saying which tells us that Persons who are not wise [enough] to find out what they ought to do, are seldome sufficiently wise to follow the advice of those who are able to tell them.

When I was of your age I was not, I can assure you, sufferd to go to Bath, or to any such hazardous Place unless in the Company & under the Protection of near Relations or friends of experience in the world. In all such places as Bath, where a variety of Persons assemble together, there are always an Abundance who maintain themselves by inveighing young men into Play, & sometimes by Leading them into still worse scrapes. These persons are always on the watch for unprotected youth. I have frequently been assaild by such when I was several years Older than you are, & I only escapd because I was kept out of their way till I had experience enough to avoid them. Many, very many of my Contemporaries were not so lucky as I have been, & sufferd severely in consequence of having made acquaintance with Persons unknown to them, who were Civil and Obliging to them at such Public Places.

I therefore seriously advise you to desist from all intention of visiting Bath at Present, unless it is [in the] Company of some steady friend much Older than yourself, & well acquainted with the world. I am ready to admit that you Cannot keep properer Company than the young gentleman you associate with at Corpus, but, my dear Sam, if three of you, Equaly inexperiencd in the ways of the world, go to Bath together without someone to point out to you the Snares which Sharpers set for young men, & the wiles they use to Entrap them, it will be like the blind Leading the blind. You will all fall into the Pit together.

Those who mean to profit by travelling to see their native Country, should first have made themselves able to profit by what they will see in their Travels.[3] I was myself a great Traveller, but I did not begin till I was more than 20 years old. In a year or two I may possibly advise you to travel. At present, my advice is that you stick to your business closely, & are Content to spend your vacation under the Protection of your guardian. I shall be happy to see you when you Come to Town whenever you Choose to Call upon me, & ready to assist you in procuring for you any Rational amusement you may wish for.

[S.L. Banks MS. TE 1:57.]

[1]Samuel Tyssen Jnr. was at Corpus Christi, Oxford, as a gentleman commoner. He matriculated in 1803, aged 17.
[2]Banks was a trustee of the Tyssen estates in Norfolk after the death of Samuel Tyssen (1756–1800). Tyssen Jnr. wrote to inform Banks of his plan to visit Bath with some university friends: Tyssen to Banks, 16/3/1803, S.L. Banks MS. TE 1:56.
[3]Banks travelled to Newfoundland and Labrador, 1766, on HMS *Niger* under Captain Sir Thomas Adams. His objective was to botanize and explore. Banks also made excursions to Chatham, western England, the Midlands and Wales immediately prior to his great voyage to the South Seas: to Chatham, Kent, from 21 February to 4 March 1767; then to Eastbury and Bristol from 15 May to 20 June 1767; finally on to Wales and the Midlands from 13 August 1767 to 29 January 1768.

Letter 91

1803

To Robert Brown F.R.S.

Soho Square
8 April 1803

Dear Sir,

It gave me sincere pleasure to learn by yours[1] from Port Jackson that you was in health, & had been So fortunate in the first part of the interesting business in which you have so hansomely volunteer'd yourself.

Your Commander deserves, in my opinion, great credit from the Public for the pains he must have taken to give you a variety of opportunities of Landing & Botanising. Had Cooke[2] paid the same attention to the Naturalists as *he* seems to have done, we should have done much more at that time. However, the bias of the public mind had not so decidedly marked Natural History for a favorite pursuit as it now has. Cook might have met with reproof for sacrificing a day's fair wind to the accommodation of the Naturalists. Capt. Flinders will meet with thanks & praise for every sacrifice he makes to the improvement of natural knowledge which is compatable with the execution of his orders.

The seeds you sent by the Ship that brought your Letters came safe & in good order to hand. They are all sown in Kew Gardens, & much hopes built on the success of them, which we expect will create a new Epoch in the prosperity of that magnificent establishment by the introduction of so large a number of new Plants as will certainly be obtained from them. We have waited in expectation of a second parcel, but, as none have yet arrived, we are in fear that some accident has happened to the Vessel in which they have been forwarded.

Pray give my best wishes to your able & quiet Assistant, Peter Good,[3] & Mr. Aiton's also.[4] His diligence & docility have been before tried. I have no doubt therefore of your being satisfied with his Assistance.

Pray tell John Allen,[5] my Derbyshire friend, that all his relations in that Country are in good health, & that on the receipt of the Certificate he sent me home I immediately remitted £50 to Mr. Milnes[6] of Ashover as he desired me to do.

Tho' our Peace with France proves no more than a turbulent & quarrelsome Truce,[7] Botany advances rapidly. I am in hopes just at this time of sending a Gardiner to China[8] to reside there some years, & send over to Kew annual supplies of the beautifull & valuable produce of that Country.

I very much approve of your employing yourself when in Harbor rather in making descriptions & enlarging your observations, than in attempting to make Copies or prepare any thing to be sent home. Without you the Specimens or descriptions you might dispatch would lose much of their value. You have, I see, nevertheless made a deposit with Govr. King:[9] a very good measure certainly.

Many thanks for your news respecting the French discovery Ships.[10] They do not, by what I have been able to learn, appear likely to prove formidable rivals to you as Investigators. They seem too much afraid of the Land.

Mr. Dryander, Mr. Bauer, Mr. Dickson, Dr. Smith, Mr. Turner of Yarmouth, Mr. Koenig,[11] & indeed all your friends here, who, thanks to God, appear to continue in very good preservation, desire their best remembrances may be made acceptable to you; & I beg, my dear Sir, that you will believe me,

Very sincerely yours,

Jos: Banks.

Pray deliver my best wishes for all success in his merotorious labour to Mr. Bauer, who I know is indefatigable.

[N.H.M. B.L. D.T.C. XIV. 43–45; B.L. Add. MS. 32439, f. 95.]

[1]Brown was the naturalist on the voyage of HMS *Investigator* under Matthew Flinders, 1801–1803. His last letter to Banks was: Brown to Banks, 17/10/1802, N.H.M. B.L. D.T.C. XVIII 277–278. Brown sent detailed accounts of the progress being made on the expedition: Brown to Banks, 30/5/1802, N.H.M. B.L. D.T.C. XIII 141–146.
[2]Banks referred back to his voyage on HMS *Endeavour* with James Cook, 1768–1771.
[3]Peter Good was the gardener on board HMS *Investigator*.
[4]William Townsend Aiton.
[5]John Allen, a Derbyshire miner. Banks had arranged for him to be the miner on board HMS *Investigator* through William Milnes.
[6]William Milnes (d.1814): estate steward for Banks at Overton, Derbyshire. In February 1801 Milnes suggested John Allen might be a suitable person to join the *Investigator* mission. By April Banks was preparing Allen to go: Milnes to Banks, 8/4/1801, M.L./D.L. Banks Papers Series 63.44.
[7]The Peace of Amiens, 1802–1803. War was declared on France on 17 May 1803.
[8]William Kerr (d.1814): plant collector. Kerr was a plant collector for the Royal Botanic Gardens at Kew from 1803 to 1814. He was sent to Canton in 1803, and worked under David Lance (fl.1772–1803), superintendent of the Honourable East India Company's Canton factory.
[9]Philip Gidley King, Governor of New South Wales.
[10]This was the French expedition under Nicolas Thomas Baudin in the ships *Le Géographe* and *Le Naturaliste*, 1800–1803.
[11]Some of this select group have already been mentioned: Jonas Dryander; James Dickson; Dr. James Edward Smith. Those who have not been mentioned are: Franz Andreas Bauer FRS (1758–1840): botanical artist, and older brother of Ferdinand Lucas Bauer. Franz was

based at the Royal Gardens, Kew, 1790–1840. His skill as an artist was outstanding, like his brother's; Dawson Turner FRS (1775–1858): banker, botanist, antiquary and bibliophile. In 1830 he accepted the role of Banks's biographer, and produced 23 volumes of copies made from his friend's correspondence; Charles Konig (or Carl Dietrich Eberhard König) FRS (1774–1851): mineralogist. He worked as Keeper of the Natural History Department at the British Museum, 1813–1837, and from 1837–1851 as Keeper of the Mineralogical and Geological Branch.

Letter 92

1803

To Arthur Young F.R.S.

[Soho Square]
27 May 1803

The Comm[itt]ee[1] do not expect in Agricultural Analysis the same degree of precise accuracy as is necessary in that intended to illustrate Philosophical experiments. It will be enough for them if the Component parts of substances, & their respective proportions to each other, are marked with sufficient precision to demonstrate their probable effects on vegetables.[2]

The Comm[itt]ee are aware that at present the science of agricultural Chemistry is in its infancy, & that, till it has been more matured, each analysis will take up a considerable portion of time. They trust, however, that it will not be long before Mr. Davy himself, or some one named by him & acting under his superintendance, will undertake the business of analizing soils and manures for individuals at a moderate fixed price for each substance that will be brought them.

The Royal Institution wish to have Mr. Davy's Lectures repeated at their house, & have desired me to ask whether the B.[oard] of A.[griculture] have any objection to a measure which appears to them likely to extend still farther [blank].

Having been requested to suggest what I think [is] a proper recompence to Mr. Davy on account of his 6 Lectures delivered at the Board, & also a plan for securing his services in Future to the B. of A., I beg leave to propose that 60 guineas be given by the board to Mr. Davy as remuneration for his 6 Lectures, being at the rate of 10 guineas for each Lecture, & that the office of Professor of Chemical Agriculture to the board with a Salary of £100 a year be offered to his acceptance. The duty of the Professorship [will] be to Read Lectures in the Spring, at such time as shall be fixed by the Board, on the application of Chemistry to the improvement of the art of agriculture, & in making an analysis of such substances as shall be put into his hands by the Comm[itt]ee in Case he is of [the] opinion that the results of their analysis is likely to throw light on the Theory & Practice of that useful art.

[N.H.M. B.L. D.T.C. XIV. 86–87; De Beer Collection 16.]

[1]A note to this letter reads 'Sir Joseph Banks, in the name of the Committee of the Royal Institution.' As an improving landowner, a member of the Board of Agriculture, and a

founder of the Royal Institution, Banks was well placed to arrange for Davy's work on agricultural chemistry.

[2]Davy delivered his first lectures for the Board of Agriculture at the Royal Institution in this year. His subject was 'The Connection between Chemistry and Vegetable Physiology.' Davy continued to lecture annually until 1812, and at the request of the Board of Agriculture he published *Elements of Agricultural Chemistry, in a course of lectures for the Board of Agriculture...*, London (1813). He showed how useful chemical science is to agriculture through soil analysis, an understanding of plant growth, the use of vegetable and plant manures, and the benefits of certain fertilizers. This was all in keeping with the Board of Agriculture's attempts to raise output by improved farming methods, and Davy's own brilliant field and laboratory work.

Letter 93

1803

To Admiral Horatio Nelson, 1st Viscount Nelson, and Duke of Bronté

Soho Square
8 August 1803

My dear Lord,

To be intrusted with a Commission[1] by a Person whose invaluable services to his Country have so highly distinguishd him, I conclude as a Favor of great importance. To Participate in The Friendship of such a one I consider as an honor highly gratefull to my feelings. Your Lordship may therefore Rest satisfied that I Shall undertake your Commission with Pleasure, & Execute it with Zeal. I will take Care to offer the Sculpture you have Capturd to government in a Proper manner, to State the Value of it with Justice & Correctness, &, in Case of their Chusing to Purchase [it] for the advantage of the Arts in Britain, I will see that your brave & meritorious Tars are not deprivd of any part of their Rights, for it would not be fit that the Public in such a Case Should Pay Less than opulent individuals are willing to do.

We are here Arming with a degree of Spirit & Enthusiasm as puts all chance of a successfull invasion out of the question. If some bounds are not set to the alacrity of all Ranks of People, we shall soon have as many millions of men in training here as Bonaparte has 100,000s of Troops. We are told also that the Emperor of Russia[2] Grins Ghastly at the Corsican Tyrant. Thanks to your Lordships exertions, and the animation of the People, we all now Sleep sound in our beds. We look forward, however, to the Time which will bring your Lordship home Crownd with Fresh Laurels, [and] when our Military neighbors will have repented of their insolent aggression sufficiently to be inclind to allow us an honorable, a safe & an Amicable Peace.

Beleive me, my Dear Lord,
with sincere Esteem
& affectionate Regard,
Your Lordships
Faithfull Hble Servt,

Jos: Banks.

[P.S.] If other Commanders would, like your Lordship, look to the benefit of Public Taste & the improvement of Science when their Captures give them an opportunity of doing Service to Literature, we should Profit materialy in Every war, Especialy from your Lordships Station.

[N.M.M. CRK/2.]

[1]Nelson to Banks, 9/7/1803, N.L.A. MS. 9/114; N.H.M. B.L. D.T.C. XIV 97–98. Nelson was in the Mediterranean on board HMS *Victory* keeping watch on the French fleet at Toulon. He informed Banks that a British frigate had captured a French corvette at Athens, and that this ship carried 'some Cases of I know not what but I suppose things as choice as Lord Elgins'. He asked Banks to advise the government of the option to purchase this cargo, but stressed that everything was prize and as such 'the property of the lowest seaman'.
[2]Alexander I (1777–1825), Emperor of Russia. Relations between Russia and France were indeed deteriorating as French activity in the Near East appeared more widespread and threatening. By late 1803 there were fears in St. Petersburg that France might overrun the Ottoman Empire, threatening Russia's Black Sea trade, and even her southern provinces. As a result, Russian forces were strengthened in the Mediterranean, and especially in Greece. There were 11,000 Russian troops stationed in the Ionian Islands by the autumn of 1804, the same year that consuls were sent to Salonika, Évvoia, Arta and Préveza. In August 1804 the ruler of Montenegro placed himself under Russian protection, and Russian co-operation with the British, already at war with France, seemed very likely.

Letter 94

1803

To Thomas Coutts

Soho Square
24 December 1803

My dear Sir,

I have receiv'd your Favor of the 21st Instant, &, as it is not in my power to comply with your wishes, I fear I must trouble you with a longer Letter than I shou'd otherwise wish to do.

During the last French War I kept up, at no small Personal hazard, an uninterrupted Communication with my Literary Correspondents in France, & was fortunate enough to do them several Sevices, which other People were either unwilling to undertake, or unable to perform.

During the short Peace I found I had more Friends in France than an Englishman ought to have. The Chief Consul[1] spoke of me frequently with apparent Respect, & the Institut chose me their first Foreign Member.

I proceeded to do all kinds of good Offices for my literary Friends, even after the present War broke out, but I was not a little astonish'd to learn, as I had carefully kept myself free from all sorts of political intermeddling, that the Senator Fourcroy[2] had denounc'd me as having under the Mask of a literary Correspondence maintain'd Spies in different parts of France, & that on some of my Friends having remonstrated with the Chief Consul on the impossibility of this Charge being true, he had answer'd that he had the Proofs in his Pocket.

After so infamous an Attack upon my Character, wholly unmerited on my Part, & which both the Chief Consul and Fourcroy must have known to be so when they accus'd me, is it likely that I shou'd humble myself before these infamous Men by making a Request to them? Or that the Chief Consul shou'd grant it if I did so?

It is true that two of my Friends have been releas'd, but it was not at my Request, nor have I acknowledg'd their liberation as in any degree owing to me.

I sincerely wish it was in my power to obtain Mr. Fergusson's liberation,[3] as I know & respect his Character as a Lover of Science, but I'm sure, Sir, You wou'd not advise me to humble myself before the vilainous Corsican Consul, or expect I shou'd have any Success was I to be so foolish. Lady Banks & my Sister desire to join in best Compts. to you & Mrs. Coutts,[4] & are very thankful for your obliging Offer of your Box, which they will occasionally profit by.[5]

I beg, my dear Sir, you
will believe me
Your faithful humble Servt

Joseph Banks.

P.S. Mr Ferguson's Letter is enclos'd.

[Sotheby's catalogue: sale, 24 June 1975, lot 261 for £240. Robert E. Levitt, Durban, South Africa was the owner. Copy at The Natural History Museum, London, in the Banks Archive.]

[1]Napoleon Bonaparte.
[2]Antoine François Fourcroy (1755–1809): chemist; educationalist; senator.
[3]Robert Ferguson FRS (d. 1841). In fact, Banks did take action on Ferguson's behalf, and the Scot was freed: Banks to Delambre, 30/1/1804, de Beer Collection 10, N.H.M. B.L. D.T.C. XIV 196–198. Ferguson wrote to Coutts to announce his release. He had decided to travel to Germany: Ferguson to Coutts, 9/5/1804, N.H.M. B.L. XIV 269–270. Coutts therefore wrote to thank Banks, and he enclosed Ferguson's letter too: Coutts to Banks, 27/5/1804, N.H.M. B.L. D.T.C. XIV 267–268.
[4]Susan Coutts (*née* Starkie), (d.1815): formerly a servant.
[5]This letter was probably dictated by Banks during an attack of gout, and is in Lady Banks's hand.

1804

To John Maitland M.P.[1]

Soho Square
31 March 1804

My dear Sir,

I am sorry I was out when you were so good as to call upon me a few days ago. I have of late stayd cheifly at home, which makes me consider myself the more unfortunate.[2]

As you & the Gentleman[3] concernd with you seem determind to persevere in your New South Wales Sheep adventure, & as I am aware that its success will be of infinite importance to the Manufacture of England, & that its failure will not happen without much previous advantage to the infant Colony, I should be glad to know whether the Adventurers would be contented with a Grant of a large quantity of Land as Sheep Walks; only resumeable by the Government in any Parcels in which it shall be found convenient to grant it as private property on condition of an equal quantity of land being granted in recompence as Sheep Walk; the Lands to be chosen by your Agents in Lots of 100,000 Acres each, & a New Lot granted as soon as the former has been occupied as far as 100,000 of Acres.

Such a Grant appears to me likely to answer your purpose to its utmost extent. You may upon it institute a Code of Laws like the Mesta of Spain, making all the inferior Shepherds dependant on the superior, & by granting to the Shepherds a small share of the Profits make honesty their best policy.

I doubt much the Propriety of Granting Land in Perpetuity to a Scheme which embraces nothing but the feeding of Sheep. The run over land suited for that purpose is all that can be wanted, & as your Sheep will retire as the Colony increases no inconvenience can possibly derive from such an arrangement; &, at all events, I think you may be certain that the tops of the Hills[4] represented by Capt. Waterhouse[5] as such excellent sheep Pasture cannot be wanted for the Plough in less than 4 or 5 Centuries.

[M.L./D.L. Banks Papers Series 23.31.]

A number of Respectable merchants of London and other parts appear much inclind to Associate, under the advice of Capt. Macarthur[6] of his Majesties N.S.

Wales Rangers and the opinion of Capt. Waterhouse of H.M. Navy, for the purpose of Cultivating on a large Scale the breed of Fine wool'd Sheep in H.M. Colony of N. South Wales; & it seems very probable that if due encouragement is given /to the undertaking/ that a Capital of £10,000 at the Least will be raisd by Subscription for this Patriotic Purpose.

The Success of this enterprise will manifestly be an advantage of no inconsiderable importance to the manufacturing interest of this Kingdom, & Even in the Event of its Failure much benefit must arise to the infant Colony by the money that will be Sent there for the purpose of trying the experiment.

As Pasture */for Sheep/* is the only thing Requird for this experiment, and as it is Alledgd that Abundance of Pasture fit for Sheep is to be met with among the uncultivated wilds of New South Wales, it would not be a judicious or a wise measure to make Absolute Grants of Land to the undertakers of it.

A Privilege Somewhat similar to that which the Proprietors of the Merino Flocks in Spain Enjoy would fully answer Every possible purpose of such an undertaking.

Suppose, then, that on Condition of £10,000 being Raisd in the first instance, & the Subscribers made into a body corporate with power to increase their Capital, that orders be transmitted to the Governor /of N.S. Wales/ to allot to the new Corporation a million of acres of Land for Sheep walks *in Lots of 100,000 acres each*; & at such places as their agents Shall fix upon, not being within 5 miles of any Settlement or intended Settlement, one Lot may be immediately put into their Possession, & another granted as soon as they have 50,000 sheep on the first, & so on in succession.

Provided, however, that Every part of these grants Shall be resumable by the Governor for the time being in Such parcels as he shall find expedient for the purpose of Locating /them/ to individuals as their Absolute property, on Condition */of an/* that a Quantity of Land Equal to that which has been resumd being in all cases granted to the Corporation for Sheep walk. /No Land, however, on which buildings of the Value of £100 or up Should have been erected by the corporation, or within one furlong of such buildings, shall be resumable by the Governor within 5 years notice & Payment of the Value of the buildings./

The Corporation to be Restrictd to the Number of one Sheep an acre on their Sheep walks.

[Related notes written in Banks's hand: M.L./D.L. Banks Papers Series 23.37.]

[1]Banks marked the letter in his own hand 'John Maitland' at the end of the first side. He made the comments which follow in his own hand too, apparently as a draft note to be appended to the letter. The text of the letter was copied neatly by William Cartlich, Banks's clerk, c.1780–1815.

[2]Banks was suffering from severe gout.

[3]The 'Adventurers' included John Macarthur (1767–1834), who had arrived in London from New South Wales in 1802. He was a strong man, who was willing to promote the fine-wool breeds in his colony, and he looked to merchants and the government in Britain for encouragement. Macarthur brought wool specimens from his own flock, and in 1803

produced a 'Statement of the Improvements and Progress of the Breed of Fine-woolled Sheep in New South Wales'. Banks read and commented on this document for officials. Though cautious, he saw clearly the benefits of self-reliance in wool production for Britain and the settlers alike if the project could be made to work.

[4]The Blue Mountains, as noted by Banks in relation to a letter from Waterhouse to Macarthur, 12/3/1804: 'Letter from Capt. Waterhouse to Mr. Macarthur March 12 1804', (S.C.) Banks Collection I. 22. 19. A copy of Macarthur's 'Proposal for establishing a Company to encourage the Increase of Fine woolled Sheep in New South Wales' immediately precedes this document, and is accompanied by Banks's opinion of that proposal: (S.C.) Banks Collection I. 22. 17–18.

[5]Henry Waterhouse joined HMS *Sirius* as a midshipman in 1786, and sailed to New South Wales. In December 1789 he took HMS *Sirius* to Norfolk Island, where it was wrecked in the following year. Waterhouse returned to England in 1791, and was promoted to lieutenant. In July 1794 he was appointed to HMS *Reliance*, and sailed again for New South Wales, arriving in Port Jackson in September 1795. He was sent to the Cape in 1796, and returned in June 1797 with what is thought to be the first merino sheep landed in Australia. Waterhouse owned a farm near Parramatta. He kept his own small flock of merinos there. He returned to England for the final time in 1800, where he eventually settled near Rochester.

[6]A conditional grant of lands for sheep pasturage was made to Macarthur by the government: Cottrell to Cooke, 14/7/1804, *Historical Records of New South Wales*, F.M. Bladen (Ed.), Sydney (1897), vol. V, pp. 398–400. Macarthur purchased six shearling rams, one 4-tooth ram and three old ewes at first public sales of His Majesty's Spanish sheep on 15 August 1804. He arrived back in New South Wales with his merino stock on 29 November 1804. Only five rams and one ewe survived to reach Elizabeth Farm, Parramatta.

Letter 96

1804

To Sir Benjamin Thompson F.R.S., Count Rumford

[Soho Square]
April 1804

My dear Count,

I am glad to find by your letter that you are well & happy where you now are. In truth, you seem so much so that your friends here begin to suppose you will take root in the [soil] where you now grow.[1] I cannot, however, disguise that your not appearing in England last year, as I had reason to expect you would have done, has been a material disappointment to me, and a great detriment to the R[oyal] Institution.

It is now Intirely in the hands of the prophane. I have declared my dissatisfaction at the mode in which it is carried on, & my resolution not to attend in future.[2] Had my health & spirits not failed me, I could have kept matters in their proper level, but sick, alone & unsupported I have given up what can not now easily be recovered.

The Royal Society, however, goes on extremely well. Our members are industrious, especially Mr. Hatchett,[3] whose chymical discoveries do him every day more & more credit. We shall not now, I trust, go astray as I think we have not one attending member who is at all addicted to politicks.

All our subordinate Societies also seem to Prosper, & labour diligently in their respective departments. We have newly formed one for the improvement of Horticulture, which promises to become very numerous.[4]

I send you a Book[5] on the subject of heat &c., which certainly contains many interesting experiments, & much bad reasoning. Upon the whole, however, it appears to me that he will not be an unlikely Candidate for your medal[6] of the next year. Pray let me have your opinion on the subject, &, if you disapprove him, who you think a more successfull promoter of the Science you wish to encourage.

I send your medals by Mr. [Levyth]. This is the first opportunity I have had, or you would sooner have received them.

[N.H.M. B.L. D.T.C. XIV. 249–250; U.Y. Beinecke Banks MS.; de Beer Collection 23.]

[1]Rumford was indeed happy, and with Marie Anne Lavoisier, widow of the famous French chemist. They acquired a house on the Rue d'Anjou in Paris, to which many alterations were made. Rumford did not return to England again, and the pair married the following year on 24 October.
[2]The lectures offered at the Royal Institution had veered towards fashionable topics such as poetry, the arts and belles lettres. The Reverend Sydney Smith (1771–1845) spoke on moral philosophy, and even promised to consider the best way to take sugar in tea. These talks had popular appeal, but some felt they were not in keeping with the original scientific and social objectives of the Royal Institution. It seems that managers such as Thomas Bernard (1750–1818) favoured greater emphasis on philanthropic pursuits as ends in themselves, while Rumford and Banks wanted to foster science as the best way of achieving benefits for society.
[3]By May Charles Hatchett had read three papers to the Royal Society: *Philosophical Transactions*, vol. 94 (1804), pp. 63–69, 'Analysis of a triple sulphuret, of Lead, Antimony, and Copper, from Cornwall. By Charles Hatchett, Esq. F.R.S.' Read 26/1/1804; pp. 191–218, 'Analytical Experiments and Observations on Lac. By Charles Hatchett, Esq. F.R.S.' Read 2/4/1804; pp. 315–345, 'An Analysis of the magnetical Pyrites; with Remarks on some of the other Sulphurets of Iron. By Charles Hatchett, Esq. F.R.S.' Read 17/5/1804.
[4]The Horticultural Society.
[5]J. Leslie, *An Experimental Inquiry into the Nature and Propagation of Heat*, London (1804). This book caused controversy in Paris when it arrived. Professor John Leslie had evidently repeated Rumford's experiments on light and heat, coming to similar conclusions. Professor John Leslie was based in Edinburgh, and said that the idea for his research had come in 1801. At this time Rumford had been busy in London organizing the Royal Institution, but he had also visited Edinburgh. Such an unusual coincidence made some people suspicious. Rumford responded with a long 'Historical Review of the Various Experiments of the author on the Subject of Heat', *Mémoires sur la Chaleur*, Paris (1804). However, it seems that both men worked separately, achieving the same results almost at the same time.
[6]In 1796 Rumford gave a large sum of money for encouraging research in the particular areas of physics which interested him most, heat and light. He donated one thousand pounds in stock to the Royal Society so that every two years a medal might be awarded for work of outstanding merit: Rumford to Banks, 12/7/1796, C.K.S. U951Z32/37. He also arranged for a medal to be presented by the American Academy of Arts and Sciences, Boston. The Rumford prizes are still valued as one of the highest scientific honours a physicist can receive.

Letter 97

1804

To Robert Brown F.R.S.

Soho Square
30 August 1804

Dear Sir,

Poor Flinders is a Prisoner, & I fear not very well treated.[1] He put in the Little Cumberland into l'Isle de France for Water, Provision & some Repairs, wholly ignorant of the War. The Governor, unwilling I suppose to beleive that any Person would venture upon so long a Voyage in so small a Vessel, accusd him of being a Spy, & maltreated him.[2] His letters, sent by stealth to the Admiralty, state all this, but conclude in saying that he was when he wrote them rather better treated. All means possible have been taken here to promote his Release which, as the French are great favorers of Science, & as the Ship had a Passport, will I hope in time be effected.

I conclude that if you have not proceeded on your return to Europe before this comes to hand that yourself & Mr. Bauer[3] will get a Passage in the first Government Vessel that returns. You Will have exhausted all your neighborhood, & the Southern Isle, where I conclude the Plants are very different from those of Sydney.

12 Kaigs of your Specimens sent home by the Calcutta have been Sent to my House by the Admiralty a few days ago for safe Custody. I shall take care of them, & be particularly cautious that no one shall inspect or handle them till your return. 4 boxes of Seeds were also on board for Kew Gardens, which have been forwarded to Mr. Aiton.[4]

No mention is made in any letter that I have heard of poor Peter Good.[5] We are told by Mr. Allen,[6] the Miner who returnd in an Indiaman, that he died at Port Jackson, which I fear has been the case. The Seeds sent home by you of his Collecting have producd some very curious Plants, particularly one of the Lobelia Tribe with an irritable Stigma. There are two species of it, which differ in this particular from all others of the Vegetable Kingdom.[7]

Botany flourishes much, & Kew Gardens have been greatly improvd since you saw them. The King has a Gardener in China, who has begun to send home very fine things, & will, I have no doubt, continue to do so.[8]

Dryander, Aiton, Dickson, Dr. Smith, Dawson Turner & Koenig,[9] all are in good preservation, & desire their best wishes may be acceptable to you. Corea[10] went to France two years ago under a slight degree of mental derangement. He is still resident at Paris, but does not appear to have quite recoverd.

Mr. Westhall has taken his option to leave the Ship & seek his Fortune in India.[11] He has sent home several scetches, some of them interesting. His finishd Drawings were all spoild in the wreck.

Pray give my best respects to Mr. Bauer. His Brother[12] is quite well, & very busy in Microscopic Drawings just at present.

I am, Sir,
With much esteem & regard,
your faithfull Servt,

Jos: Banks.

Your Plants are in good Condition except one Parcel, & this but only the uppermost Plant. I have not thought it necessary to open one Parcel.

[M.L./D.L. Banks Papers Series 66.06; B.L. Add. MS. 32439, f. 95; N.H.M. B.L. D.T.C. XV. 84–86.]

[1]On 8 June 1803 Matthew Flinders arrived in Port Jackson with a sick boat. Disease had spread among the crew, claiming the lives of some of his best men. Futhermore, HMS *Investigator* was soon found to be unseaworthy. There was much evidence of rotten timber. The Governor of New South Wales, Philip Gidley King, provided the armed vessel *Porpoise* for Flinders to return to England in. On 9 August it departed in a flotilla, with Flinders travelling as a passenger. Eight days later the *Porpoise* sailed onto a reef, as did one of the two accompanying ships, *Cato*. A third, the *Bridgewater*, somehow avoided disaster, but left Flinders and his companions to fend for themselves. They bravely salvaged one of the cutters, which was christened 'Hope', and used it to return to Port Jackson for help. Meanwhile, most of the crews remained on a nearby sand bank, using provisions from the wrecks to stay alive. Three small vessels were mustered in Port Jackson, and they left to rescue the castaways on 20 September. One of them, a schooner called *Cumberland*, was commanded by Flinders. They arrived at Wreck Reef on 7 October, and in a few days had got off the men and stores. Those who wanted to sailed for Canton or Port Jackson, but Flinders selected a few men with the intention of returning to England in the *Cumberland*.
[2]Flinders was taken prisoner when the *Cumberland* arrived in Port Louis, Mauritius, on 17 December 1803. He was quite unaware that the Peace of Amiens had failed, and that war had been declared in May. He carried, anyway, a passport issued by the French government for a scientific mission in HMS *Investigator*: Otto to Hawkesbury, 23/6/1801, P.R.O. FO. 27/58. Nevertheless, Flinders could not convince General Charles-Mathieu-Isidore Decaen (1769–1832), who heard his case, that this was not an attempt by the British to spy on the French. The general would not believe that Flinders now led such an expedition in a vessel of just twenty-nine tons, and his suspicions were further aroused because the passport referred only to HMS *Investigator*. Both men quickly fell out, and Decaen took little pity on the unfortunate Flinders.

[3]Brown and Ferdinand Lucas Bauer decided not to depart on the *Porpoise*. Both remained at Sydney to continue their field work, descriptive taxonomy and drawings. The living plants, dried specimens and seeds Brown sent with Flinders were lost in the *Porpoise* wreck. However, Brown spent nine months from November 1803 to August 1804 collecting in Tasmania. He also travelled to Hunter River, and the Hawkesbury and Grose rivers. Bauer visited Norfolk Island from September 1804 to February 1805. Both men returned to England in October 1805. Ironically, they sailed home in HMS *Investigator*. The ship had been repaired.
[4]William Townsend Aiton.
[5]Peter Good was too ill even to be moved ashore when HMS *Investigator* arrived at Port Jackson. On 12 June he died of dysentery, which he had contracted in Timor.
[6]John Allen reached England in the Indiaman *Henry Addington* on 8 August 1804.
[7]The 'Lobelia Tribe', Trigger plants (*Stylidium* spp.).
[8]William Kerr, who was mentioned in Letter 91, Note 8.
[9]This group was described in Letter 91, Note 11.
[10]Abbé José Francisco Correia de Serra FRS (1750–1823): Portuguese botanist.
[11]Westall did not hurry back to England from Canton after the *Investigator* expedition foundered. He travelled to India, and arrived in London in 1805. He then explored Mauritius and Jamaica, returning in 1806.
[12]Franz Andreas Bauer, who was recording the effects of blight on wheat. His work formed part of a pamphlet which Banks produced on the subject for the benefit of farmers and landowners: 'A Short Account of the Cause of the Disease in Corn, called by the Farmers the Blight, the Mildew, and the Rust' (1805). This account was reproduced in more than one journal so that by the end of the year Banks's views had gained a wide circulation.

Letter 98

1805

To Captain William Bligh F.R.S.

Soho Square
15 March 1805

My Dear Sir,

An opportunity has occurred this day, which seems to me to lay open an opportunity of being of service to you;[1] and as I hope I never omit any chance of being useful to a friend whom I esteem, as I do you, I lose not a minute in apprising you of it.

I have always, since the first institution of the new colony at New South Wales, taken a deep interest in its success, and have been constantly consulted by His Majesty's Ministers through all the changes there have been in the department which directs it relative to the more important concerns of the colonists.

At present, King,[2] the Governor, is tired of his station, and well he may be so. He has carried into effect a reform of great extent, which militated much with the interest of the soldiers and settlers there. He is, consequently, disliked and much opposed, and has asked leave to return.

In conversation, I was this day asked if I knew a man proper to be sent out in his stead — one who has integrity unimpeached, a mind capable of providing its own resources in difficulties without leaning on others for advice, firm in discipline, civil in deportment, and not subject to whimper and whine when severity of discipline is wanted to meet [opposition]. I immediately answered: 'As this man must be chosen from the post captains, I know of no one but Captain Bligh who will suit, but whether it will meet his views is another question.'

I can, therefore, if you chuse it, place you in the government of the new colony with an income of £2,000 a year, and with the whole of the Government power and stores at your disposal, I do not see how it is possible for you to spend £1,000. In truth King, who is now there, receives only £1,000 with some deductions, and yet lives like a prince, and, I believe, saves some money; but I could not undertake to recommend any one unless £2,000 clear was given as I think that a man who undertakes so great a trust as the management of an important colony should be certain of living well, and laying up a provision for his family.

I apprehend that you are about 55 years old.[3] If so, you have by the tables an expectation of 15 years' life, and in a climate like that, which is the best that I know, a still better expectation. In 15 years £1,000 a year will, at compound interest of 5 per cent, have produced more than £30,000, and, in case you should not like to spend your life there, you will have a fair claim on your return to a pension of £1,000 a year.

Besides, if your family goes out with you, as I conclude they would, your daughters will have a better chance of marrying suitably there than they can have here,[4] for as the colony grows richer every year, and something of trade seems to improve, I can have no doubt but that in a few years there will be men there very capable of supporting wives in a creditable manner, and very desirous of taking them from a respectable and good family.

Tell me, my dear Sir, when you have consulted your pillow, what you think of this.[5] To me, I confess, it appears a promising place for a man who has entered late into the status of a post-captain; and the more so as your rank will go on, for Phillip, the first Governor, is now an admiral, holding a pension for his services in the country.

I have, etc.,

Joseph Banks.

[G. Mackaness, *The Life of Vice-Admiral William Bligh*, Sydney (1931), vol. II, pp. 96–97.]

[1]Lord Camden, John Jeffreys Pratt (1759–1840), approached Banks in his search for a successor to Philip Gidley King as Governor of New South Wales: Banks to Chapman, 19/4/1805, P.R.O CO201/38 44274. Whoever was appointed would need to be strong in character and firm in discipline, because the colony suffered at the hands of powerful men who monopolized the rum trade, and controlled the economy. They used the New South Wales Corps for their own purposes, and exploited the relative weakness of the Governor. Some had become extremely rich this way, and strict curbs were needed to prevent further abuse. Banks was in no doubt that Bligh would be able to impose his authority on the unruly factions.

He may also have considered the return of John Macarthur to the colony. Macarthur had come to England under arrest for a duel he fought in 1801. While in England he disagreed with Banks, but still managed to obtain government support for his proposal to develop the wool industry in New South Wales. By 1804 Macarthur was travelling back to Australia with grants of land, sheep and a reputation for trouble to which more than one Governor could attest.

[2]Philip Gidley King had struggled with many difficulties in his six years as Governor, among them the best way to establish the colony's economy on a stable basis.

[3]Bligh was born on 9 September 1754, and was 51 years old when Banks wrote this letter.

[4]Bligh's family did not leave with him. His wife, Elizabeth (c.1752–1812), was afraid of the sea, and since her health had been poor she remained in England. A consequence of this was that she maintained a close contact with Banks, and actively defended her husband's reputation when events in New South Wales deteriorated.

Bligh secured a place for his son-in-law, Lieutenant John Putland (d.1808), and so Bligh's daughter, Mary (1783–1864), accompanied her husband. John died of consumption in 1808,

and Mary subsequently married Maurice Charles O'Connell (1766–1848). He was Lieutenant-Colonel of the 73rd Regiment, which replaced the New South Wales Corps. Bligh was not pleased at the match initially, but O'Connell went on to command his regiment, and became Lieutenant-Governor of New South Wales. He was also knighted.
[5]Bligh did not accept this offer immediately because it was not clear that his family would leave for the colony. Bligh also felt he lacked sufficient knowledge about the colony and the duties of a Governor. Most of all, he was worried that his career in the Navy would be jeopardized. However, in April he accepted the post, and on 29 of the month his half-salary as Governor of New South Wales commenced.

Letter 99

1805

To Professor John Leslie

Soho Square
19 April 1805

Sir,

I have lately receiv'd from Paris Count Rumford's "Memoires sur la Chaleur".[1] As I do not know that any other copy of it has reached London I cannot venture to send it to you, lest my friends here who wish to peruse it might be disappointed. You will, however, no doubt wish to know in what manner you are spoken of in it. I have therefore enclosd a Copy of all that in my opinion is immediately necessary for your information.

I am sorry to learn by a letter from Dr. Wright[2] to Dr. Garthshore[3] that the Clergy of Edinburgh, a set of Men hitherto honored for their mild & moral conduct as well as for the purity of their religious proceedings, have instituted what in my humble opinion much resembles a persecution of you for Tenets, which, if not strictly within their notions of Orthodoxy, are surely such as wise & well informed Men should not select & hold forth to light as proper objects of Ecclesiastical censure.[4] They would surely have acted more properly, & in a manner better becoming their Station in the community, by suffering your Book to remain quietly on the Philosopher's shelf without dragging forward into public notice a few passages, which, at best, can only lead to a controversy in which Orthodoxy will be put to the hazard, & more of it probably lost than gain'd.

No Church surely has ever increased its influence over the minds of Men by unwholesome severity. Mildness of demeanor carries, or rather forces, Religion into the hearts of Men, & it soon finds it[s] way from their hearts to their souls; while the opposite extreme renders its Teachers odious in the eyes of all Mankind, & puts Religion itself in continual risk of being abandoned. Unprejudicd & strong headed men have always thought for themselves, & will continue to do so in spite of all the Bishops, Priests & Deacons that ever have or, ever will be ordaind; tho they may conscientiously abstain from commenting & enlarging upon them, lest weak brethren should be bewilderd.

Surely a Man may fulfill his duty to his Creator, & render his Redeemer propitious, without assenting unconditionaly to every indigested Tenet which our half-informed Predecessors have left behind as a legacy to their more enlightened successors. If it has pleas'd God to permit his creatures to increase in wisdom, he will not condemn them for assenting to new Opinions which their reason demonstrates to be just.

I am, Sir,
Your obedient Hble Servt

Jos. Banks.

P.S. If the Clergy wish to retain that respect in which their Cloth is, & ought to be held by the Mass of Mankind, they should avoid instead of courting Controversy, as it is evident to a moderate understanding that they must lose more than they can gain by every public altercation in which they are by any means involved.

[N.H.M. B.L. D.T.C. XV. 349–350.]

[1]Leslie's book, *An Experimental Inquiry into the Nature and Propagation of Heat*, caused controversy in Edinburgh and Paris alike. See Letter 96, Note 5.
[2]Dr. William Wright FRS (1735–1819): physician; botanist.
[3]Dr. Maxwell Garthshore FRS (1732–1812): physician; author on obstetrics.
[4]Church ministers in Edinburgh were critical of Leslie for quoting David Hume (1711–1776) with approval in his publication. They opposed his candidature for the chair of mathematics at Edinburgh when it became vacant. In May a General Assembly of the National Church intervened to end the dispute, and by March, Leslie had been elected to the chair. In another letter, Leslie explained why he thought the clergy of Edinburgh were against him, and also gave an account of the scientific work which had clashed with Rumford's: Leslie to Banks, 28/4/1805, N.H.M. B.L. D.T.C. XV 367–369.

1805

To Ann Flinders

Soho Square
29 April 1805

Madam,

I wish it was in my power to give you more satisfactory accounts of the Situation of my worthy friend Capt. Flinders, but such as I have, madam, you may be assurd I Shall always be happy to Communicate.

As the governments of England & France, owing to the extreme ill Temper & Savage disposition of the Latter, have not since the Last war broke out held any Kind of Communication with Each other, it was impossible for our government to Complain with any Effect of their Abominable Breach of the Law of Nations in Confining Capt. Flinders.

Under these Circumstances I applied for Permission to write to my Literary friends in his Favor, which, as I had already procurd the Enlargement of 4 or 5 English Prisoners by their means, our government Readily Consented to.[1]

Early in the month of August this Letter was sent. It Enterd fully into Capt. Flinders Case, and no argument I could make use of to induce the Tyranical government of France to Consent to his Liberation was omitted.[2]

This Letter was intrusted with others to the Care of a Prince Pignatelli,[3] a Spaniard of high Rank, who undertook to deliver them at Paris without delay. Indeed, this was the only mode of Conveyance that was at that time to be Obtaind.

It was not till the 18 April an answer to this Letter was Receivd. The Prince, as it appeard, Left his baggage & the Letters he was intrusted with at Rotterdam in Holland, & diverse accidents prevented their arrival at Paris till the month of February.

So long a delay made me almost inclind to give up all hopes of any Effect from my Letter, & I felt much anziety on account of my unfortunate friend, whose destiny, I concluded, was decided by the Abominable Bonaparte, & who, of Course, I Could not expect to be Liberated till at Some distant period peace and a good understanding might again take place between the two Countries.

I was, however, much Satisfied in Learning by this Letter of the date of March 5th[4] that the delay of the Princes Baggage was the Real Reason why no answer had Come to my hands, & that the National Institution of Paris had unanimously recommended to the Minister of Marine[5] to Procure Capt. Flinder's liberation in Strong & Pointed terms.[6]

My hopes, therefore, are once more Alive on the subject, &, as I flatter myself that all has been done that Can be done, I feel a Satisfaction in hoping that it will not be Long before Bonaparte's determination on the subject is sent to me, & that it will be favorable, in which Case, madam, you may depend on an Early Communication.[7]

I beg, madam, that you will beleive
Your Obedient Hble Servant,

Jos: Banks.[8]

[N.M.M. FLI/1&26.]

[1]Camden to Banks, 10/8/1804, M.L./D.L. Banks Papers Series 66.04, N.H.M. B.L. D.T.C. XIV 20.
[2]Banks to Delambre, 22/8/1804, M.L./D.L. Banks Papers Series 66.05, N.H.M. B.L. D.T.C. XIV 54–57.
[3]Francesco Pignatelli (1732–1812): Neopolitan General.
[4]Delambre to Banks, 5/3/1805, C.K.S. U951Z32/46.
[5]Denis Decrès (1761–1820), Comte de L'Empire and Duke: Vice-Admiral; General Inspector of the Mediterranean Coast; Minister of Marine and the Colonies, 1801–1814, and for 100 days in 1815. It seems he was burned to death by a servant.
[6]Delambre assured Banks that the Institut National would always try to assist men of science, and that Decrés had been approached on the matter of Matthew Flinders. Delambre to Banks, 5/3/1805, C.K.S. U951Z32/46: 'A la lecture de votre lettre la classe a unaninmement arrêté de recommander au Ministre de la Marine l'affaire du Capitaine Flinders...Nous attendons la réponse de Ministre, et nous vous prisons de croire que dans toute occasion l'institut cherchera avec empressement les moyens d'obliger les savans vos compatriotes et à diminuer autant qu'il sera en son pouvoir les inconvéniens qui résultent pour eux des querelles qui divisent les nations.'
[7]Ann wrote to Banks regarding her husband, Matthew, more than once. Banks responded with details of the long negotiations for his release from Mauritius, usually in a tone designed to keep Ann's spirits up: Letters 107, 109, 111 and 113.
[8]Many of the sentiments here were also explained in a fine letter drafted by Banks to Matthew: Banks to Flinders, 18/6/1805, *Historical Records of New South Wales*, F.M. Bladen (Ed.), Sydney, vol. V, pp. 646–647, 'Mrs. Flinders I heard of very lately, as full of anxiety for your return as possible. I have heard many times from her on the subject, and always done my utmost to quiet her mind and sooth her apprehensions.' p. 647.

Letter 101

1805

To William Smith M.P.

Soho Square
14 May 1805

My dear Sir,[1]

I am Sorry I cannot attend the meeting at Mr. Wests[2] this morn, being oblig'd to go into the Country. Another time I hope I Shall be more fortunate, & I hope also that that the next meeting will be held at some house not of an Academician, Lest we be supposd to associate with the Less popular part of the R[oyal] Acad[emy].

Will it not be right to mention in our Prospectus that the three great schools of Painting were each of them the Produce of the Moneyd Prosperity of the three Countries in which they flourishd? The Venetian school arose while that Town was the Emporium of the East; the Flemish when Antwerp was that of the Western World; & the Roman when appeals to the Roman Ecclesiastical Courts made their Lawyers almost as rich as our Civilians are now, & when Benefices held by Roman Bishops &c. in foreign Countries renderd their clergy almost as rich as our merchants now are. In this Point of view the time is come when England has the means through her Commercial prosperity to Foster a fourth school, & has every prospect of excelling France more Effectualy in Patronage than France can excell her in ancient models of sculpture. If half of the money that has of Late years been Lavishd upon Repainted originals & Copies impudently foisted on the Public as originals had been divided among our artists, the business would by this time have been done. Foreigners have become as anzious to posess modern English Pictures as Englishmen are to Obtain Counterfeit antiques. The arts always will flourish in Proportion to the Patronage given them by the Rich. We have the means which no other nation Posesses of giving that Patronage, but we have for years past squanderd away upon strangers the bread that we ought to have bestowd on our children.

Excuse this Long Rigmarole. Allow me only to add that we ought, I think, in no case to admit a Picture which has not Previously be[en] exhibited at the R. Academy. This appears to me a Sine qua non.

Nothing has preventd England from Producing a School of Excellent Painters of its own. The Crowds of old pictures, fictitious & Real, that have Pesterd us for years will Continue to do so till we give a due preference to the merit of our own. Neither of the Three old schools were plagued with foreign Pictures to abate the Love of their Countrymens work as we are.

Adieu my dear Sir,

Jos: Banks.

[H.B.D. G.A.C. A.S.L.: 14 May 1805.]

[1]There is a note at the head of this letter: 'Sr. Jos. Banks on the British Institution'. Another at the foot reads: 'His Portrait hanging up in the Study'. Neither is in Banks's hand.
[2]The British Institution for Promoting the Fine Arts in the United Kingdom, which was an association founded in London in June 1805. It offered prizes and bought modern works, and held 2 annual exhibitions, one of contemporary British painting and one of old masters. The British Institution's aims were to "encourage and reward the talents of the Artists of the United Kingdom; so as to extend and improve our manufactures, by that degree of taste and excellence which are to be exclusively derived of the Fine Arts; and thereby to increase the general prosperity and resoures of the Empire."

The meeting to discuss the new body appears to have been at the house of Benjamin West PRA (1738–1820). West painted the portrait of Banks wrapped in a Maori cloak, 1771–1772, now at the Usher Gallery, Lincoln. See Illustration I.

Letter 102

1805

To Robert Fulke Greville F.R.S.

Soho Square
20 August 1805

My dear Sir,

I Enclose you a justification[1] of my conduct, drawn up as concisely as I am able to do it. I hope you will not think it too Long. Whether you use it in the whole or in Part, & the manner in which you introduce it, I Leave wholly to your Prudence & Friendship.[2]

This Frett,[3] as you Call it, appears to me to Lie deeper than you suppose. The Explosion was delayd from the time when you first mentiond the circumstance till the Catalogue[4] gave opportunity for a Vent. It was not the Ebullition of a moment, but the Consequence of a Concerted & Contrivd measure, or I am much mistaken.

We have both Observed for some years past that H.M. mind is more irritable & Less placable than it usd to be, & since this change has taken place I do not recollect an instance of any one of whom so hard a thing has been said as he has said of me being restord to Confidential favor. It will be far better for me to be dismissd than to Remain under sufference only. I feel a personal friendship for the King, & if it is returnd as it usd to be I can never forego it, but coldness from a friend, tho' a King, I can not support.

I might have added to my Justification that no one chief in the Agricultural Line attended the Sale except Hugh Hoare, & he refusd to give the Prices the Sheep fetchd; & I have no idea of Mr. Coke having bought had he been there, & a firm opinion that he will consider the average price of £24.3. which he must pay as a very exhorbitant one.

It was impossible for me to get a List with the names [of the buyers] copied in time for the Coach on Friday. I Should have sent one by Saturdays Coach, had not so many of them been the mere agents of proprietors who were not able to be present. I have not yet made out such a one as is fit for the King to See; nor do I think I Shall be able to make a perfect one at any time, tho I have written Several Letters for information.

Your Favor did not Reach my hands till after three yesterday, & but for an accident would have been Left behind till this Evening. I fear this is the nature of the Weymouth mail. The other Gen[eral] Post Letters were on my table before 11.

beleive me, my dear Sir,
with Sincere Gratitude & Esteem,
Most Faithfully yours,

Jos: Banks.

[N.H.M. B.L. D.T.C. XVI. 112–113; B.L. Add. MS. 42072, ff. 79–80; B.L. Add. MS. 33981, f. 218.]

[1]Banks to Greville, 20/8/1805, B.L. Add. MS. 40272, ff. 81–82, N.H.M. B.L. D.T.C. XVI 110–111, M.L./D.L. Banks Papers Series 73.091.

[2]Banks had managed the Royal flock for a number of years, paying particular attention to the introduction of Spanish merino sheep with their fine wool. Public sales of surplus sheep were organized in fields and barns near the Pagoda at Kew Gardens, and these served to disperse the Spanish breed. One such was held in August 1805, before which Banks decided to sell three ewes privately to an eminent agriculturist, Thomas William Coke, 1st Earl of Leicester (1752–1842). He told Greville of this: Banks to Greville, 4/8/1805, B.L. Add. MS. 42072, ff. 71–72. Banks hoped to increase the popularity of the sheep through the patronage of such men.

George III was unwell prior to the sale, and so did not know of the arrangement, but was furious when informed. Despite a friendly acquaintance of over twenty years, he accused Banks of betraying public trust that all the sheep would be sold at auction: Greville to Banks, 18/8/1805, (S.C.) Banks Collection II. 11. 129–130, S.L. Banks MSS 'Wool' by box and folder, N.H.M. B.L. D.T.C. XVI 105–107. Banks was subsequently reassured that the King's anger was not lasting, but gave up his duties anyway: Letter 103.

[3]Greville to Banks, 19/8/1805, (S.C.) Banks Collection II. 11. 129–130, S.L. Banks MSS 'Wool' by box and folder, N.H.M. B.L. D.T.C. XVI 108–109.

[4]A sale catalogue, which Banks forwarded on Saturday 17 August: Banks to Greville, 17/8/1805, B.L. Add. MS. 42072, ff. 75–76.

Letter 103

1805

To Robert Fulke Greville F.R.S.

Overton
31 August 1805

My dear Sir,

You have certainly Liberated my mind from a Load of unpleasant sensations by your Friendly Letter, Receivd Last night.[1] I feel grateful, as I ought to do, for the Pardon H.M. has been graciously Pleasd to bestow upon me, & am releivd from the fear I felt of the Consequences of Royal displeasure.

I must, however, Frankly Confess that the whole business honestly & Fairly Stated does not much flatter those feelings in which I have for some years been usd to indulge myself. I fancied, humble as I am, that the King felt a Friendship for me, & I am certain that I felt for him a Sentiment which Crowns & Sceptres Cannot Command from a man of honest & independent Spirit.

I did not therefore Expect to become a Victim of anger originating in a want of information which might have been Easily procurd; nor that, after I had fully explained the Cause & the Real motives of my Conduct, I should be dismissed by a declaration that my intentions were good, which clearly implies that my Conduct was bad.

My intention of Resigning, if I can do it with perfect decency & decorum, the mangement of the Flock, which has for so many years been my delight, is not at all alterd. To be bound in a matter, the Conduct of which is an untrodden path, by the Actual Letter of a Rule Laid down by myself as a general one, subject to many exceptions, because H.M. has approvd them will at once paralyse all my attempts to carry the business to greater perfection. If the discretion I have hitherto exercisd is at an End, Every improvement deriveable from Circumstances of which H.M. has not been previously acquainted will be rendered impracticable.

Besides, when I consider that [in] the detail of this most distressing business the defence of my character was necessarily intrusted to the discretion of a man who has not a drop of gentle blood in his veins, & who has not even had the benefit of a gentlemans Education, & of course scarcely acknowledges the existence of that sentiment which men of blood & descent almost Idolise under the name of honor, it makes my Soul shudder.[2] Thanks to god & to your exemplary Friendship I have surmounted the misfortune with which I have been threatened of Labouring under, H.M. displeasure. How soon I may again incur it if I continue his servant, no man Can guess as it did not in this Case originate in wisdom or discretion. I Shall therefore Labor silently but unremittingly till I can say 'me uvida Suspendisse potenti vestimenta maris deo'.[3]

Your active friendship on this occasion Can never be forgotten or Ever recompensd by me. But for that I Should have continued under the Lash of Royal displeasure in aggravation of my present feeling. You have lifted that mountain from my breast, & I thank you. The Remainder must Continue to oppress me. Neither you, nor I, nor any one Living can in any degree alleviate [it].[4]

The Loss of Friendship, like disappointment in Love, can derive Consolation from no other source than the Lapse of time.

Beleive me, my dear Sir,
Sincerely & Faithfully yours,

Jos: Banks.

[B.L. Add. MS. 42072, ff. 88–89; M.L./D.L. Banks Papers Series 73.093.]

[1]Greville to Banks, 25/8/1805, (S.C.) Banks Collection II. 11. 131–132, S.L. Banks MSS 'Wool' by box and folder, B.L. Add. MS. 42072, ff. 85–86, N.H.M. B.L. D.T.C. XVI 117–119.
[2]Richard Snart (fl.1802–1832): superintendent of the royal parks and gardens, 1802–1813.
[3]Horace, *Odes*, 1, 5, 13–16, 'Me tabula sacer/ Votiva paries indicat uvida/ Suspendisse potenti/ Vestimenta maris deo'. This is the complaint of a rejected lover, declaring 'I have hung in yonder temple/ My dripping garments, vowed/ to the god who curbs the main'.
[4]Some words are heavily crossed out here.

Letter 104

1806

To Thomas Manning

Soho Square
20 April 1806

My dear Sir,

I Should most willingly have Obeyd your Commands in Soliciting a Passport[1] for you from the Emperor of the French, had it not been renderd impossible by the Present State of the two Countries.[2] I do not Know that any intercourse upon which the Least dependence can be placd now exists between them. Of this, at Least, I am sure, that I have not had an opportunity of writing to any of my Friends in the National Institut for many months. It is true there is said to be a negociation on Foot, & that our Minister Fox[3] corresponds with M Taleyrand[4] by the Flags of Truce which board our vessels in the Downs, but you may Easily Conceive that it would not be a Prudent request in me to charge their Courier with a Private Letter soliciting a Favor for an individual.

I cannot, however, help adding on this occasion that I do not doubt from the Knowledge I have of the French nation, whose respect for Science & Scientific men was never for a moment abated during the most Horrible parts of their Revolution, that you will Easily gain your Personal liberty in Case of Capture, unless it should be your Fortune to be the Prisoner of an ignorant & illiterate man, which Can hardly happen. Your Project is in itself so very worthy of protection, & so extremely interesting to the inhabitants of all Civilisd nations, that no one Can doubt of being Rewarded instead of blamed by a Nation & a Government so liberal in all matters of Science as the French are for having done his Endeavour to Promote it; & more particularly by the Emperor,[5] who, I have no doubt, would himself notice the Conduct of a man who broke through some rules of discipline in order to Liberate you in the Case of Capture. In Fact, the Conduct of the Emperor himself in ordering your Liberation from among the detain'd English, solely on account of the Interest taken by himself & the Learned men of France in your hazardous & most Laudable Enterprise, holds out an Example which Cannot Fail of producing its Full Effect on the mind of Every man Capable of Commanding a French Vessel. His Plain understanding of it as absolutely void of Politick, will inevitably inform him that to Liberate again the Person, who has once been liberated by his Emperor, Cannot but be an acceptable act in the Eyes of his Sovereign; & that, on the Contrary, to detain a man & prevent his Proceeding on the very business which induced Napoleon to give him Liberty, Cannot but be a Step that would be condemnd by Every good French man, from the Emperor to the Lowest Pretender to Literature in the vast dominions of France.

I have argued this opinion of mine Rather more at Large than I possibly should have done in hopes to free your mind from the apprehensions you are under of molestation from Capture, & to Console you as well as I am able under the impossibilty which the Present State of affairs between the two nations Presents to the Obtaining [of] the Passport which you so very much wish for. I shall be happy if this Letter Places your mind in a State of Ease, & decides you to trust yourself without delay to the Hazards of the Ocean, & the Less Formidable terror of being interruptd by a nation too Civilisd, in my opinion, to be suspected of any thing tending in Ever so small a degree to Barbarism & the undervalueing of the progress of human Knowledge.

Beleive me, my dear Sir,
with the most sincere Esteem,
& the most ardent wishes for the
success of your Most Laudable Project,
Your Most Faithfull
& most Hble Servt

Jos: Banks.

[N.H.M. B.C. 95–96.]

[1]Manning was a great scholar of Chinese culture. He studied the Chinese language in Paris from 1800, but when war broke out in 1803 he left France for England. Napoleon himself signed Manning's passport home, the only one the Emperor did sign. Subsequently, Manning wanted another passport in order to make his famous journey to China and Tibet. Banks arranged free passage to Canton by writing on Manning's behalf to the Court of Directors of the Honourable East India Company: Court of Directors to Banks, 11/4/1806, N.H.M. B.C. 91–92; Banks to Manning, 12/4/1806, N.H.M. B.C. 93–94. However, even Banks could not secure a passport from the French for this trip. Manning departed in June 1806, travelling widely in the East until he returned in 1817. During this time he sent Banks information about such things as plants and china-ware. He was also part of William Pitt Amherst's (1773–1857) Embassy to Peking, 1816, acting as a junior secretary and interpreter.
[2]Nevertheless, Banks wrote to France with news of Manning's journey: Banks to Lacépède, 5/7/1806, N.H.M. B.L. D.T.C. XVI 278–279.
[3]Charles James Fox, Foreign Secretary. Fox had contacted Talleyrand in the hope of negotiating peace between France and England, but no agreement was reached.
[4]Charles-Maurice de Talleyrand Périgord (1754–1838): French statesman and diplomat.
[5]A reference to Napoleon's treatment of Manning in 1803.

Letter 105

1806

To Captain Pierre Bernard Milius

Soho Square
24 May 1806

Sir,

It gives me inexpressible pleasure to feel myself, under the favor of my Government, the fortunate instrument of restoring to his Friends a Man so well spoken of by every person who had the pleasure of his acquaintance in New South Wales as Capt. Milius certainly is. To contribute in any shape to the liberation of a Man who has been employed by his Country in a Voyage of Discovery will always be an important object in my estimation; but in your case, when the character of a Discoverer is joined to that of a man who defended the Ship in which he was taken with more than usual Galantry, & whose amiable manners have attach'd to him a number of my friends, the pleasure of restoring to him his liberty is trebly valuable.[1]

It must, however, occur to you that the gratitude, whatever its amount may be, which you feel on this occasion may be exerted by you in a manner not a little pleasing to yourself, & at the same time infinitely interesting to me. Capt. Flinders, who put into the Island of Reunion[2] with Letters of recommendation from Capt. Baudin,[3] who was sailing homewards in a Vessel of not more than 30 Tons,[4] who was ignorant of the war, & who was in actual want of water & Provisions, has been detained there under a groundless pretence of having come in as a Spy, & is still detaind there as a Malefactor.

You, Sir, who know the man, & who are aware of the nature of his Expedition, one of the most hazardous that has succeeded, will readily give him credit for his honesty. His business was to Solicit another Ship in the place of the Investigator, that was condemn'd. He had no view whatever of doing any other thing, & it was quite impossible for him to take a Survey of the Harbor he was entering half so good as those already publishd, had he been desirous, which he certainly was not, of so doing.

To my friends of the Institut I must request you to state such facts as you know, & such opinions as you yourself can form of the State of Capt. Flinder's[5] detention. Your Emperor is, I know, of a most honourable disposition; always inclined to justice; never erring from it unless under the influence of misinformation; & a determined friend to Science. He must have been misled by false accusations, & may, I am sure, be induced to enquire more particularly into the matter. If he does, the result is certain to be the liberation of our friend, to the honor of the French Nation, & of those who direct its national conduct.

I have also a private request to make of you in which I am confident my friends at Paris, who are not a few, will assist. I have a young friend, the son of a merchant at Boston, near my Estate, who was detained with the other English while quite a boy sent to France for his Education. He has at various times attempted to procure leave to go home, but has never been able to obtain the Passport of the Grand Judge, without which he cannot pass the Frontier.

He is now only 18 years of age. His name is Thomas Gee.[6] He resides at Rostrenen[7] in the department of the North, under the care of M. Verdier. If the passport of the Grand Judge can be obtained for him, which surely cannot be very difficult, it will lay me under no small obligation.

Wishing you, Sir, a prosperous Voyage, a safe return to your Native Country, & a happy meeting with your family & friends, I beg you to beleive me,

Your very faithful,
& very humble Servant

Jos: Banks.

[N.H.M. B.L. D.T.C. XVI. 269–271; M.L./D.L. Banks Papers Series 73.096.]

[1]Milius had commanded the French frigate *La Didon*, which was captured following an engagement with *HMS Phoenix* in 1804. He became unwell and asked Banks to help obtain parole for him to take the waters at Bath: Milius to Banks, 3/9/1804, B.L. Add. MS. 8099, ff. 401–403; Milius to Banks, 22/9/1805, B.L. Add. MS. 8099, ff. 399–400; Banks to Milius, 24/5/1806, N.H.M. B.L. D.T.C. 269–271. Milius's release brought Banks into conflict with the First Lord of the Admiralty, Lord Howick (1764–1865): Marsden to Banks, 8/8/1806, N.H.M. B.L. D.T.C. XVI 306–307.
[2]Réunion. Mauritius lies 65 miles to its north-east, and was the island where Matthew Flinders was detained.
[3]Nicolas Thomas Baudin died on Mauritius in 1803. His ship, *Le Géographe*, was sailed back to France under Milius's command. *Le Naturaliste* was the second ship for Baudin's mission, and carried all the natural history specimens. It was captured in the English Channel by HMS *Minerva* on 27 May 1803, and taken to Portsmouth. Banks then intervened to release the vessel: Banks to Hamelin, 1/6/1803, R.B.G. Kew Archives B.C. 2 276.

The letter of recommendation which Banks referred to was written by Baudin in gratitude for the assistance he and his crew received from Philip Gidley King, the Governor of New South Wales. On 20 June 1802 Baudin arrived in Port Jackson in *Le Géographe* with a crew suffering badly from scurvy. Medical help and supplies were immediately provided by the colony. Baudin described these events in an open letter, and explained that Flinders might visit Mauritius. He advised the French authorities there to treat the Englisman well: *Historical Records of New South Wales*, F.M. Bladen (Ed.), Sydney (1896), vol. IV, p. 968–969. However, Flinders did not take a copy of Baudin's recommendations, and was made a captive when he eventually reached Mauritius.

[4]The schooner, *Cumberland.*

[5]Banks clearly hoped to free Matthew Flinders in return for the release of French prisoners like Milius.

[6]Thomas Gee: son to a merchant at Boston. A passport was obtained for him: Lacépède to Banks, 1/7/1806, B.L. Add. MS. 8100, f. 16. Gee was duly released: Rivaud St. Germain to Banks, 4/7/1806, B.L. Add. MS. 33981, ff. 237–238.

[7]Rostrenen, Côtes-du-Nord, France. Some 25 miles south-west of Guingamp.

Letter 106

1806

To Jacob Pleydell Bouverie, 3rd Earl of Radnor

[Revesby Abbey]
[6 October 1806]

My Lord,[1]

It gives me satisfaction to learn from your Lordship's letter[2] that Lord Howick considers the supposition of his having discountenanced my endeavors to obtain, through the medium of my Literary friends in France, the liberation upon parole of Englishmen detained there as having originated in a mistake. After that declaration his Lordship will, no doubt, join his endeavours to mine for the rectifying that part of the mistake which still remains unaltered, & [which] constitutes the chief obstacle to my future proceedings on this subject.

I some time ago requested, at the desire of Persons high in office, the liberation of Mr. Egerton[3] & of Capt. Flinders,[4] & had been indulged, as I thought, with the credit of having obtained here the freedom of Capt. Milius & a French Surgeon.[5] Mr. Egerton & Capt. Flinders were immediately set free, & a third person, in whose fate my friends knew I took a personal interest, was liberated at the same time;[6] but I was requested to obtain in return the freedom of the French purser, a prisoner here.

My request for the release of this purser, which I concluded Lord Howick would have granted to me with pleasure, was the subject on which the mistake in Question originated.[7] It was refused to me in unpleasant terms. No hopes were given me that I should be able to obtain his freedom in future, & in point of fact he still remains a prisoner here.

Unless this request, & more than this, is acceded to on the part of Government, I can see no hopes that my Literary friends, who have now obtained for me Ten English prisoners at the least, will continue to perform the unpleasant task of soliciting the restoration of Prisoners there with any Energy. I wrote to Mr. Marsden[8] as soon as your Lordship's first Letter came to my hands, & requested he wd sound Mr. T. Grenville[9] on this head, but I have not yet received his answer. It is likely that Mr. Grenville, tho' unacquainted with the particulars of the Transaction between Lord H. & me, will adhere to Lord Howick's measure unless his Lordship will request him to alter it.

Whether your Lordship will choose to interfere farther on this delicate subject must be left to your own discretion. I can only add that I feel myself infinitely desirous of assisting Lord Shaftesbury;[10] that I have little doubt of being able to procure his Lordship's liberation if Government will give me their decided countenance; but that, from the tenor of letters I have very lately received, I entertain no hopes of being able to effect anything more in France unless I have the full & compleat cooperation of his Majesty's Ministers in the business, which, as I have no favor of any sort or description to ask for in return, or any view but mere motives of good will, I do not myself see any reason why they should refuse to grant to me.

[N.H.M. B.L. D.T.C. XVI. 336–337.]

[1]Lord Radnor had sought Banks's assistance in gaining the release of Lord and Lady Shaftesbury: Radnor to Banks, 23/9/1806, N.H.M. B.L. D.T.C. XVI 326–327. However, Banks had been involved in a disagreement with Lord Howick. The First Lord of the Admiralty opposed Banks's attempts to intervene on behalf of prisoners. Banks therefore told Radnor he would not act 'without the direct consent of Lord Howick': Banks to Radnor, 28/9/1806, N.H.M. B.L. D.T.C. XVI 328. For manuscript originals from Radnor, and draft replies by Banks: N.L.A. 9/136.

[2]Radnor wrote to Howick to obtain consent for Banks to assist Lord and Lady Shaftesbury. Howick replied that 'I can have no objection to the use of any influence Sir Joseph may possess for the purpose of obtaining the release of any of our Countrymen': Howick to Radnor, 2/10/1806, N.H.M. B.L. D.T.C. XVI 332. However, since the British would no longer release French prisoners, it seemed unlikely that the French would co-operate by releasing any of theirs.

[3]Reverend Francis Egerton FRS (1756–1829), afterwards 8th Earl of Bridgewater. Banks had sought Egerton's release earlier in the year: Banks to Lacépède, 24/5/1806, N.H.M. B.L. D.T.C. XVI 272–273. The French co-operated: Rivaud St Germain to Banks, 4/7/1806, B.L. Add. MS. 33981, ff. 237–238.

[4]Matthew Flinders, who had not been released from captivity in Mauritius. Orders for his liberation were sent from France in March 1806, but they were ignored.

[5]Rivaud St. Germain: a French doctor, whom Banks had freed: Banks to Rivaud St Germain, 24/5/1806, N.H.M. B.L. D.T.C. XVI 268. Rivaud carried home Banks's pleas for British prisoners: Banks to Lacépède, 24/5/1806, N.H.M. D.T.C. XVI 272–273. Actions like this gained Banks useful allies in France, but not at the Admiralty.

[6]Thomas Gee.

[7]Banks requested the liberation of 'Mr. Julien Jouneau agent Comptable de la Frégate, *La Didon*': Banks to Marsden, 7/8/1806, N.H.M. B.L. D.T.C. XVI 304–305.

[8]William Marsden, at this time First Secretary at the Admiralty.

[9]Thomas Grenville (1755–1846): bibliophile; President of the Board of Control, 1806; following Howick, First Lord of the Admiralty.

[10]Anthony Ashley Cooper FRS (1761–1811), 5th Earl of Shaftesbury, and the Countess of Shaftesbury (*née* Barbara Webb).

Letter 107

1807

To Ann Flinders

Soho Square
22 May 1807

Madam,

Your Letter unfortunately has been in search of me in Lincolnshire, where I was for a few days on Election business.[1] It Came to my hands only yesterday, or it would have been Sooner acknowledgd.

It Greives me to hear that Capt. Flinders, after having for so long supported with manly fortitude the very disagracefull treatment he has met with from those Enemies to humanity, the French, has at last given way to oppression, & Sufferd his Spirits to Flag.[2] I can not, however, have a doubt from the well Known Energies of his mind that his Low Spirits are only a temporary Attack of depression, which will not be Lasting.

Respecting the Possibility of his having Receivd the news of his Liberation at this time, it Certainly depended on the opportunities the French may have had of sending off the orders Given in Capt. Flinders' favor. The order for his Liberation issued in March, & three Copies of it, were sent to the Coast to be forwarded by neutrals.[3] It is not, however, probable that many Neutrals at that time were to be met with who Saild from France to the Isle of Reunion. The Americans who frequent that Island usualy sail from America to India without touching at France. It always, therefore, appeard Probable that the dispatch forwarded by the Admiralty to S[i]r Edwd Pellew[4] in December Last would bring the first news. This has no doubt before this time Reachd the Island, & I have reason also to beleive that it has been, or Soon will be, attackd by an English Force[5] so that we have now Every degree of hope that your husband either already is, or soon will be, set Free.

I am, madam,
your Obedient Hble Servt,

Jos: Banks.

[N.M.M. FLI/1&26.]

[1]Lord Grenville's 'Ministry of All the Talents' collapsed in March, and was replaced by a troubled administration under the Duke of Portland. Banks avoided meddling in party politics at a national level, and so was able to advise successive governments as a 'neutral'.

However, in Lincolnshire his influence was strongly felt at election time, when he ensured that 'the peace of the county' remained undisturbed. In other words, the sitting Members of Parliament were returned with as little dispute as possible, especially if they represented the landed interest. Banks was principally concerned with elections for the county and for the boroughs of Boston and Lincoln.

[2]Flinders suffered bouts of illness and unhappiness during his captivity. He described them to Ann: Matthew to Ann, October 1804, P.R.O. Adm. 7/707. A bout of depression followed the departure in August 1806 of his friend Thomas Larkins, formerly commander of the Honourable East India Company's ship *Warren Hastings*.

[3]The order to release Flinders was signed by Denis Decrès on 21 March 1806. Copies were then sent to Mauritius. Banks received confirmation 'that orders have been sent to the Isle de France [Mauritius] for the Liberation of Capt. Flinders' early in August 1806: Banks to Marsden, 7/8/1806, N.H.M. B.L. D.T.C. XVI 304–305. The Governor of Mauritius, General Decaen, decided not to carry them out. They were vague at best.

[4]Sir Edward Pellew (1757–1833), 1st Baronet and 1st Viscount Exmouth: admiral, RN.

[5]British maritime power grew after the French Fleet was defeated at Trafalgar in 1805. Decaen was isolated when Britain took possession of the Cape of Good Hope in January 1806. His supplies and communications were at risk. Indeed, the British Navy was so successful at intercepting orders from France that at the Admiralty some worried Decaen might not receive his instructions to release Flinders.

Pellew — then a rear-admiral and commander-in-chief in the East Indies — was told to deliver a copy of those which had been captured. This was done under a flag of truce by the English frigate *Greyhound* in July 1807. In addition, a French version got through against the odds. However, Decaen had little intention of releasing a British officer who was well acquainted with the vulnerability of Mauritius.

Letter 108

1807

To Henry Francis Greville

Soho Square
June 1807

My dear Sir,[1]

I am thankful to you for your letter of the sixth instant,[2] which I should have answered sooner had the printed paper which came to my house some days ago been at hand. From the tenour of the printed paper I conceived that I was invited to become a member of an intended Society, and that the Room in which the Society was to meet was to be occasionally employed for other purposes. Your letter has shown me my mistake. I find from it that new rooms are to be built and exclusively reserved for the use of the intended Society, and that the Members of it are chosen by a Committee.

I must beg on any terms to exculpate myself from all idea of my having ever considered the amusements of the great world as frivolous. I respect, I assure you, Sir, the recreations of the public, and look up to those who direct them with taste and judgement. I see no objection to a room being occupied one day by a Society of Philosophers, and on the next day a company of masqueraders; and the less so as it will probably on both occasions be frequented by the same persons. I was myself an attendant on plays, operas, concerts, masquerades etc. till prevented by infirmities; and was my health now restored to me, I should again be a partaker in the declines of life of those gaieties which added so much pleasure to the commencement of it.

I am sorry I cannot accept your obliging invitation to dine on the 28th as I shall then be in the country. But this is of no moment, and I know myself too well to suppose myself a proper member of a Society for Belles Lettres. I am scarce able to write my own language with correctness, and never presumed to attempt elegant composition either in verse or in prose in that or in any other tongue. It is fitting, therefore, that I continue to confine myself, as I have hitherto done, to the dry pursuits of Natural History.

[B.L. Add. MS. 33981, f. 256v.]

[1]Banks made two drafts of this letter.
[2]Greville to Banks, 6/6/1807, B.L. Add. MS. 33981, ff. 255–256. The proposal was for a Belles-Lettres Society. Banks was invited by Greville to join a group of men of 'rank,

fortune, & Literary attainments'. These included Richard Payne Knight and John James Hamilton, Lord Abercorn (1756–1818). One question was whether such a society — for the 'noblest of purposes' — should meet in a building used for public entertainment — 'the fashionable world' of 'frivolous pursuits'.

1808

To Ann Flinders

Soho Square
19 July 1808

Dear Madam,

I conclude that you have receivd a Letter from Capt. Flinders by the way of America. The one alluded to in the French Letter [is] inclosd in yours, which I return. I have one which Came to hand by yesterdays Post, & States my Friend, your husband, to be in good health, tho no time has yet been fixd upon by the General[1] for obeying the orders he has receivd Respecting his Liberation.

I cannot, I confess, attribute the disobedience to orders which the General has been guilty of to any Source but that of Cowardice. He suspects that your husband is able to give Such information to his brother officers in India as might enable them to attack the Island with advantage, & has therefore resolvd not to Let him proceed to any place where he has any Likelyhood of Seeing any of them Soon.[2]

The authenticity of the order has been acknowledgd by a Copy of it being sent to your husband. The Reason given by the Cheif of the Etat Major for delaying its execution is the want of a Proper opportunity; that is such a one as they may think a proper one. The French Frigate La Som[e]illante was, when he wrote, fitting for a Voyage to France, & her departure Cannot I think but furnish a proper opportunity of Sending him home as She will Land him in France, & by so doing put him as much as possible out of the Condition of managing an attack upon the Island. I Trust, therefore, that he is before this time on his passage home in this French Frigate, & I most Sincerely hope that She may be Capturd & soon brought in to a British Port.

I have Solicited Lord Mulgrave,[3] as I have done other First Lords, in the hope of Obtaining post for him during his Absence, but have always met with the Same answer that no instance occurs of giving an officer preferment while he is a Prisoner.[4] I have no doubt, however, of a favorable a reception for him from Ld. Mulgrave as we can wish for if we Can get him home.

I am, Madam,
your Faithfull Hble Servt.,

Jos: Banks.

[N.M.M. FLI/1&26.]

[1]General Decaen, the captain's chief antagonist.
[2]This impression was one Matthew Flinders had suggested in a letter to Banks earlier in the year: Flinders to Banks, 24/1/1808, *Historical Records of New South Wales*, F.M. Bladen (Ed.), Sydney (1898), vol. VI, pp. 420–422. In the same letter Flinders confirmed that he had seen the order for his release, but that excuses were now being used to detain him: 'By giving me a copy of the Marine Minister's letter which ordered my liberation, the General avowed the authenticity and sufficiency of the order, and the letters I have from the Chef D'etat Major announce that the only cause of this order not being put in execution is the want of a convenient opportunity.' p. 421.
[3]Sir Henry Phipps (1755–1831), 1st Earl of Mulgrave, 1st Viscount Normanby and 3rd Baron Mulgrave: 1st Lord of the Admiralty, 1807; master of the Ordnance, 1810–1818.
[4]Banks failed to gain Flinders a promotion because strict Admiralty practice did not allow this for prisoners.

1808

To Thomas Phillips R.A., F.R.S.

Revesby Abbey
12 September 1808

Dear Sir,

I cannot say that I Feel myself able to give you advice on the subject of your Letter. That you have Painted a Picture of me that does honor to your Talents as an Artist, I am Quite sure;[1] but whether a Print Taken from that Picture will be a Profitable Concern is a matter of which I, who never was Concernd in the Science of money Getting, have not the means of Forming a Proper Judgement.

The Sale of Such a Print must, I conceive, depend upon three things: the Excellence of the Painter; the Talents of the Engraver, & the notoriety of the Person it Represents. The Two first I readily admit to be of the very best description. Relative to the Third, I have many doubts. There are already three prints of your Hble Servant to be met with in the Print Shops: one from a Picture of Sir Joshua, an admirable mezzotinto;[2] another from the Pencil of the President,[3] & a third, a most decided Likeness, from a Crayon Picture of Russell.[4] How these three have fard in the world, the Printsellers will tell you. I doubt, however, whether any adequate Reward was Obtaind by the artist for Either of the Large ones. A man like me, who has never medled in Politics, & who Cannot, of Course, possess a Squadron of Enthusiastic Friends, is not likely to Sell a dear Print. A Cheap one will answer better among the men of Science, many of whom have honord Russels Print with a Place in their apartments.

For me at my age to Solicit a Subscription for the Publication of my Portrait would, I think, be interpreted by Cool headed men into an act of Vanity. I feel inclind, it is true, to do it as an act of gratitude to you for the Genius you have exercisd, & the Pains you have taken, but will any body Know that this is an active motive in my mind, & will not the most alert & prattling part of the Community be more inclind to discover a Trait of Vanity in the actions of my mind than one of a more amiable nature? I dare not for these reasons Subscribe, nor should I venture, on account of the high Price put upon proofs, & Very little real superiority they have over Prints, to purchase privately more than a Few of them. Some prints I certainly should try to Lay by [for my] Family, in hopes that they may become usefull to some one sometime hence as Presents, when dificult to obtain in the Shops

More on this Subject I Know not how to Say. I beg, however, to hint that, much as I admire the Burin of Schiavanetti,[5] I Should, was I to direct, Employ Sharp.[6] He has Engravd Boulton[7] & John Hunter[8] admirably, & is now Employd on Home[9] as a Fourth in the Set of which there never Can be many. Some Sale will arise, but I very much doubt, in any Case, whether a Print of me will Ever Satisfy the just Claims of a good Engraver, & Confess therefore that I See no hopes of any profit whatever being Reapd by the Father artist, the Painter.

I am, Sir, with much regard
& Esteem, your very Hble Servt,

Jos: Banks.

[P.M.L.]

[1]See Illustration XIV, with explanatory Note. The portrait Banks referred to was painted by Phillips in 1808 for the Spanish astronomer, Jose Mendoza y Rios FRS (1762–1816).
[2]See Illustration IV, with explanatory Note. This portrait was painted in 1772–1773 by Sir Joshua Reynolds PRA. A mezzotint engraving by W. Dickinson was made later in 1773, probably for Banks's uncle, Robert Banks-Hodgkinson.
[3]See Illustration I, with explanatory Note. Benjamin West was President of the Royal Academy for most of 1792–1820. This portrait was painted by him in 1773, and a mezzotint engraving of it was made by John Raphael Smith (1752–1812), 1773.
[4]An excellent pastel crayon portrait by John Russell RA (1745–1806) drawn in 1788. In it, Banks was shown holding 'Carte de la Lune par J. Russell', a reference to Russell's work depicting the Moon using Dr. (later Sir) William Herschel's reflecting telescopes. The original is owned by Lord Brabourne. A stipple engraving of the portrait was made by Joseph Collyer (1748–1827) in 1789.
[5]Niccolo Schiavanetti (1771–1813): Italian engraver from Bassano. Phillips's contribution to the Royal Academy exhibition of 1809 included his first portrait of Banks, which was engraved by Schiavanetti, 1812.
[6]William Sharp (1749–1824): engraver.
[7]Matthew Boulton.
[8]Dr. John Hunter.
[9]Dr. Everard Home.

xv. **Sir Joseph Banks P.C., K.B., F.R.S.**
Portrait, 1815, by Thomas Phillips R.A. The Royal Society, London.

Note: This portrait was commissioned by the Royal Society in 1815 as a copy of one painted in 1808 by Phillips. The original was painted for the Spanish astronomer, Jose Mendoza y Rios F.R.S. In the later 1815 version Banks is depicted as the formidable President of the Royal Society, with the mace of office in the foreground, and above his head the Society's motto: *Nullius in verba* ('not bound to swear on the word — or the text — of any master'). He also wears the star and sash of the Order of the Bath. The paper in Banks's hand is Humphry Davy's Bakerian Lecture for 1808: *Philosophical Transactions*, vol. 99 (1809), pp. 39–104, 'An Account of some new analytical Researches on the Nature of certain Bodies, particularly the Alkalies, Phosphorum, Sulphur, Carbonaceous Matter, and the Acids hitherto undecompounded; with some general Observations on Chemical Theory. By Humphry Davy, Esq. Sec. R.S., F.R.S., Ed. and M.R.I.A.' Read 15/12/1808. To show his respect for Mendoza's work, Banks requested that the paper on the table should be, once again, from the *Philosophical Transactions*: vol. 91 (1801), pp. 363-374, 'On an improved Reflecting Circle. By Joseph de Mendoza Rios, Esq. F.R.S.' Read 4/6/1801. This imposing portrait currently hangs behind the President's chair in the Royal Society's Council Room, 6 Carlton House Terrace, London.

Letter 111

1810

To Ann Flinders

Soho Square
12 June 1810

Madam,

I wish I could give you any Comfortable or Encouraging news respecting my worthy Friend Capt. Flinders. Government can do nothing in his Favor under the Capricious & insolent Government of the Tyrant of France. I have Left no means untried, but whether or not I have done good, or may have done harm, I have no means of Learning. I Sent word above a year ago, by a way very Likely to Reach the Ears of Bonaparte, that nothing but the Liberation of Capt. Flinders would Ever induce me to beleive that The Letters I receivd from Paris were not a deceipt, as I thought it impossible that Gen. de Caen[1] would venture to disobey the orders of such a sovereign as Bonaparte unless he had secret instructions from France.

There is, however, a hope from another Point.[2] The Trade of India has been of Late so Seriously injurd by vessels from the Isle of France that I am told vigorous measures are to be adopted for seizing The Island.[3] If this Succeeds, & if it is Plannd with wisdom & executed with Spirit it must succeed, we may have the hapiness of Seeing the Gallant Capt. return before we expect him.

I am, madam,
Your Obedient Hble Servt,

Jos: Banks.

[N.M.M. FLI/1&26.]

[1]General Decaen.
[2]Banks was cautious when discussing the possibility of release, as earlier hopes of liberation had proved false. He was correct in the view that British naval power would be more effective than correspondence in gaining Flinders his freedom.
[3]The British squadron in the Indian Ocean blockaded Mauritius from June 1809. The island was captured following an attack on 3 December 1810.

1810

To William Scoresby F.R.S.

Soho Square
8 September 1810

Dear Sir,

I thank you for your Letter & for the Observations Containd in it, which prove that you Continue to exercise your Talents & your industry in a way that Cannot Fail in due time to make you usefull to the Public, & Respected by them in proportion as they find your application to Study has improvd the Knowledge of Science.

I am Sorry to have disappointed you Last year in Failing to provide for you a proper apparatus for Obtaining the temperature of the Sea at Considerable depths.[1] I myself was disappointed in procuring it by the unexpected death of Mr. Cavendish[2] who had undertaken to Superintend the Contrivance, & afterwards also by the Loss of Mr. Gilpin[3] who overlookd the execution. The Loss of these two admirable men, for such they were both of them, made me at the time too negligent of & indeed unfit for my usual pursuits. I Trust, however, that the instrument is now Ready for delivery as I Saw it a few days before I Left town, & gave then the final directions to Mr. Cary,[4] instrument maker in the Strand, to Finish it for me. He promisd to have it done without delay. I Shall be glad if you Come to London to have the pleasure of Seeing you, that I may instruct you personaly in the use of it. If you do not, I will have it deliverd to your order, wherever you please, with the best account I can provide of the management of it, which is by no means difficult. I confess I am at a Loss, however, to guess how your Bucket Could have been broken by the force of water if the water was able freely to Pass through it. Nor do I comprehend how the Pores of the wood Could be filld with Congeald water, when the temperature of the Sea was only 28, & that of the water in the Bucket warmer.

I have often noticd the variations[5] you mention in the Color of the Sea, but as I have been sometimes able to ascertain that they have depended on accidental circumstances, I have my doubts of any usefull conclusions arising from them. Perhaps, however, it may be different in the Chilly Latitudes you have visited.

I gave your Letter to Mr. Home,[6] who will be glad to see you if you Come to London, & to Thank you for your attention to your [his] wishes.[7]

I beg my Compts to your Father, & am, Sir,
your Very Hble Servt,

Jos: Banks.

[Warren R. Dawson, *The Banks Letters: A Calendar of the Manuscript Correspondence*, London (1958), p. 739, gives the reference 'Whitby, < B. 1 >'. However, the original was not found at Whitby Museum. A Warren Dawson transcript and a photostat copy of this letter are at the Natural History Museum, London, in the Banks Archive. The letter was published in: T. and C. Stamp, *William Scoresby: Arctic Scientist*, Whitby (1975), pp. 49–50.]

[1]Scoresby had been investigating the temperature of the Arctic Sea. Initially, he had used a device of his own. This was a ten-gallon cask with home-made valves on both ends. Banks intervened to provide a better instrument made of wood bound with brass. In time, these early experiments led Scoresby to the conclusion that the Arctic Sea is warmer near the bottom than it is at the surface.

[2]Henry Cavendish, who died in March.

[3]George Gilpin (d.1810): assistant astronomer on Cook's 2nd voyage, 1772–1776; assistant, Royal Observatory, 1776–781; clerk of the Royal Society, 1785–1809; secretary, Board of Longitude, 1801–1810.

[4]William Cary (1759–1825): instrument maker, 182 Strand, London. Cary was formerly an apprentice to Jesse Ramsden.

[5]Scoresby found that the 'variations' were caused by 'minute animalcules', or plankton.

[6]Dr. Everard Home.

[7]Following this letter, Scoresby wrote to Banks to say that the instrument failed in deep waters: Scoresby to Banks, 15/9/1811, W.M. < 7 >. At three hundred fathoms the wood swelled and the glass broke, but Scoresby tried again using another version made of brass, which proved more successful. This was his 'marine diver'.

Letter 113

1810

To Ann Flinders

Revesby Abbey
25 September 1810

Madam,

I have infinite Satisfaction in informing you that Capt. Flinders has at Last Obtaind his Release, & is expected in England in a few weeks, & that on his arrival he will be immediately made a Post Captain.[1]

I am, Madam,
your most Faithfull Servt,

Jos: Banks.[2]

[N.M.M. FLI/1&26.]

[1]Matthew Flinders was released on parole, which he signed on 7 June 1810. He had been held captive for six and a half years. He sailed from Port Louis six days later, and reached Spithead on 24 October. Flinders was made Post Captain of HMS *Ramillies* on his return.
[2]Flinders wrote a joyous letter to Banks on reaching England: Flinders to Banks, 25/10/1810, *Historical Records of New South Wales*, F.M. Bladen (Ed.), Sydney (1901), vol. VII, pp. 436–437.

xvi. **Captain Matthew Flinders F.R.S.**
Portrait, 1808, by T.A. de Chazal at Mauritius. Reproduced from E. Scott, *The Life of Matthew Flinders*, Sydney (1914), p. 366.

Letter 114

1810

To Dr. Everard Home F.R.S.

Revesby Abbey
22 October 1810

My dear Sir,

I was so stunned by the unlooked for Blow I received from the Perusal of your Letter[1] that I was utterly unable to answer by return of Post. I have lost my right hand, & can never hope to provide any thing as a Substitute that can at all make amends to me. My chief pleasure, that of my Library, is reduced almost to a Shadow when I consider the many points from whence I derived satisfaction in the possession of it.

I always hoped that Dryander[2] would outlive me. He was younger & less afflicted with disease than myself. Probably it is better as things now are for him. He would have lost more in surviving me than I lose in his Death. I had arranged for him the use of my Library as long as he lived in the same manner as he has been used to enjoy it, a Breakfast for him & my Friends every morning &c. &c., but that might have Failed.

I attribute his loss intirely to his indiscretion in not disclosing his disease long ago, & console myself in his having been under your care. If an unerring hand guided by Skill, Practice & what is best of all, Superior Talents & a warm heart, could have Saved him, he would still have been with us, but heaven's will be done!

For me, I console myself in Feeling that my departure cannot be long delayed, & that, should I linger, I must Soon be deprived of all those enjoyments to which he used to administer while he participated in them. My hearing begins to Fail. One eye, as I told you before, is of little use. In short, the infirmities of age overtake me fast. My life has not been Idle, & I feel that Exertion has acted on my Constitution, & hastened that Consumation which all of us must look forward to. I see it without Fear, & I think I can bear The loss of my Sight & hearing. I have Patience, & I pride myself in Profiting by it as a Remedy to every ill in Life.

I wrote to Brown[3] to thank him for his kind & important services on this occasion, & to beg of him to see the Remains of my Friend properly laid in the Ground. My library cannot, till I return & have made Some arrangements, be open to all comers as it was.[4] I shall Soon be in Town, & I hope to make some arrangement.

My Pen almost refuses its office. Adieu, then, my dear Home. My obligations to you I cannot easily express, but be assured I Feel them most sensibly.

Sincerely & Faithfully yours,

Jos: Banks.

[N.H.M. B.L. D.T.C. XVIII. 88–89.]

[1]Banks knew that Dryander had been unwell, and gave instructions that he be cared for at Soho Square: Banks to Home, 11/10/1810, B.P.L.M.A. Ch. I. 5. 32–32a, N.H.M. B.L. D.T.C. XVIII 84–87. However, the seriousness of Dryander's condition was not fully understood.
[2]Jonas Dryander had been botanical curator and librarian to Banks, 1777–1810. He did a great deal in assisting Banks to organize the herbarium and library at Soho Square. See Letter 34, Note 2.
[3]Robert Brown took up many of the duties Dryander had performed, especially in the herbarium. Banks arranged in his will for Brown to receive £200 a year, and the use of the library, herbarium and other resources at Soho Square. Banks also decided that the British Museum should receive his collections when Brown died.
[4]Banks's collections were available to outside visitors on introduction.

Letter 115

[1810]

To John Barrow F.R.S., 2nd Secretary of the Admiralty

Revesby Abbey, Boston
24 October [1810]

My dear Sir,

Many thanks for your Letter. It has Raisd the Spirits of our Lincolnshire Lads very much. L[or]d Wellington seems to have Seizd upon the most vulnerable Part of the Present French Character, their Rashness, & by his Cool & wise Conduct Led Massena[1] into a scrape from which, God send, he may not be able to Extricate himself, which, if God in his infinite mercy to us is Pleasd to send Plenteous Rains, it seems Clear he never Can do.[2]

I am Gratefull for Mr. Yorkes[3] kindness to Flinders as far as it goes, & Shall always feel for his kindness a proportionate degree of Gratitude; but as I am of opinion he has not gone nearly So far as he might have done, I must attribute the Late date of Flinders's Commission to his General Conduct being appreciated at the Admiralty at a Lower rate than that his Friends put upon it.[4] In fact, the Admiralty are the best judges, & if they Rate his merits & his Sufferings no higher than an antedate of his Commission to his arrival at the Cape, we must Submit. The most plain Line would have been to date it from the day of his being deliverd from French Custody, & Placd under English Charge.

I Call it *Custody* because I do not admit that Flinders was ever a Prisoner of War. He was, it is true, detaind as the Detainees are in France, but the Council of War in Paris Long ago declard it their opinion that his imprisonment was improper, & of Course their Cheif, the French Emperor,[5] Sent orders for his Liberation. These were forwarded to me, Sent by me to the Admiralty, & from the Admiralty forwarded to the Isle de France. Surely this act must be Considerd as a termination of his imprisonment, & of Course an Eminently Proper one for the date of his Commission. If he is Estimated at the Admiralty as deserving of one, it would mark our attention to our officers when under misfortune; & Shew how little the office inclind to take advantage of that harsh Rule which interdicts the Promotion of a Brave man, however well he has Fought, if the Fortune of war has placd him in the hands of his Enemies till his Enemies are Pleasd on their own mere motions to Restore his Liberty. Surely this Rule is more like a French one than an English one.

These are the arguments I mean to offer to Mr. Yorke when Flinders arrives. If you Should think fit to Read this Letter to him, it will give him the means of being prepard with answers when I urge my request. He will, however, be pleasd to Remember that I am Gratefull for what I get, & not Sulky when refusd.[6]

Adieu, my dear Sir,
Faithfully & Sincerely yours,

Jos: Banks.

[M.L. Banks MSS. 743/3.*]

[1]André Masséna (1759–1817), Duc de Rivoli and Prince d'Essling: perhaps the greatest of Napoleon Bonaparte's marshals. Banks referred to the Peninsular War in which Sir Arthur Wellesley, Viscount Wellington FRS (1769–1852), confronted heavily reinforced French armies, fresh from victories in Spain, and led by Marshal Masséna. Wellington retreated into Portugal in 1809, where he decided to remain and fight a defensive campaign. The French advanced into Portugal in September 1810 to meet him, but Wellington had already ensured that all supplies in their path were removed or destroyed. After one sharp engagement he waited in Lisbon, protected by the impregnable lines at Torres Vedras. October rain drizzled on Masséna's enormous army as he stared in astonishment at these defences, of which he had not been warned. Thwarted by them, Masséna fell back in March 1811 as his troops starved, and Wellington harried him on the way. Napoleon replaced Masséna later in 1811.
[2]Banks's confidence at this stage of the Peninsular War was not shared by many politicians in London. Indeed, Banks must have been privileged with information about Wellington's determined strategy, or a clearer assessment of morale among the commanders in Portugal than was generally available. His prediction was, anyway, accurate enough.
[3]Charles Philip Yorke FRS (1764–1834): politician; Secretary at War, 1801–1803; Home Secretary, 1803–1804; First Lord of the Admiralty, 1810–1811. Yorke became First Lord on 1 May 1810.
[4]Flinders was made Post Captain of HMS *Ramillies*. Both he and Banks argued that the commision should be back-dated to 1804 as this was the year Flinders might have returned had he not been detained by General Decaen: Flinders to Banks, 25/10/1810, B.L. Add. MS. 32439, f. 332. However, no time earlier than Yorke's accession to the Board of Admiralty was permitted, and so 7 May was offered as a compromise.
[5]Napoleon Bonaparte.
[6]Banks wrote to accept the compromise: Banks to Barrow, 20/10/1810, R.S. M.M. 6 71. Banks had recently lost his good friend, Jonas Dryander, and in his grief lacked the spirit to confront an inflexible Admiralty.

1813

To William Jackson Hooker F.R.S.

Spring Grove
19 June 1813

My dear Sir,

Tho I realy Cannot think it Possible that your Relatives & Friends in Norfolk Can Consider an Island[1] half as large as England to be of a deadly & unwholesome nature because one Town upon it is notoriously So, I See their Objections are urgd with So much determination and Eagerness that I am far indeed from advising you to despise them. I have, however, no doubt that arguments or injunctions Equaly Strong will be urged by them if you attempt to extend your views Further than the Exhausted Azores, Originaly Scarce worth the notice of a Botanist, & now almost intirely Transferd to Kew Gardens by the indefatigable Ma[s]son.[2]

From the Complexions of all that has Passd of Late in the Conduct of your Friends, I have no doubt that they wish to force you to adopt Sardinapalus's[3] advice to his Citizens to Eat, drink & propagate. How you will Like to be married and Settled in the countrey, as Joe Miller[4] wishd the dog had been who flew at him to bite him, I know not, but that this fate is prepard for you Somewhat Earlier than the natural period of Renouncing an active Life is a matter of which I have no doubt; but Pressd as you are, I advise you to Submit & sacrafice, if you Can, your wish for Travelling to the importunities of those who think they Can Guide you to a more Serene, Quiet, Calm & Sober mode of slumbering away Life Than that you proposd for yourself.

Let me hear from you how you feel inclind to prefer Ease & indulgence to Hardship and activity. I was About 23 when I began my Peregrinations. You are Somewhat older, but you may be assured that if I had Listend to a multitude of voices that were Raisd up to dissuade me from my Enterprise, I Should have been now a Quiet Countrey Gentleman, ignorant of a multitude of matters I am now acquainted with, & probably have attaind to no higher Rank in Life than that of a Countrey Justice of the Peace.

Adieu, my dear Sir.
Very Faithfully yours,

Jos: Banks.

The Ladies beg to be Rememberd to you. I get better daily, but very Slowly.

[R.B.G. Kew Archives, Director's Correspondence 1. 39.]

[1]Java. Banks recalled here the visit made by HMS *Endeavour* to Batavia from October 1770 to January 1771. Batavia was a notoriously unhealthy place. Many of the *Endeavour* crew and Banks's party died from diseases contracted there. By the time of this letter the Dutch East Indies were under British control, and so there was an opportunity for a mission rich with botanical promise. Hooker expressed an interest in going, but then withdrew on the advice of people like Dawson Turner, his father-in-law.

A former naval surgeon, Dr. Joseph Arnold (1782–1818), went instead. He was appointed personal physician to Sir Thomas Stamford Bingley Raffles FRS (1781–1826), Governor of the Honourable East India Company's establishments in Sumatra. Arnold discovered the magnificent *Rafflesia arnoldii* on a trip with Raffles into Sumatra in May 1818. However, Arnold also caught a fever from which he died.

[2]Francis Masson was a tough, hard-working Scot, who collected tirelessly for the Botanic Gardens at Kew. Banks selected an example of industry and adventure designed to inspire Hooker.

[3]The name given to the last king of Nineveh. Sardanapalus was known for his luxurious and effeminate behaviour.

[4]Possibly a reference to do with Joseph Miller (1684–1738), a popular comedian, as described in *Brewer's Dictionary of Phrase and Fable*.

Letter 117

1813

To Dr. James Edward Smith F.R.S.

Spring Grove
6 July 1813

My dear Dr.,

I cannot help feeling anxious in the Extreme for the success of your present undertaking,[1] and this anxiety is founded not alone in friendship for you, but in the Love I have always felt for the Science of Botany, & the ardent wish that a Science in itself so usefull, So Engaging & so Eminently suited to cherish the best Feelings of Gratitude to a benevolent creator, Should be offerd to the Consideration of the Youth of this Countrey at a time when their minds are not Engagd in any Particular pursuit. That you will Read Lectures Regularly I well know, & by these Lectures I expect that Recruits to the Study of natural history will be Raised in abundance. It is grevious that neither of our Universities Should offer to those whom they Educate the choice of that Science which I judge to act more Strongly in Regulating the Passions of mankind, & in Demonstrating the wisdom of the works of Creation. May success attend you, & Cambridge be Releivd from the Scandal of Receiving a Royal Bounty without enforcing the application of it to the Purpose it was intended to answer.

Let me hear from you my dear Dr., for I am anxious till I do, & the more So Since the very han[d]some Conduct of the University in the Case of the Chemistry professor[2] gives me much Solid grounds to Expect your Success.

Beleive me, my dear Dr.,
always Faithfully yours,

Jos: Banks.

I have heard nothing yet from the Vice chancellor,[3] a bad omen for Biggs, & a Strong mark of my inefficiency in offering advice to alma mater.

[L.S. Smith MS. 301. 7. 47.]

[1]Smith hoped to succeed Professor Thomas Martyn (1735–1825) as the Cambridge Professor of Botany. He believed Martyn would soon die. In fact, Martyn survived another twelve

years, and Smith never obtained the chair of botany because some university members objected to his unitarian views.

[2]Dr. Smithson Tennant FRS (1761–1815): chemist; Professor of Chemistry at Cambridge, 1813.

[3]Dr. John Davie (1777–1813): theologian; Vice-Chancellor of Cambridge University, 1812–1813.

Letter 118

1813

To Charlotte Seymour, 11th Duchess of Somerset

Spring Grove
21 July 1813

I hope your Grace will not write again to me in the Stile of Friendly Kindness which pervaded your Last Letter. Another Such would make me Repine at what I ought not to Regret.[1]

I have Enjoyd a Long Life, Abundance of health, & an unvaried Continuance of what People Call Good Fortune. I must not, now I have So nearly Reachd my appointed Time, Regret that the Pleasures of this Life have decreasd, & the pains of it have increasd. Such a change is Fitting to Prepare me to meet a Greater Change without Fear or Regret.

Tho I am Every day Sensibly better than the day before, my Legs Still Refuse to do any Part of their usual Duty. I have not yet ventured to get into a Carriage, nor am I likely to do So Soon. My weight is too heavy to be Easily managed. The Going into or Coming out of a Carriage must therefore always be a hazard, & a Fall in the helpless State I am in could Scarcely happen without the Consequence of a Broken Limb.

Was I able to go out, I Should instantly Fly to obey your Graces Commands as the most Gratifying indulgence I could bestow upon myself. If I amend So as to be able to go out, Bulstrode[2] will be the First Place I Shall Solicit Leave to visit, but the [situation] in which I am accustomd to indulge myself does not allow me at present to think of the Time when that is Likely to Happen. Hope is indeed the Cheif Comfort of the afflicted, but it ought to be that Regulated & Patient hope which does not hazard the intermixture of Disappointment.

May all that is Powerful & benificent preserve your Grace, the Duke[3] & your beautifull Children from all the Calamities of this Life is the Sincere wish of,

Your Graces Most Hble,
& Most Faithfull Servt,

Jos: Banks.

[B.R.O. Bulstrode (Additional) D/RA/A/1B/4/1.]

[1]Banks was very unwell for most of this year, and was unable to visit his estates. He had suffered intermittently from gout since November 1787, but 1813 saw his ailments increase. He was passing an occasional calculus, there were signs of uræmia, and he appeared anæmic too. From time to time Banks was barely able to use his right hand. His legs also swelled and were bandaged. He was therefore confined to his bed or a chair, but used a wheel-chair when he could move more freely.
[2]Bulstrode, Buckinghamshire. Bulstrode Park belonged to the Dukes of Somerset from 1810 to 1885. The estate contained lands in a number of parishes, including Hedgerley. The family house is now in the Gerrard's Cross area.
[3]Edward Adolphus Seymour FRS (1755–1855), 11th Duke of Somerset: mathematician; antiquarian; President of the Royal Institution, 1801–1838; President of the Linnean Society, 1834–1837. He was a landowner in Lincolnshire, who therefore shared many interests with Banks relating to county affairs.

Letter 119

1813

To Sir Everard Home F.R.S.

Spring Grove
22 August 1813

My dear Sir Everard,

Many thanks for your Letter.[1] I wish it may be in my Power to obtain for you any information about muscles [mussels] & Whelks during your Stay at Lenning /?/;[2] but I have Little hopes as Rennie,[3] the only person by whom I can obtain it, is in the Country.

I have a Favorite here that amuses me, & I think may instruct me. It is a Spider who presented himself on my Table with a fly in his mouth, & on examination was found to have only 3 legs instead of 6, & those three all on one side. Whether they were lost by accident or never developed I know not, but no Trace of either of them is visible.

What Struck me most was the Change in the Animal's manners suited to his present Cripled State. Instead of spinning a web & entraping Flies, he sought in the evening, for I found him by Candle light, & Seized them while at rest.

He now lives in a Small Glass & seems contented. He eats a fly every Day & Grows fat. I shall learn from him something. If his legs grow again, *that* will be a Piece of usefull information. If not, I shall conclude that he never had any. In truth it is curious to see how well he walks upon the half of his natural legs, & all of them on one side.

I am going on Prosperously. I am in the 27th day of Freedom from the Gout. 3 days ago I felt in the night Some acid in my Stomach. It alarmed me, & the next night I increased my dose of Magnesia to 20 Grains. Yesterday I was Quite well, & this morn I voided a little Sand. Should it prove that my Gout is inclined to change itself into [a] Stone, & that magnesia Protects me from the Evils of the alternative, it will be an additional Triumph for you & for Brand.[4]

Adieu, my dear Sir Everard,
always Faithfully yours,

Jos: Banks.

[N.H.M. B.L. D.T.C. XVIII. 276–277.]

[1]Home's last letter to Banks has not been located, but the subject of mussels and whelks was raised in: Banks to Home, 5/9/1811, N.H.M. B.L. D.T.C. XVIII 104–105.
[2]The inserted question mark appears to refer to 'Lenning' as a place name which cannot be identified. It may be Lerrin, Cornwall.
[3]Possibly John Rennie FRS (1761–1821): civil engineer. The remaining correspondence between Banks and Home has no mention of Rennie in it.
[4]Dr. William Thomas Brande FRS (1788–1866): a chemist with a range of scientific interests who, along with Home, had analysed calculi.

Letter 120

1813

To Henry Goulburn

Soho Square
10 November 1813

My Dear Sir,

As the Period at which the final destiny of our friend Admiral Bligh is on the point of arriving, it being intended by Lord Bathurst to include his Pension in the next Estimates, I take the liberty to request of you to peruse the Memorial enclosd, which was laid before Lord Bathurst in the Spring.[1]

Lord Bathursts[2] intentions appear to be to Estimate the amount of the Pension by the actual time which he held the Government, without the least regard to the Sufferings he experiencd from the rebellious proceedings which displacd him.[3]

Now, if I am right in beleiving that his ill treatment originated in his strict obedience to the Orders he receivd from home, & that no improper Act originating in his own mind tended even to accelerate the catastrophe to which he was the sacrifice, it appears to me no more than justice that his Pension should be, if not quite as much as that given for the regular time of Residence, at least not much below it. I have conversd with Lord Bathurst on the subject, who, tho always kind to me, does not seem to think of the Admirals services, & of the privations he has endurd, & the sacrifices he has made, as I do. If you & my Lord Castlereagh[4] think of Bligh as I do, I am sure his Lordship will speak a word in his favor to Ld. Bathurst, which would be decisive. Allow me, therefore, to request you to lay the matter before Lord Castlereagh, & to assure his Lordship that if he will do me the favor to hear your Representation of the Admirals services, I shall think the amount of the Pension his Lordship & you will think right a perfectly proper one, & shall venture to hope that Lord Castlereagh, actuated more by the sentiments of Justice he is so well known to posess, than by his Lordships friendship to me, which I know how to value, will be kind enough to say a word on the subject to Ld. Bathurst.

Adieu, my dear Sir, excuse this trouble from
Your very faithful
Old friend & Hble Servt.

Jos: Banks.

[S.L. Banks MS. SS. 1:15.]

[1]Bligh submitted a memorial to the Admiralty on 19 April 1813. He wanted to retire and take up the pension Banks had promised some years earlier: Banks to Bligh, 15/3/1805, G. Mackaness, *The Life of Vice-Admiral William Bligh*, Sydney (1931), vol. II, pp. 96–97, Letter 97.

Goulburn sent Banks a copy of a letter he had sent to Bligh. In the letter to Bligh, Goulburn explained that a pension of £200 a year was to be proposed to parliament: Goulburn to Banks, 31/5/1813, S.L. Banks MS. SS 1:15 (enclosed with this was a copy of Goulburn to Bligh, 31/5/1813).

[2]Henry Bathurst (1762–1834), 3rd Earl Bathurst: Master of the Mint, 1804–1812; Foreign Secretary, 1809; President of the Board of Trade, 1807–1809; Secretary for War and the Colonies, 1812.

[3]Banks offered Bligh the governsorship of New South Wales in 1805, which Bligh accepted. Bligh was determined to follow instructions strictly once in post. In particular, he sought to curb the illegal trade in spirits, which was crippling the colony's economy. This brought him into conflict with some influential figures, such as John Macarthur. He also antagonized the New South Wales Corps, which had a monopoly on the importation and sale of strong liquor. On 26 January 1808 Bligh was arrested by Major George Johnston (1760–1823) of the New South Wales Corps, who assumed the title of Lieutenant-Governor.

Johnston was eventually recalled to England for this, and tried by courts martial at Chelsea Hospital. His trial commenced on 7 May 1811, and on 3 July he was found guilty of mutiny. He was therefore cashiered. Macarthur appeared as a witness for the defence, but fared no better than Johnston. He was barred from returning to his home and family in New South Wales for eight years. Bligh kept Banks informed of proceedings throughout.

[4]Robert Stewart FRS (1769–1822), 2nd Marquis of Londonderry and 1st Viscount Castlereagh: statesman; Secretary for War and the Colonies, 1805–1806 and 1807–1809; Foreign Secretary, 1812–1822.

1813

To Charlotte Seymour, 11th Duchess of Somerset

Soho Square
10 December 1813

I Return with Abundant Gratitude the Book your Grace was so good as to Send to me by the Duke. I trust your Grace Receivd 'The Curse',[1] in which Poor Pallas[2] is made to talk like a Naughty [child], as I Sent it Some Days ago. Had it not been for your Graces Kindness to me I Should not have seen it, or Received the amusement I did Receive from some of the well directed & severe applications of the Scourge with which it Abounds. I cannot, however, Easily forgive my Lord[3] for Finishing his Poem with a satyre on his Native Country, & for telling the world because we Bombarded Copenhagen that we have Lighted up Fires from the Baltic to the Mediterranean.[4] To exceed the bounds of Truth is not Good Even in a Politician; & to be blind to the humane character of his Countrymen is to mingle them in the Atrocities of the French nation; & to do away almost all Remembrance of the best Pride of human Nature is Surely illaudable.

The 'Bride of Abydos'[5] has Flights in it that Reachd the innermost recesses of my Feelings, but so many Careless Points in its Construction as to Put it, in my Judgement, in much hazard Should any Keen Critic choose to set his Teeth into it. The Frequent & Sudden Changes of metre are often too Evidently Causd by the dificulty of Restraining his words in the Compas prescribd by his Plan. I will touch his First Page Gently, but will not Torment your Grace with any other Part of my acidity.

the Cypress & myrtle — Are Emblems of Deeds. The one is the Emblem of Death, the other of Love, but neither of them are deeds.

The Cedar — No Cedar grows on the Banks of the Hellespont. This Cannot be Called the Land of the *Vine* where wine is Forbidden.

The beams Ever Shine — Sunbeams are the implements which the Sun uses in Shining. It is not our Legs that walk, tis we who walk with our Legs.

The Light wings of the Zephyr — My Lord must Furnish his wings with noses before they Can wax oppressed by Perfume. This surely is a Lamentable metaphor. For giving wings to winds my Lord has the Authority of Sternhold & Hopkins,[6] but it must be bad Taste to give Corporeal Limbs to an incorporeal Energy, for so I may almost Call a Zephyr.

Gul in her Bloom[7] — Sure it is bad Taste to Call a Rose a Gull, & merely because my Lord has read a Few words of Hafiz[8] to Compell the Reader to Refer to the notes to Find that Gul is Persian for a Rose, when the word Rose is as Capable of Personification, & as fitting to its Place in the Verse as Gul is.

The Citrus & olive are Fairest of Fruit — Are Either of these Fair in appearance or Fitting for the Desert? The names fit the Verse better, & are Less Familiar to the Ear than the Peach & the Apricot or melon.

The Voice of the *Nightingale* never is mute — This implies a Dayingale, if such there is in my Lords Elysium.[9] In Fact, between Nightingales & Bulbuls[10] my Lord has Left us in a wood.

The *Purple* of Ocean — I have Read of black Seas & Red Seas, and I have Seen Green Seas. I must Conclude the Purple Sea to be in consequence of its approach to the Black Sea.

End of Page 1, & of my Remarks.

Deprivd as I unfortunately am of the Power of Enquiring in Person after your Graces health, your Situation Continualy presents itself to my mind. Your Grace is Soon to undergo that Trial[11] which Providence has assignd to the weaker Sex for the benefit & advantage of the Stronger, who are permitted to Enjoy the Pleasures, & Exempted from Enduring the Pains necessary to the Continuance of the Human Race.

Inscrutable in this Case to the wisdom of man are the ways of Providence, but as in all Cases we are Permitted to understand they are the ways of Justice wisdom & Benevolence, we Cannot doubt that they are Equaly So in those we are yet Forbidden to Comprehend.

May the Gracious & divine Bounty which Sustains us all Grant to your Grace the inestimable Gift of Patience, which alone Can blunt the arrows of Pain, & deprive Torture of its Sting. May it also Grant to your Grace an Exemption from those Perils which Sometimes, but (thanks to God) not always, surround the Couch of Trava[i]ll. Till your Grace is Restord in Safety to your Relations and Friends, I Shall be incapable of Enjoying Pleasures unalloyd by anziety.

I have the honor to be,
with unfeignd Esteem & Regard,
your Graces most Hble
& most Faithfull Servant,

Jos: Banks.

[B.R.O. Bulstrode (Additional) D/RA/A/1B/4/3.]

[1] *The Curse of Minerva* (1811), by Lord Byron. Pallas appears to the speaker in the poem as he is admiring the Parthenon, her temple. Pallas complains about antiquarian collectors who damage the remains, and mentions Lord Elgin (1766–1841) in particular. Elgin employed

artists to make drawings of the sculptures and ruins at Athens. He then arranged for the Parthenon frieze to be conveyed to England, 1803–1812. Known as the 'Elgin marbles', the frieze was eventually sold to the nation amid controversy in 1816. Such collectors are cursed by Pallas in an exaggerated but amusing tirade. In the last section of the poem Pallas makes more serious references to the horror of war, and especially British naval attacks on Copenhagen.

[2]Of Greek legend, Pallas or Pallas Athena, goddess of war, and of many crafts and skills. She was a daughter of Zeus. The Romans identified her with Minerva.

[3]George Gordon Byron FRS (1788–1824), 6th Baron: poet. Byron gained an early reputation for atheism, radicalism and scandal. In 1810 he travelled across Europe to Greece, living the life of a romantic adventurer. On his return Byron published Cantos I and II of *Childe Harold's Pilgimage* (1812). They were an immediate success, and Byron hastily wrote other poems such as *The Bride of Abydos* (1813). His work featured the moody, haunted and glamorous Byronic hero. In 1816 Byron left England for the last time, but remained a controversial figure, both for his promiscuous behaviour and for his remarkable verse.

[4]In 1807 the British government was alarmed at the prospect of the Danish fleet falling into French hands. The Danes were under pressure from Napoleon Bonaparte to join the hostile alliance against Britain, and thereby tighten his embargo on trade, possibly by closing the Baltic to British shipping. When Britain's demand that the Danes surrender their fleet was refused, a squadron was dispatched. It landed an armed force, which attacked Copenhagen. Some 33 ships were taken, including the entire Danish battle line. Neutral Denmark joined France as a consequence, and Russia declared war on Britain as well.

[5]*The Bride of Abydos: A Turkish Tale* (1813) by Lord Byron. Byron took less than two weeks to write this poem. It featured the heroine, Zuleika, whose erotic remarks to Selim provoked criticism. However, by early 1814 some 125,000 copies had been sold. Banks wittily referred to the first thirteen lines of the poem.

[6]Thomas Sternhold (d.1549): translator; versifier of the Psalms with John Hopkins. John Hopkins (d.1570): translator; versifier of the Psalms with Thomas Sternhold.

In the mid-sixteenth century many believed that metrical versions of the psalms were closer to the Hebrew than those in prose, which could not be sung. Sternhold and Hopkins therefore set the Psalms to simple ballad measures, and their metre was widely used in religious worship. Numerous editions based on Sternhold's early work, *Certayne Psalmes chosen out of the Psalter of David, and Drawn into English metre...*, Edward Whitchurch (c.1549), were published, even into the nineteenth century.

[7]'Gul' is the first note given for this poem by Byron. It means 'rose' in Persian and Turkish. Byron's use of the root term here is indeed unusual. However, the image occurs frequently in *The Bride of Abydos* as a symbol of love and feminine beauty in keeping with the Eastern traditions which inspired Byron so much.

[8]Hafiz was one of the greatest lyric poets of fourteenth century Persia.

[9]Of Greek legend, Elysium was the abode of the blessed after death.

[10]'Bulbul' is Persian for nightingale. In Eastern legend the bulbul is generally associated with love and melodious song. Byron links it to the rose in his *Turkish Tales*. The attachment of the nightingale to the rose was well known in Persian fable.

[11]The Duchess was pregnant, and due to give birth.

Letter 122

1814

To Charlotte Seymour, 11th Duchess of Somerset

Soho Square
16 February 1814

Nothing but a Condescention unexampled in the annalls of Nobility Could have induced your Grace to ask that as a Favor of me which must Ever Constitute my Cheeifest Pride. To Call a Seymour my Godson is an honor I Could never have aspird to, or Ever hopd to Attain. To your Graces Boundless Favor alone I am indebted for it. I was disappointed. I Felt the disappointment Severely. Your Graces Goodness alone was Able to Replace me on the Proud Eminence I had before almost attaind, & to have Receivd Such a Favor from your Grace must be to me a Subject of Perpetual Gratitude.

I would not for the world be Absent at a Ceremony so interesting to me as that of the Baptism of this Precious infant.[1] Whenever your Grace Fixes the day I Shall attend it.

I have the honor to be,
with inexpressible Gratitude &
with due Respect,

Your Graces Most Hble
& most Faithfull Servt,

Jos: Banks.

[B.R.O. Bulstrode (Additional) D/RA/A/1B/4/5.]

[1]Algernon Percy Banks Seymour (1813–1894), 14th Duke of Somerset. Banks's godson was born on 22 December 1813

1814

To William Townsend Aiton

Soho Square
7 June 1814

Dear Sir,

Among the innumerable indulgencies which I have for a long time enjoyed, derived from the Gracious kindness of our beloved & afflicted Monarch, the connection[1] I have been permitted to form with the Royal Gardens at Kew is among those the most grateful to my feelings; & I beg you to be assured that as long as I shall be permitted to continue it, I shall cherish & improve it to the best of my power.

Among other indulgencies allowed to me I was permitted, as you, Sir, know, to point out destinations, & to draw instructions for those persons whom you recommended as properly qualified to travel as Collectors for the R[oyal] Botanical Gardens. I think I may venture to affirm that, until that arrangement was interrupted by the almost impossibility of sending home living Plants in Ships liable to the detention of waiting for Convoy, His Majesty's Botanic Gardens stood unrivaled in the whole of Europe for the extent of its Collections as well as for the Beauty & interest of the Plants it consisted of.

The arrival of the definitive Treaty with France,[2] & the Certainty that before any Collections can be sent home Ships will sail as they were used to do, without being subjected to any uncertain delays, makes me anxious to see the establishment of Foreign Collections resumed; & the more So as the Emperor of Austria,[3] who has formerly freighted Ships at an immense expence, & sent well educated Botanists to collect for his Garden at Schönbrun[n],[4] the only rival to Kew that I am acquainted with, will, no doubt, resume the business of improving it.

The Climate best suited for Collecting is, as you know, the Southern Temperate Zone; & in that part of the World no places are so productive as the Cape of Good Hope & New South Wales. If H.R.H. the Prince Regent should permit you to engage Two Collectors,[5] these are the places to which I should wish them to be sent. The Plants of both these Countries are beautiful in the extreme, and are easily managed as they suit the Conservatory & have no occasion for the unnatural heat required by the intertropical vegetables. I should wish also to send a man to Buenos Ayres, but at present, & until Spain has reconquered her disobedient Colonies, this cannot be thought of.[6]

Should you be allowed to send to the Cape & to New South Wales, I have no doubt of being able to give such instructions to the Governors of those Places as will enable H.R.H. Collectors to visit at a very easy charge Countries that are hitherto unexplored, & that will add Riches to the Royal Collection beyond the Sanguine expectations of those who have had less experience in the produce of those Countries than I have.

[N.H.M. B.L. D.T.C. XIX. 40–41.]

[1]Banks had been unofficial Director of the Royal Botanic Gardens at Kew from 1773. Under his direction the gardens had flourished. He had been able to obtain much royal support from George III, and had used his considerable influence to obtain seeds and plants from around the world.

[2]The first Treaty of Paris was signed on 30 May 1814, following Napoleon's defeat by the armies of the Grand Alliance, and his removal to the island realm of Elba. However, he returned to France on 1 March 1815, and was received with much popular enthusiasm in Paris.

[3]Francis I (1792–1835): he was Holy Roman Emperor until 1806. From 1804 he was Francis I, Emperor of Austria.

[4]The Imperial Garden was attached to the splendid palace at Schönbrunn, Vienna.

[5]Banks wrote a detailed report on the state of the Botanic Gardens, which he sent to George Harrison (d.1841), Assistant Secretary to the Treasury: Banks to Harrison, 1/9/1814, B.L. Add. MS. 56299, ff. 41–45v, N.H.M. B.L. D.T.C. XIX 56–63. In it, he recommended James Bowie (c.1789–1869) and Allan Cunningham (1791–1839) as collectors who ought to be sent to the Cape of Good Hope and New South Wales. His plan was accepted shortly afterwards: Harrison to Banks, 13/9/1814, *Bulletin of Miscellaneous Information*, Royal Botanic Gardens, Kew (1891), p. 308. With other arrangements for the mission completed, the two men departed on 3 October in HMS *Duncan*. Their destinations were Rio de Janeiro, the Cape of Good Hope and then New South Wales.

[6]Napoleon Bonaparte's subjection of Spain and Portugal in 1810 greatly weakened their control over their colonies in South America. Although the French were eventually forced out of both countries, in South America nationalists started to rebel against their European masters. The struggle continued with popular figures such as Simon Bolivar (1783–1830) and José de San Martin (1778–1850) leading it. In July 1816, Buenos Aires declared independence.

Letter 124

1815

To Charlotte Seymour, 11th Duchess of Somerset

Soho Square
27 April 1815

My Old heart is Certainly as active as Ever. It has now been at work more than 72 years Day & night without Ceasing, & will Continue to work on without one interval of Rest as Long as I Continue in this Transitory world. A Proof to me that Sleep is not given to mortals as a necessary Refreshment, is their bodies might have been Created of the same indefatigable Stuff as their hearts, had not Providence found it advisable to Restrain the Natural Enmity of man to man into one half of the time he has to Carry it on. Bonaparte,[1] who Extends, it is Said, this one half into $\frac{3}{4}$, & Sleeps only one Quarter of his Life, is more mischevous than ordinary men.

Is the heart the Seat of Love? I doubt it. The Islanders of the South Sea Say that the Soft Passions originate in the intestines. Dr. Gell says in the Back of the head. I beleive it to belong to the Nervous System, & that the heart is wholly destind to be the Seat of Friendship. If it was otherwise, Love would have Remaind in me as unchangd as my heart is. Friendship, however, Remains in Full & undiminishd Force, & it Felt, I can assure your Grace, more Satisfaction than Pen Can Express at the Contents of the Last Letter I Recd from Bradley.[2]

I had yesterday the Great Pleasure & honor of a visit from the Dear Ld. Seymour,[3] who Called upon me in his way to School. He gave me a Prosperous account of his Little Relatives at Bulstrode, & Flatterd me much by a kind message from Lady Jane.[4] He did not seem so well Pleasd with the Return to discipline as I could wish him to be. Unless the native wildness of man is broken down into Civilizations by the discipline of a School, the Fathers hand finds it dificult to Restrain it at a later Period.

Your Graces Ever Devoted,

JB.

[B.R.O. Bulstrode (Additional) D/RA/A/1B/4/15.]

[1]Napoleon Bonaparte.
[2]Maiden Bradley, a family seat at Warminster, Wiltshire.
[3]Edward Adolphus Seymour (1804–1885), 12th Duke of Somerset.
[4]Lady Charlotte Jane Seymour.

Letter 125

1815

To Sir Humphry Davy F.R.S.

[Revesby Abbey]
30 October 1815

My dear Sir Humphrey,

Many thanks for your Kind Letter, which has given me unspeakable Pleasure. Much as by the more brilliant discoveries you have made The Reputation of the R.S. has been exalted in the Opinion of the Scientific world, I am of Opinion that the Solid & Effective Reputation of that Body will be more advancd among our Contemporaries of all Ranks by your Present discovery than it has been by all the Rest.[1]

To have Come forward when Calld upon because no one Else Could discover the means of defending Society from a Tremendous Scourge of humanity, & to have by the application of Enlightened Philosophy found the means of Providing a Certain Precautionary measure Effectual to guard mankind for the future against this alarming & increasing Evil, Cannot fail to Recommend the discoverer to much Public Gratitude, & to place the Royal Society in a more Popular Point of view than all the abstruse discoveries beyond the understanding of unlearned People Could do. I Shall most Certainly direct your Paper to be Read at the very first day of our meeting.[2]

We Should have been happy to have seen you here, but I am Still happier in the Recollection of the Excellent Fruit which was Ripend & Perfected by the very means of my disappointment,[3] your early return to London. I trust I shall arrive there on the morning of our first meeting, & that it will not be long after before I have the Pleasure of Seeing you.

My dear Sir H.,
always faithfully yours,

Jos: Banks.

[R.I. MS. 27D/2.]

[1]Davy had been working on the problem of explosions in deep mines caused by fire-damp. This was being ignited by miners' lamps. Some attempts had been made to put a stop to the rapidly increasing number of underground disasters, but without complete success.

In 1815 Davy was asked to find a solution. He visited mines in Newcastle, obtaining samples of fire-damp for analysis in the chemistry laboratory. Together with Michael Faraday (1791–1867), he confirmed that the main constituent of the gas involved was light carburetted hydrogen (methane). Davy looked at the explosive characteristics of the gas when mixed with air, and the degree of heat that ignited it.

The breakthough came when he discovered that apertures of less than one eighth of an inch prevented flames from igniting fire-damp. He also found that flames passed more easily through glass tubes than through metal ones, and that metal troughs and slots of less than one-seventh of an inch diameter would stop ignition. Fine wire gauze had the same effect. Armed with such information Davy was in a position to design a new lamp that would not cause explosions in mines, a safety lamp.

[2]Davy's work was initially described in a series of papers to the Royal Society: *Philosophical Transactions*, vol. 106 (1816), pp. 1–22, 'On the fire-damp of coal mines, and on methods of lighting the mines so as to prevent its explosion. By Sir H. Davy, LL.D. V.P.R.I.' Read 9/11/1815; *Philosophical Transactions*, vol. 106 (1816), pp. 23–24, 'An account of an invention for giving light in explosive mixtures of fire-damp in coal mines, by consuming the fire-damp. By Sir H. Davy, LL.D. F.R.S. V.P.R.I.' Read 11/1/1816; *Philosophical Transactions*, vol. 106 (1816), pp. 115–119, 'Farther experiments on the combustion of explosive mixtures confined by wire-gauze, with some observations on flame. By Sir H. Davy, LL.D. F.R.S. V.P.R.I.' Read 25/1/1816.

In effect, Davy had found a way of protecting the coal supplies that would fuel industrial development throughout the nineteenth century, and was awarded the Society's Rumford Medal. A further paper followed: *Philosophical Transactions*, vol. 107 (1817), pp. 77–85, 'Some new experiments and observations on the combustion of gaseous mixtures, with an account of a method of preserving a contained light in mixtures of inflammable gases and air without flame. By Sir Humphry Davy, LL.D. F.R.S. V.P.R.I.' Read on 23/1/1817. Davy also published his findings in *On the Safety Lamp for Coal Mines with Some Researches on Flame*, London (1818).

[3]Davy declined a visit to Banks at Revesby Abbey in his haste to reach London to start his laboratory research.

xvii. **Sir Humphry Davy F.R.S.**
From the portrait by Sir Thomas Lawrence. The Royal Society, London.

1816

To Thomas Andrew Knight F.R.S.

Soho Square
23 January 1816

My dear Sir,

I have delayed an answer to your favor[1] too long in the full expectation of receiving your letter with the name & address of your correspondent, which I concluded would somehow reach my hands as I can without the least trouble procure it to be directed in German, which is always the best way as the German postmasters are not so good linguists as the French & Dutch, with whom English addresses pass very well. Be so good, then, as to send it with the name & address, & I will do all the rest.

I shall say but little on the subject of equivocal generation when opposed to your Brother,[2] whose strong mind has so much exhausted itself on Greek &c. as to leave little force left for the explanation of Natural Phenomena. For the Louse & woodcock argument, be good enough to tell him that nature has provided for every species of animal one or more especial lice, who fatten upon the Carcase they are destined to inhabit, but cannot exist in any other animal or by any other food.[3]

Thus, the noble creature, man, is the destined prey of the head Louse, the body Louse & the crab Louse, neither of which can exist in any other situation than on the human body. Of course, as man was the last work of Creation, he must have maintained all these three animals untill he had a wife who might release him from supporting one or two of them; but till Abel, the younger brother of Cain, was born, there were not more men than lice destined to feed upon them.

But enough of this nonsence. Untill an actual experiment has taught us that an animal can proceed from another without having been created or begotten, what inducement can we have for believing that possible from abstract reasoning which appears impossible from actual experiment? Carlisle[4] has not entered my house since the Commee of Papers of the R.S. refused to print a paper of his, &, I am told, has declared that he never will. I hear that he is employed in Hatching a Publication in which countenance will be given to those equivocal Doctrines, but I do not hear of one experiment he has to produce in favour of his doctrine.

The by-Law party of the Horticultural have triumphed, & Salisbury, who has been defeated, threatens to withdraw.[5] I dread the consequence, because the Lawgivers look up to R. Wilbraham[6] as their leader, who cannot know any thing of Horticulture, & who never was famed for being very nice in his general Conduct.

Adieu my dear Sir. May we when we meet again See better times for farming & for Horticulture.

Believe me always,
Truly & faithfully yours,

Jos: Banks.

[N.H.M. B.L. D.T.C. XIX. 233–234.]

[1]The last letter Banks received from Knight appears to have been: Knight to Banks, 29/12/1815, N.H.M. B.L. D.T.C. XIX 228–230. This was answered: Banks to Knight, 15/1/1816, H.R.O. T74/ 583/ 31.

[2]Richard Payne Knight, the numismatist and virtuoso. He was resident at 3 Soho Square from 1808 to 1824.

[3]Lice are highly host-specific, and each species of human louse specializes in infesting particular parts of the body. The lice (*Anoplura*) Banks has in mind are *Pediculus capitis*, *Pediculus humanus* and *Pthirus pubis*. The body louse is also known as the clothing louse because it lays its eggs on clothing and visits the body to feed.

[4]Sir Anthony Carlisle FRS (1768–1840): surgeon. He was resident at 12 Soho Square from 1800 to 1819. Carlisle's 'Paper' was not published in the *Philosophical Transactions*, and there is no record of it in the Royal Society archives. Of this, Banks said: 'I do not think I have been misinformed about Carlisle's pique against me. He certainly has not darkened my door since the Committee of Papers of the R.S. declined to print his work. Why I should endure personal punishment for a sin in which 12 or 14 persons participated, I do not comprehend. It proves, however, that Carlisle has a potatoe in his head, which he uses occasionaly instead of his Brains', Banks to Knight, 16/2/1816, N.H.M. B.L. D.T.C. XIX 246–248.

[5]Richard Anthony Salisbury resigned his office as Secretary of the Royal Horticultural Society in 1816 after a committee investigating the affairs of the Society produced a highly critical report of the accounts. This was issued in March 1816. Although Salisbury was not explicitly referred to, his position was untenable. Joseph Sabine FRS (1770–1837), who had been an Inspector-General of Assessed Taxes, prepared much of the report. Sabine was appointed Secretary in Salisbury's place.

[6]Roger Wilbraham FRS (1743–1829): politician; scholar.

Letter 127

1816

To Thomas Andrew Knight F.R.S.

Soho Square
23 February 1816

My dear Sir,

It gives me pain to inform you that Dr. Phillip[1] was last night rejected at the Royal Society. I had not an idea of any such intention on the part of the members present, & never felt more surprised than I did in drawing the Ballot to find the negative drawer, in which I did not expect to find 4 balls, almost full. The members were 56 in favor & 32 negatives, but our Statutes require [a] full $\frac{2}{3}$ of the Votes present to make an Election. It required, therefore, 64 balls to elect him, & this notwithstanding the Austrian Archdukes were present as fellows, & no doubt voted for the Candidate, as did those fellows who attended their imperial highnesses, who would not otherwise have been at the meeting.

The cause of this has been the stile of his paper on the experiment tried on Rabbits by severing the Nerves on which digestion &c. depend. These cruel experiments were detailed with all the particulars of the tortures experienced by the animals, exactly as if the audience were to be composed of none but Butchers & Surgeons. I was in my bed when the paper was read, but the Vice president[2] in the chair told me the next day that the disgust expressed by the audience was marked by more than half of them leaving the room in Succession one at a time. The paper was not concluded, but at the next meeting we suppressed the continuation of the Experiments & read only the results.

I am sorry for the event, though it is a victory of humanity over Science. Our comparative anatomists here are carefull in the extreme to pass over in their descriptions all that possibly can be avoided of the effects of their knives on the Sensations of their Victims, but I suspect that this check will suspend for a time at least their Communications to the Royal Society. In fact, these matters are better Suited to Medical & anatomical publications than to the audience, mixd as it certainly is, of the Royal Society.

I have told you before that the Doctor's Deduction is held in extreme doubt by our operators. Of course, I conclude his paper will not be printed in the Phil[osophical] Trans[actions]. He will, however, of course print it in some other way to vindicate his opinion, & controvert those of our Physiologists.[3]

I am sorry we have lost so good a Philosopher as Dr. Phillips certainly is as an associate, but I trust that we shall not be deprived of his future endeavours to Scrutinise the secrets of nature.[4] If you write to him, I pray you to assure him that I grieve at his disappointment, but that the Conduct of my too humane Colleagues has not abated an atom of the esteem in which I have hitherto held him. There seems to be a Spirit rising against Experiments tried upon living animals. I was astonished to hear the other day that Carlisle[5] in his lecture to the Royal Academy had introduced rather abruptly an attack upon these experiments, Concluded by an Assertion that anatomists ought to confine their researches to the Bodies of Dead animals.

Adieu, my dear Sir,
always faithfully yours,

Jos: Banks.

[N.H.M. B.L. D.T.C. XIX. 249–251.]

[1]Alexander Philip Wilson Philip FRS (c.1770–1851): physician; physiologist.
[2]Banks had been suffering from gout. The Vice-President was Sir Everard Home.
[3]Philip was upset at being rejected when he stood for fellowship of the Royal Society, and wanted to publish his work along with an attack on the Society. It was the beginning of an interesting dispute which rumbled on for years: Banks to Knight, 1/3/1816, N.H.M. B.L. D.T.C. XIX 252–254; Banks to Knight, 16/3/1816, N.H.M. B.L. D.T.C. XIX 256–257; Banks to Knight, 23/3/1816, N.H.M. B.L. D.T.C. XIX 258–260.
[4]Philip was rejected again in 1818, and complained about the Royal Society in the second edition of his *Experimental Inquiry into the Laws of the Vital Function, with some observations on the nature and treatment of internal disease*, London (1818). Banks continued to support Philip as a candidate for Royal Society fellowship throughout, but regretted the physician's angry public comments: Banks to Knight, 20/6/1818, N.H.M. B.L. D.T.C. XX 99–101; Banks to Knight, 30/6/1818, N.H.M. B.L. D.T.C. XX 102–104; Banks to Knight, 18/1/1819, H.R.O. T74/ 583/ 33; Banks to Knight, 25/1/1819, H.R.O. T74/583/34; Banks to Knight, 18/8/ 1819, N.H.M. B.L. D.T.C. XX 202–204. Philip was eventually elected a Fellow of the Royal Society in 1826.
[5]Sir Anthony Carlisle.

Letter 128

1816

To William Körte

[Soho Square]
[4 October] 1816[1]

Sir,

There are, as you State in your Letter, in the British Museum Two Volumes of the Drawings of Maria Sybylla Merian,[2] which Cost Sir Hans Sloane,[3] as Tradition informs us, for Each Drawing five Guineas. She was Certainly in her Time an Excellent Paintress,[4] but as The modern improvements in the Science of Botany were wholly unknown to her, She has of a necessity Omitted to Represent Those parts of Flowers which were uninteresting in her Time, but are now Essentialy necessary to make a drawing usefull to a Botanist.

For this Reason, Sir, I feel no inclination to become a Purchaser of the Drawings you offer to me. As, however, their merit to an amateur of Printing who is not a Botanist Cannot be all diminishd by the omissions I have Stated Above, I think I may hope that you will find a Purchaser at the moderate value you have set upon them. If I meet with any Person of that description likely to Purchase them, I will not Fail to acquaint him in what manner to Proceed in order to Obtain the Refusal of them.

I am, Sir,
your most Obedient Servt,

Jos: Banks.

[H.C. This letter is inserted in the extra-illustrated edition of: *Letters of Samuel Johnson*, G.B. Hill (Ed.), London (1892), vol III., facing p.188.]

[1]The date here is when the letter was received.
[2]Maria Sybille Merian (1647–1717): Swiss painter; entomologist. She was among the first to describe, with illustrations, the life-histories of certain butterflies.
[3]Sir Hans Sloane PRS (1660–1753), 1st Baronet: collector; scientist; antiquary; Secretary of the Royal Society, 1693–1712; President of the Royal Society, 1727–1741.
[4]Some of the most important drawings by Merian were made when she journeyed to Surinam in 1699. She stayed in South America until 1701 studying insects, and produced very impressive compositions of plants and insects using watercolours and body-colours on vellum.

This work was published as part of *Metamorphosis insectorum Surinamensium...*, Amsterdam (1705). Two volumes held at the Department of Prints and Drawings in the British Museum are simply inscribed 'Merian's Drawings of Surinam insects, &c', MSS. Sloane 5275–6.

Letter 129

1816

To Davies Gilbert F.R.S.

Spring Grove, Near Hounslow
16 October 1816

My dear Sir,

Many thanks for your Letter, & the very ingenious hint it Contains Relative to our operations on the Leng[t]h of the Pendulum.[1] It Requires a Person versd in Studies I have never Pursued to ascertain the Correctness of the Idea. To me, I Confess, it appears unquestionable, & it gives me the Pleasing hope that we shall again Triumph over our Rivals in Science as we have done in arms.

M. Arrago,[2] who was here not long ago with Gay Lussac,[3] has Convinced himself Satisfactorily of the Superiority of our Tools, as well as the manager of them, over those of the Paris Observatory, & has most Liberaly Reported so much in their favor to his countreymen that an order has been sent from France for a Circle by Troughton[4] in all Respects Similar to that of Greenwich.[5] In the Article of Flint Glass, however, we have Sacraficed to the arbitrary and Clumsey methods of Collecting the Duty our Former Superiority. Reichenbach[6] has made an achromatic Object Glass 7 $\frac{1}{2}$ inches aperture without a shade of Color or a blemish of any kind. Our ministers ought to know this, & devise Some method of allowing us to Retrieve the Superiority we have been deprived of.

I greive over the depressd State of the mining interest which, like Every other branch of Commerce, Sinks under the Poverty of all Europe caused by the Rapine of Bonaparte.[7] That a wretch who has Entaild Such a mass of misery upon a Larger Extent of this Globe & a greater Number of his Fellow men than any Former Tyrant has ever held in Subjection Should be Sufferd to Live, while his dishonorable & dishonest Conduct is Felt So far & So wide, is indeed an instance of honorable Feelings haveing Gone mad. Could not an act of Parliament passd here, & a decree of the French Representative body Constitute a /Jurisdiction & a/ Judicature on board an English & a French Ship by which he might be Tried & Executed? It would yeild Some Consolation to those who Still Suffer, & will Suffer a long time from the Consequences of his Crimes, & a Solace to the Souls of the multitudes whose blood he has Spilld, that his Execrable Carcase was Sunk in the bottomless ocean, & Establish an Example of Terror to Future Tyrants.

I wish I could agree with you that a Coinage of Copper Could at this time be Tolerated.[8] Copper Tokens at present do, & will for Some months Longer, Supply the place of the Silver Sixpences. When the new Coinage of Silver Appears I conclude it more likely that a diminuition of the Copper Currency should be found necessary than that any increase Should be Tolerated by the Public. In any case, I Fear that 1000 Tons of Copper Removed in [any] way from the Stock in hand would be a very insufficient Remedy, & it would do nothing for Tin, which is Still more in want of assistance.

Adieu, my dear Sir,
Most Faithfully Yours,

Jos: Banks.

I am Still in bed most part of the day, but I hope to be able to do my Duty for one year more at Least. Excuse me for Troubling you with another Letter for Mr. Holt.

[MS. Gunther 14/17; B.L. Add. MS. 56301(1), ff. 38–39.]

[1]The length of a pendulum vibrating seconds, which both Banks and Davies Giddy worked on for some time. Fellows of the Royal Society had been considering a uniform system of weights and measures since at least 1742. Although committees were appointed from both the Royal Society and the House of Commons to investigate the subject, no firm conclusions had been reached. In 1814 Gilbert (as Giddy was known after 1808) was one of 23 Members of Parliament included on a committee which failed to settle the matter.

However, the next year Gilbert and two others were asked to prepare a bill for establishing and preserving a standard system of weights and measures, and in 1816 the Government formally requested the Royal Society Council to assemble a committee. Its purpose was to assist in 'ascertaining the length of the pendulum vibrating seconds of time in the latitude of London, as compared with the standard Measure in possession of this House; and for determining the variation in length of the said pendulum at the principal stations of the Trigonometrical Survey extended through Great Britain, and also for comparing the said Standard Measure with the ten-millionth part of the Quadrant of the meridian, now used as the basis of linear measure on the Continent of Europe.' House of Commons, 15 March 1816.

Banks was part of this committee, as was Gilbert, and Henry Kater FRS (1777–1835) undertook the observations. The results were not decisive, General Roy's Scale being recommended as the Standard Yard, and later dropped in favour of the Parliamentary Standard. Banks chaired a Commission of Inquiry in 1819, which met at Soho Square to report on the pendulum experiments made by Kater. Gilbert was one of the commissioners. The Commission continued to produce reports after Banks's death. Thus, he did not live to see standards of length, weight and geodetic measurement finally determined, despite the efforts of some of the best scientists in Britain and France.

[2]Dominique François Jean Arago FRS (1786–1853): French astronomer; physicist. He was a republican, who went to Spain in 1806 for the Bureau des Longitudes with his countryman, the astronomer and mathematician Jean Baptiste Biot FRS (1774–1862).

In 1816 Arago was in London making arrangements for the northern extension of the west European arc of meridian, to be made by Biot with William Mudge FRS (1762–1820) and Thomas Colby FRS (1784–1852) of the Ordnance Survey in 1817.

[3]Joseph Louis Gay-Lussac FRS (1778–1850): French chemist; physicist. Gay-Lussac made a balloon ascent with Biot, 24 August 1804, to determine whether the magnetic intensity of the Earth decreased at great altitudes. They found that there was no change up to 4,000 metres.

[4]Edward Troughton FRS (1756–1835): London instrument maker. He supplied a transit circle to the Observatory at Greenwich in 1812, along with other instruments. Troughton had experience in making brass scales for measuring the standard yard. He was also one of the first to use the micrometer microscope for measuring and laying down distances on a straight line.

[5]The Royal Observatory, Greenwich.

[6]Georg Friedrich von Reichenbach (1771–1826): Bavarian artillery officer. He set up a firm in Munich producing precision instruments. With his partners he made theodolites and transit circles of great rigidity and accuracy.

This was not the first time Banks complained about the scientific progress Reichenbach had made while taxes prevented the British from competing in the manufacture of flint glass for telescopes: Banks to Blagden, 22/12/1814, R.S. BLA. B 56, 'I have just now Learnd from a Letter of Hardings to Dr. Tiarks in my Library that Reichenbach has Retrievd the art of making Flint Glass, driven from This Countrey by the Severity of the Laws for Collecting the Tax upon Glass. He has Finishd an Object glass of 8. inches, & is now at work on one of Twelve inches. The Telescope with this Object glass he values at £600. He makes also small instruments for seeking after Comets & Planets, which we Call Sweepers, for £10. These Harding Says are incapable of improvement. The aperture is 4 inches.' Although British optical instrument makers were recognized as the best in the world in the eighteenth century, Reichenbach and his associates were able to surpass them.

[7]Napoleon Bonaparte was defeated at the Battle of Waterloo on 18 June 1815. He abdicated for a second time on 22 June, and surrendered on 15 July to the commander of the British ship HMS *Bellerophon* in the Breton port of Rochefort. In October Napoleon was taken to St. Helena, where he remained until 1821 when he died.

[8]Banks was appointed to the Coin Committee when he became a Privy Councillor in 1797: Letter 66. In the letter above Banks expressed his opinion on old silver coinage, especially sixpences. There was a shortage in circulation, and what remained had become so worn that some silver coins were being refused. Gilbert proposed an issue of copper coins as a way of supplying the want of small change and of diminishing copper stocks. However, a new issue of £2.5 million of silver coins was planned for February 1817. Although copper could be used in place of the lower denomination silver coins, Banks judged that less not more copper would eventually be needed. He was correct.

Letter 130

1817

To William Scoresby F.R.S.

Soho Square
22 September 1817

Dear Sir,

I have Read in the Liverpool Paper that you have this year Seen the Coast of West Greenland free from ice, & have Sailed along it for a Considerable distance.

You will Oblige me much if you will Give me a Letter Stating Such particulars as you have Observed Relative to the decrease of the Polar Ice;[1] a matter in my judgement of Great importance to the Prosperity of this Countrey if, as I conceive to be the Case, the Frosty Springs & Chilly Summers we have been Subject to for many years Past, so much so that it is now 16 or 17 years Since we have had a full Crop of apples for Cyder, are Caused by the increase of Ice which seems to have accumulated for Many years past.

It is Said that Islands of Ice have been This Summer Seen in far Greater Numbers than usual, & the Floods in all the mountainous parts of Europe, of which our Newspapers have given us Continued accounts, seem to Prove a diminuition of Cold in the upper Regions of the air.

I hope your Good Father[2] is alive & well. Lady Banks has not Forgot his Obliging Present to her of White Bear Skins, which She uses in the winter to her great Comfort.

An answer to this directed to me at Revesby Abbey near Boston will be Thankfully Receivd by,

your Faithfull Hble Servt,

Jos: Banks.

[W.M. < 2 >.]

[1] Scoresby replied in a long and fascinating letter describing the disappearance of some 2000 square leagues of Greenland ice, which he believed had drifted into the warmer south and melted. He stressed his interest in discovery over fishing, and suggested undertaking a survey of East Greenland or Spitzbergen. Scoresby also argued that exploration of Baffin Bay was needed, suggesting that 'it is simply an island or an assemblage of islands'. Scoresby

to Banks, 2/10/1817, U.Y. Beinecke Banks MS., N.H.M. B.L. D.T.C. XX 42–47. Banks shortly afterwards advised the Admiralty on the need for an Arctic expedition with the North-West Passage as one objective, and in 1818 two naval missions were mounted. See Letter 132.
[2]William Scoresby Snr. (1760–1829): whaler captain; Arctic navigator.

1817

To Thomas Andrew Knight F.R.S.

Soho Square
26 September 1817

My dear Sir,

The sight of your hand gave me great pleasure. I well knew that your Silence was owing to the employment you find for yourself in the pursuits that most interest me, but I looked forward to the moment when the recollection of your Absent Friends would call upon you to remind them of your Employment.

I am, I thank God, in a much better state of health than I have been. I have till now resided at Spring Grove, where I have been much contented with my new Gardiner,[1] who is [a] master-man in his Profession. He sometimes grins & brags by saying, 'I think, Sir, I may say that you have more stuff in your two acres of Land than all there is in Kew Gardens.' I forgive his exaggeration, as he seldom indeed brags; but, in fact, no market man in our neighbourhood crops &c. so close as he does, & this they all allow.

I have introduced a steam pipe among my houses. It is at present in its infancy, but must become the Main Spring of all forcing houses as it puts the degree of heat & its duration more into the power of the gardiner than any other mode can do. Tomorrow I set out for Lincolnshire,[2] where I shall this year spend one month.

The newspapers during the whole summer & autumn have been full of inundations in all Countries where Rivers are fed from Icy mountains, & have told us that the Coast of West Greenland, which has not been free from Ice for 150 years, has been this year seen & found clear of all obstruction for many leagues. The Atlantic has been unusually clogged with Islands of Ice.

Possibly I am too sanguine, but as I have always attributed the increasing Coldness of our Climate to the increase of Polar Ice, I feel a hope that we shall be indulged with better Springs than have lately been provided for us. It is now 16 years, I am told, since the Cyder harvest was too large for the Casks provided to contain it. In France there has been no good Vintage since the Comet year, & they are now all drinking Comet-wine.

If Galvanism & Electricity[3] are the same thing, many have been the attempts to apply the fluid to the advantage of Vegetation. Those made by men of no literary credit have succeeded to admiration, but, when repeated by cool headed performers, have intirely failed so that Electricity has been set aside from gardening for many years.[4]

I admire your sagacity in judging that the Peach is a child of the Almond, & have now no doubt of the fact. The French state in their gardining books that some almonds have Pulp like that of a Peach, but not of a good Flavor, but these blockheads never thought of drawing the Conclusion from the fact that you have now verified.

Your Currants are most interesting. I fear that, like almonds, they will seldom produce Peaches, but surely the attempt is worth all the Pains that can be bestowed upon it. My Gardiner has brought Currants from Russia, which, as he tells me, bear bunches of fruit as large as those of the cluster grape, & like those branched. Each berry [is] as large as those of the cluster Grape. Their misfortune is that they blossom earlier than ours, & are therefore more subject to early Frosts. They flowered this Spring, but every blossom was destroyed.[5]

I beg to state to your son[6] for his information the result of many adventures of mine in storming wasps nests. I always choose the day for the business, & strong sunshine I consider as the best.

My weapon is a Squib or Serpent of Gunpowder. I cut a Turf, &, having lighted the fuse of my Squib, push it into the hole of the wasps' nest, & instantly cover it with turf, & press it down with my foot. I soon hear the explosion of the firework, after which I keep down the turf for one or two minutes. I am then certain that every inhabitant of the nest is so stupified as to allow me to dig without interruption.

While digging the labourers will be surrounded by hundreds of working wasps, but not one soldier wasp, for those never go from home, & not one of the working wasps will sting unless he is ill used. Thus, in a few minutes the nest is dug out, the brood & all the wasps in the nest crushed & killed, & the hole filled up. The wasps who were abroad will haunt the Place for some days, & seldom leave it till they die for want of shelter.

The Community of wasps consists of two Trades — Soldiers & Labourers. The Soldiers always remain as sentinels near the mouth of the hole, but, within it, they are a 'genus irritabili'. [They] attack any one who disturbs them, & pursue some hundred yards, descending rapidly from the air, & striking forcibly on the object they attack. These, I apprehend, are the vindictive wasps your Son has met with, but I have never seen them in separate communities.[7]

Always faithfully yours,

Jos: Banks.

[N.H.M. B.L. D.T.C. XX. 38–41.]

[1]Isaac Oldacre (Oldaker), (1772–1852): gardener. Oldacre was employed at Spring Grove from a period late in 1813 or early in 1814. He had previously been in the service of the Emperor of Russia. January 1816 saw him succeed Thomas Fairbairn (bailiff and head

gardener at Spring Grove, 1803–1814) as head gardener. This was a position he held until 1828.

From the middle of the 1780s Spring Grove was increasingly a center for Banks's interest in plant and animal breeding. Crossing sheep, farming and gardening, cultivating exotics under glass were all carefully pursued at his Middlesex retreat.

[2]Banks inherited family estates situated in Lincolnshire, and visited his country home, Revesby Abbey, each September and October, whenever he was well enough to make the journey from London.

[3]'Galvanism' comes from the work of Luigi Galvani (1737–1798) who investigated current electricity. He was an Italian anatomist, and his most important research was on the electrical stimulation of nerves and muscles in frogs. Galvani supposed that electricity residing in the muscles and tissue caused movement when dissected frog legs were placed in contact with iron. He found that contractions occurred even when there was no electrical storm.

In fact, the source of electricity in Galvani's experiments had come from the two different metals used, and contact with the animal's body fluids. Nevertheless, current electricity was called Galvanic electricity for many years afterwards.

[4]This assessment was correct. The Committee of Papers at the Royal Society rejected accounts of the effect of galvanism on plants: Banks to Knight, 14/11/1817, H.R.O. T74/ 583/32; Banks to Knight, 16/3/1818, N.H.M. B.L. D.T.C. XX 86–87; Banks to Knight, 20/6/1818, N.H.M. B.L. D.T.C. XX 99–101.

[5]Knight published two papers related to these subjects: *Transactions of the Horticultural Society of London*, vol. III (1818), pp. 1–5, 'An Account of a Peach Tree, Produced from the Seed of the Almond Tree. In a Letter to the Secretary, from Thomas Andrew Knight. Esq. F.R.S. F.L.S. &c. President.' Read 7/10/1817; on currants in the same volume, pp. 86–90, 'On the Variation of the Red Currant (Ribes rubrum) when propogated by seed. By Thomas Andrew Knight, Esq. F.R.S. &c. President.' Read 3/2/1818.

[6]Andrew Knight (1795–1827). He was killed in a tragic shooting accident in woods near Downton Castle.

[7]There is no soldier caste in British social wasp colonies. Workers will act as soldiers when their colony is attacked, but they are all identical in appearance. They frequently spend time foraging away from the nest entrance.

Letter 132

1817

To Robert Saunders Dundas F.R.S., 2nd Viscount Melville, 1st Lord of the Admiralty

The Royal Society
20 November 1817

My Lord,

It will without doubt have come to your Lordship's knowledge that a considerable change of climate, inexplicable at present to us, must have taken place in the circumpolar regions by which the severity of the cold, that has for centuries past enclosed the seas in the high northern latitudes in an impenetrable barrier of ice, has been, during the last two years, greatly abated. Mr. Scoresby, a very intelligent young man, who commands a whaling-vessel from Whitby,[1] observed last year that 2000 square leagues of ice, with which the Greenland seas between the latitudes of 74° and 80° N[orth] have been hitherto covered, has in the last two years entirely disappeared.[2] The same person, who has never been before able to penetrate to the westward of the meridian of Greenwich in these latitudes, was this year able to proceed to 10° 30′ W[est] where he saw the coast of East Greenland, and entertained no doubt of being able to reach the land, had not his duty to his employers made it necessary for him to abandon the undertaking.

This, with information of a similar nature derived from other sources — [such as] the unusual abundance of ice islands that have, during the last two summers, been brought by currents from Davis's Straits[3] into the Atlantic; the ice which has this year surrounded the northern coast of Iceland in unusual quantity, and remained there unthawed till the middle of August; with the floods which have during the whole summer inundated all those parts of Germany, whence rivers have their sources in snowy mountains — afford ample proof that new sources of warmth have been opened, and give us leave to hope that the Arctic seas may at this time be more accessible than they have been for centuries past; and that discoveries may now be made in them, not only interesting to advancement of science, but also to the future intercourse of mankind and the commerce of distant nations.

The Prince Regent's Government has already given abundant proof of its disposition to follow up the plan of prosecuting discoveries for the extension of useful knowledge and the general benefit of mankind (which has so eminently distinguished the reign of his present Majesty).[4] We cannot therefore entertain a doubt that it will be ready to avail itself of the favourable opportunity which now appears to have occurred; to endeavour to correct and amend the very defective geography of the Arctic regions, more especially on the side of America; to attempt the circumnavigation of Old Greenland, if an island or islands, as there is reason to suppose; to prove the existence or non-existence of Baffin's Bay, and to endeavour to ascertain the practicability of a passage from the Atlantic to the Pacific coast along the northern coast of North America. These are objects which may be considered as peculiarly interesting to Great Britain, not only for their proximity and the great national advantages which they involve, but also for the marked attention they called forth, and the discoveries made in consequence thereof in the very earliest periods of our foreign navigation.

We have been induced to take the liberty of making this communication to your Lordship on finding, by a reference to our Minutes, that the voyage of discovery towards the North Pole, which was entrusted to the late Captain Phipps, afterwards Lord Mulgrave,[5] was first suggested by the President and Council of the Royal Society in a communication made to Lord Sandwich,[6] then first Lord Commissioner of the Admiralty, on the ground that such a voyage might be of service to the promotion of natural knowledge, which is the proper object of this Institution under the patronage of his Majesty. We therefore beg leave, in the event of one or more expeditions being fitted out on the present occasion, to make an offer of our services in any way that may be thought useful in the preparation thereof, or conducive to the accomplishment of the objects which such expeditions may have in view.

I am, &c.,

Joseph Banks.

[C.R. Weld, *A History of The Royal Society, with Memoirs of the Presidents*, London (1848), vol. II, p. 274–277; see also R.S. Council Minutes, IX, 137–141, 20 November 1817.]

[1]Whitby, Yorkshire. This was the port where HMS *Endeavour* was built. Prior to being selected and renamed for the voyage under Captain James Cook, 1768–1771, this ship operated as a collier vessel called the *Earl of Pembroke*. Whitby was also where Cook served his apprenticeship, learning about coastal navigation on vessels engaged in the coal trade. Scoresby was born at Cropton, near Whitby, on 5 October 1789. Whitby has been called 'a nursery of seamen'.

[2]The decrease in Polar ice was reported to Banks by William Scoresby: Letter 130. The letter above was addressed to Dundas by Banks on behalf of the Royal Society, which, like its President, had long been interested in Arctic exploration. In signing it Banks drew on his reputation and experience as a patron of science and exploration of fifty years standing. His advice could be relied on by government. Dundas replied to Banks to say that two naval expeditions would be launched in four specially strengthened whalers, Dundas to Banks, 10/12/1817, C.R. Weld, *A History of The Royal Society, with Memoirs of the Presidents*,

London (1848), vol. II, p. 279: 'Our present intention is that two of them should proceed into Davis's Straits, and from there in a north-westerly, and possibly afterwards in a westerly direction, and that the other two should proceed along the coast of East Greenland to the Northward, and if practicable, afterwards to the Westward.' The Royal Society was requested to assist with instructions, especially for the advancement of science.

One mission was to Baffin's Bay, under Captain John Ross RN (1777–1856) in HMS *Isabella*, and Lieutenant William Edward Parry RN, FRS (1790–1855) in HMS *Alexander*. The other was to the north of Greenland and Spitzbergen, under Captain David Buchan RN in HMS *Dorothea*, and Lieutenant John Franklin RN, FRS (1786–1847) in HMS *Trent*. Both departed early in 1818. They were equipped with many instruments for experiments, and personnel to conduct them, but neither revealed a great deal about the geography of the regions visited. Ice forced Franklin back despite brave efforts to reach higher latitudes than 80° 34′, this being the farthest north he travelled. Ross penetrated to 75° in Baffin Bay, ice was found. Ross wrongly concluded that there was no outlet through Lancaster Sound, and returned to England. By 21 November 1818 all four ships were back at Deptford.

There was disappointment at the inconclusive results of the voyages, but the Royal Society could console itself with the magnetic observations made by Edward Sabine FRS (1788–1883) in the north-west. These were communicated in the *Philosophical Transactions*, and in due course Sabine fulfilled a promising scientific career, which Banks had encouraged, by becoming President, 1861–1871. Furthermore, another voyage to Baffin Bay was dispatched in 1819. This was under Parry in HMS *Hecla*, accompanied by HMS *Griper*. Parry successfully navigated through Lancaster Sound as far as Melville Island, which he named after Dundas. Winter compelled Parry to seek a harbour in the island until August of the following year, when he returned to England, arriving on 3 November 1820. Banks was dead by then, and so never shared in the triumph of an enterprise he had done so much to initiate. However, his name is commemorated in Banks Island, and so, in other places on the route through to the Bering Strait, were those of a number of people drawn together here by Banks.

[3]Davis Strait, between Baffin Island and Greenland, leading to Baffin Bay.

[4]The Prince of Wales was appointed Prince Regent, subject to specified limitations, in 1811.

[5]Constantine John Phipps, 2nd Baron Mulgrave. See Letter 10.

[6]John Montagu, 4th Earl of Sandwich, who was appointed a Lord Commissioner of the Admiralty, 1744, and then First Lord of the Admiralty three times, 1747–1751, 1763–1765 and 1771–1782.

xviii. **Antoine-Laurent de Jussieu F.R.S.**
From an engraving by Ambroise Tardieu. The Linnean Society, London.

ANT.NE LAUR.T DE JUSSIEU
(Botaniste et Médecin),
Membre de l'Académie des Sciences
et Professeur de Botanique rurale au Jardin-du-Roi.
Né à Lyon (Dép.t du Rhône) le 12 Avril 1748.

1817

To Sir James Edward Smith F.R.S.

Soho Square
25 December 1817

My dear Sir James,

After thanking you & Lady Smith for your kind Commencement of our Annual Supply of Turkies, & for your Goodness to the Ladies in Procuring for them a Fresh Annual Supply of Buns, which are now become So Popular in our Family that the Ladies wish to Enter into direct Correspondence with the maker of them, my chief reason for Troubling you with this is to tell you I have paid Obdience to your mandate by reading your Article on Botany in the Scotch Encyclopaedia,[1] which, conceiving it to be an Elementary Performance, I had neglected till now to Peruse.

I was highly Gratified by the distinguishd situation in which you have Placed me, more so I fear than I ought to have been. We are all too fond of hearing ourselves well Spoken of by Persons whom we hold in high Regard. But, my dear Sir James, do not you Think it Probable that the Reader who takes the book in hand for the Purpose of seeking Botanical Knowledge, will Skip all that is Said of me as not at all tending to Enlarge his Ideas on the Subject?

I admire your defence of Linnaeus's[2] Natural Classes. It is Ingenious and Entertaining, & it Evinces a deep Skill in the Mysteries of Classification, which must, I fear, Continue to wear a mysterious Shape till a Larger Portion of the vegetables of the whole Earth Shall have been discoverd and described.

I fear you will differ from me in opinion when I Fancy Jussieu's[3] Natural Orders to be Superior to those of Linnaeus.[4] I do not, however, mean to alledge that he has even an Equal degree of merit in having Compiled them. He has taken all Linnaeus had done as his own, & having thus Possessd himself of an Elegant & Substantial Fabric, has done much towards increasing its beauty, but far Less towards any improvement in its Stability.

How immense has been the improvement of Botany since I attached myself to the study, & what immense facilities are now offered to students that had not an Existence until Lately. Your descriptions, & Sowerby's[5] drawings of British Plants, would have saved me years of Labo[u]r had they then Existed. I well Remember the Publication of Hudson,[6] which was the First Effort at well-directed Science, & the Eagerness with which I adopted its use.

The Ladies beg to join in best Compts. to you & Lady Smith,[7] & I beg, my dear Sir James, that you will beleive me,

Very Faithfully Yours,

J: Banks.

[L.S. Smith MS. 301. 20. 80.]

[1]*Encyclopaedia Britannica*, Supplement, vol. II, London (1817). The title for this account was 'A Review of the Modern State of Botany, with a particular reference to the natural systems of Linnaeus and Jussieu.' In it Smith explained: 'The names of Banks and Solander have, for nearly half a century, been in everybody's mouth. Their taste, their knowledge, their liberality, have diffused a charm and a popularity over all their pursuits; and those who never heard of botany before, have learned to consider it with respect and admiration, as the object to which a man of rank, riches, and talents, devotes his life and his fortune; who while he adds, every season, something of novelty and beauty to our gardens, has given the Bread-fruit to the West Indies, and is ever on the watch to prompt, or to further, any scheme of public advantage.'
[2]Carl von Linné (Carolus Linnaeus).
[3]Antoine Laurent de Jussieu. Jussieu's system for classifying plants influenced Brown's work on *Prodromus Florae Novae Hollandiae*..., London (1810). Smith commented at the end of his essay: 'The French school has been much flattered by our able countryman Mr. Brown, having classed his *Prodromus* of the New Holland plants after the method of Jussieu.'
[4]Jussieu distinguished relationships between plants by reference to a large number of characters. Linnaeus used a sexual system, which relied on only a few important ones. Linnaeus's system seemed artificial in comparison to Jussieu's, but it did provide a good practical system for working naturalists.
[5]James Sowerby (1757–1822): botanical artist. He produced the illustrations for the great *English Botany*, London (1790–1814), in 36 volumes with some 2592 coloured plates. Smith provided the text.
[6]William Hudson FRS (c.1730–1793): botanist. He published *Flora Anglica: exhibens plantas per Regnum Angliae sponte crescentes*..., London (1762).
[7]Lady Pleasance Smith (*née* Reeve), (1774–1877).

Letter 134

1818

To Lady Charlotte Jane Seymour

Spring Grove
10 August 1818

It Gives me Great Pleasure to hear that my Amiable Little Friend, Lady Jane Seymour, has undertaken the Study of English Botany. She will find it through Life an Agreable Amusement, which will fill up Pleasantly Many an hour that would otherwise be wasted in idleness, or renderd irksome by Ennui. The Pleasure I have Enjoyd, as long as my Eyes permitted me, of Reading in the hedges & Lanes during a journey has always Renderd that an amusement which would otherwise have been a Fatigue.

Allow me to Request your Ladyships acceptance of the Little book which accompanies this. The Present Volume,[1] the only one yet Publishd, Contains all the Plants Called by Botanists Phenogamous Plants (See Preface). The Rest, which are Mosses & other dificult Genera, I would not advise your Ladyship to undertake at present.

As soon as you begin to Examine Plants by Dr. Hulls book, I advise your Ladyship to make an herbarium by Drying the Plants you Collect between Sheets of Paper, and keeping them Separate from Each other. If you will bring this Collection to London when you visit us, I Shall have Great Pleasure in looking it over, & in Case any mistakes occur, which will always happen to beginners, in putting them Right. A Herbarium, as it is Called by Botanists, thus Collected will afford an amusement as Long as the Love for Botany Continues. Indeed, the Pleasure of adding a new Plant to it when the Collection is nearly Compleat is all but indescribable. I have Placed three or four Specimens of the Little wild sentimental Plant, Calld by the Germans Forget me not, in the book in order to show your Ladyship the mode of Drying Plants. Its name in Latin is Myosotis Scorpioides.[2]

JB

[B.R.O. Bulstrode (Additional) D/RA/A/1D/8.]

[1] J. Hull, *Elements of Botany*, 2 vols., London (1800). On page xviii of the preface; Dr. John Hull (1761–1843) introduced the 'new name of *Phanerogamia*' for plants with evident sexual organs, while retaining the 'old title of *Cryptogamia*' for those plants which have no conspicuous sexual organs.
[2] Water forget-me-not, (*Myosotis scorpioides*).

Letter 135

1818

To Sir Everard Home F.R.S.

Spring Grove
27 August 1818

My dear Sir Everard,

I enclose, at the request of Dr. Wollaston,[1] a Sample of Lint, which appears to me of an excellent quality. It is not made by breaking the texture of Old Linen, but is woven on purpose. Whether beating & bruising of flax by hammers & mills can give it that extreme softness & delicacy of feel so invaluable in the treatment of inflamed sores is for experiment to decide. I confess I doubt whether any bruising, but that which takes place between outer Garments & Flesh, can give the necessary delicacy of touch.

I was overturned two days ago by a drunken Coach man, but received no hurt. Lady B[anks], my sister[2] & me were returning home from dining with Sir A. Macdonald.[3] We are all three heavy, & I [am], as you know, quite helpless. We were obliged to lay very uneasily at the bottom of the Coach for half an hour before assistance could be got to lift us out. We all bore our misfortune without any repining, or any demonstration of the follies occasioned by fear, and we are all now quite recovered from the effects of our accident; except my Sister, who has a cut in her head filled with Lint, & [she is] doing very well. Both the ladies have gone every where since without an hour's confinement.

I have told you this long Story as the preface to the catastrophe which took place in my case. I was much pressed while I lay in the Coach with a desire of making water. As soon as I came home I passed a considerable quantity, & went to bed.

You are aware that I have been for some years vexed with a Spasm in my bladder, which very often stoped my urine, but which always gave way in a few minutes. Of late the attacks of this Spasm have been much more frequent than formerly, but its duration much shorter. At first I was obliged to restrain the action of Bladder, which always attempted to remove the obstruction, on account of the extreme pain it occasioned, & wait till I knew by a feeling in it that the obstruction was removed. Of late I was able to allow the bladder to exert itself so that I always effected the purpose of Emptying my bladder, though the urine which had run freely before the Spasm took place passed very slowly when the passage was again opened.

I believe I have before described these Symptoms to you, & have added that, frequent as these spasms were when I passed my urine sitting on a Chair, or standing on my legs when I was able, no one instance of Spasm had occurred when I lay on my side in bed.

On going to bed on the night of my overturn I felt again a desire to pass urine, which rather surprised me as I had not many minutes before passed a considerable quantity. I took the Chamber-Pot, & was surprised to experience for the first time a spasmedic stoppage as I supposed. The bladder, however, continued to act rather Strongly, & in a few seconds the water flowed again with a gush of more than usual force, attended with some pain, but not enough to make me use any kind of effort to abate it.

On receiving the Chamber-Pot from me, Phillips immediately said, '[T]here is some thing here, Sir. Could it not have been here before?' He brought me a Stone of this form & Dimentions  with abundance of jags sticking out of the sides, which make it resemble a small branch of Coral so. Had not the Stone presented itself to the urethra with its large end foremost, it must have stuck in the passage, & you must have come from Lancing to remove it, if my bladder had not burst before you could come with the accumulation of urine.

From the color of the Stone, it appears to be uric acid. From the shape of it, the possibility of its being the remains of a larger Stone is clear. [I] Suppose a stone of this shape 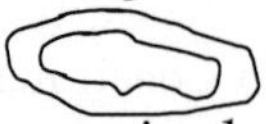 to be dissolved by a proper solvent till the size of the kernel only remained. Is it not likely, in this case, to exhibit knobs formed of the parts that were originaly formed of a Substance more hard & less easy of Solution than the general mass of the Stone? I know not how otherwise to account for the very irregular bumps on this calculus in a way so probable. If this really is the case, can the dissolution of this Stone be attributed to any thing but Wm Brande's[4] magnesia? I dare not be too Sanguine in believing it, but, for the Sake of suffering humanity as well as my regard to Wm. Brande, I cannot avoid feeling an ardent hope that subsequent observations may confirm my Theory.

Excuse, my dear Sir Everard, this tedious peace [piece] of Prattle. I have enlarged upon the subject because it is in your line. Believe me, with love to my godson,[5] & kindest respects to all your family, most faithfully yours.

Jos: Banks.

[N.H.M. B.L. D.T.C. XX. 117–120.]

[1]Dr. William Hyde Wollaston FRS (1766–1828): physiologist; chemist; physicist. He was Banks's successor as President of the Royal Society for a brief period, June to November 1820.

[2]Lady Dorothea Banks and Sarah Sophia Banks. Banks and 'the Ladies' were a close trio for many years. Sarah Sophia died a short while after this accident, much to her brother's distress.

[3]Sir Archibald Macdonald (1747–1826), 1st Baronet: judge.
[4]William Thomas Brande.
[5]Sir James Everard Home (b.1798), 2nd Baronet: captain RN.

1819

To an Unknown Correspondent

Spring Grove
21 July 1819

I Feel much indebted to your Kind intention of depositing my Picture at Oxford.[1] I am not much addicted to the Love of Posthumous Fame, but I Confess the Idea of being Rememberd[2] by those who in Future Receive intellectual nourishment from the milk of the alma mater by whom I was Fed, is an Idea that Renders the natural Fear of Dissolution Less alarming.

The Place where you chuse to deposit it must Certainly be that chosen by yourself. I confess, however, that my vanity would Receive a Gratification if it Could be admitted in to Christ Church Hall superior to what would accrue from a Place in the Picture Gallery. You, however, ought to chuse, & are able to make a better choice of a Place for it than I can do, who have now Left Oxford more than half a Century.

Beleive me, my dear Sir,
your much obligd &
Most Faithfull Sevt,

Jos: Banks.

[P.M.L. Described as 'Autograph letter signed, 21 July 1819, to an unknown correspondent.']

[1]Banks was a gentleman commoner at Christ Church, Oxford, from 1760 to 1765. In 1762 he took rooms six and eight on staircase eight in Peckwater Quadrangle.
[2]The portrait is not at Christ Church.

Letter 137

1820

To The Council of the Royal Society

Soho Square
1 June 1820

Sir Joseph Banks begs leave to inform the Council of the Royal Society that his motive for offering his resignation of the office of President was a conviction that old age had so far impaired his sight and his hearing as to render him by no means so well able to perform the duties of that respectable office as he has been. He is gratified in the extreme by finding that the Council think it possible for him to continue his services without detriment to the interests of the Society, and he begs leave to withdraw his resignation, assuring the Council that his utmost exertions shall never be wanting to conduct, so far as may be in his power, the affairs of the Society.[1]

[C.R. Weld, *A History of The Royal Society, with Memoirs of the Presidents*, London (1848), vol. II, p. 301.]

[1]Banks offered his resignation in May, having taken the chair at the Council of the Royal Society for the last time on 16 March. The Council unanimously requested that he remain President. Banks died in his bed at Spring Grove early on 19 June 1820. He was buried at the parish church of St. Leonard, Heston.

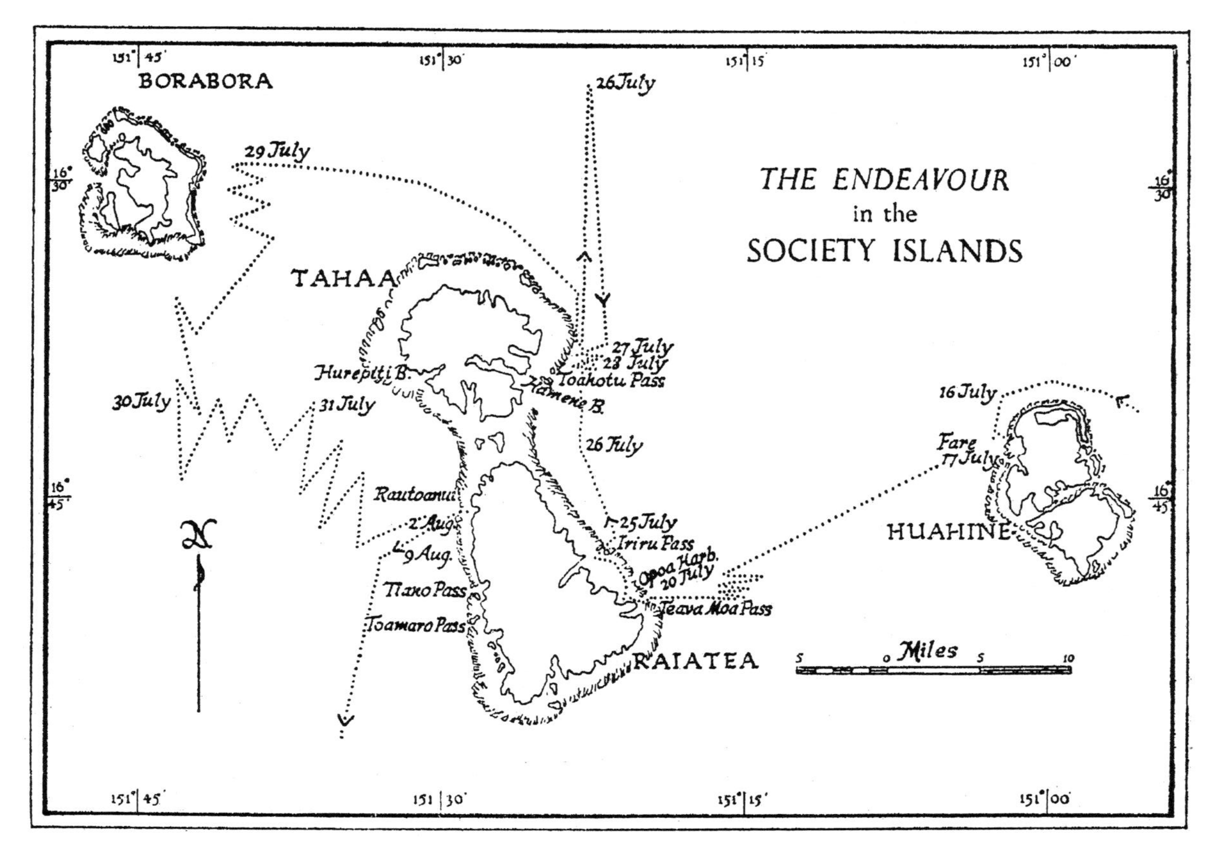

i. Track of H.M.S. *Endeavour* in the Society Islands, 16 July–9 August 1769. Reproduced from *The Endeavour Journal of Joseph Banks, 1768–1771*, J.C. Beaglehole (Ed.), Sydney (1962): vol. I, p. 315.

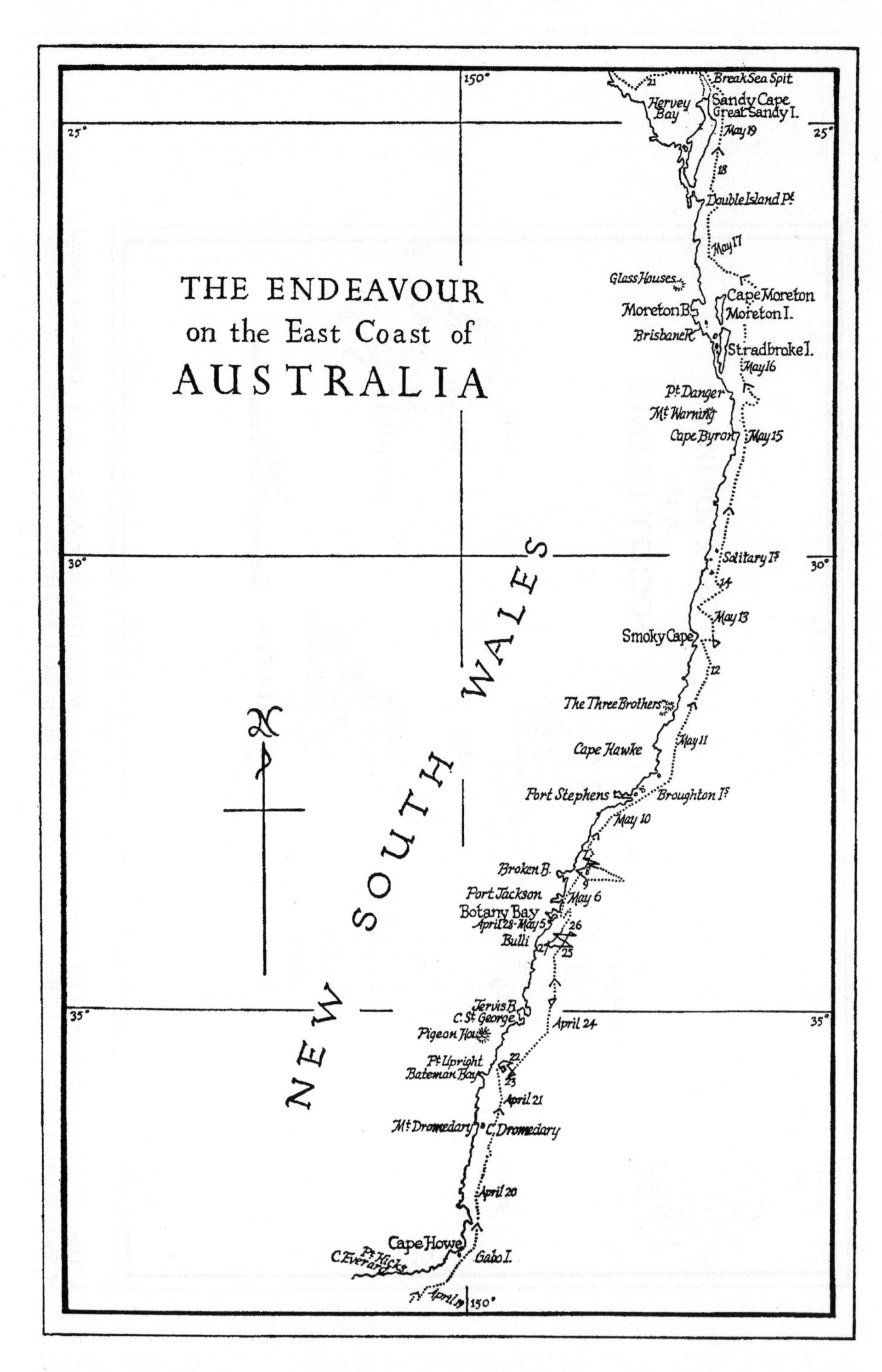

ii. Track of H.M.S. *Endeavour* exploring part of the coast of New South Wales, 19 April–21 May 1770. Reproduced from *The Endeavour Journal of Sir Joseph Banks, 1768–1771*, J.C. Beaglehole (Ed.), Sydney (1962): vol. II, p. 48.

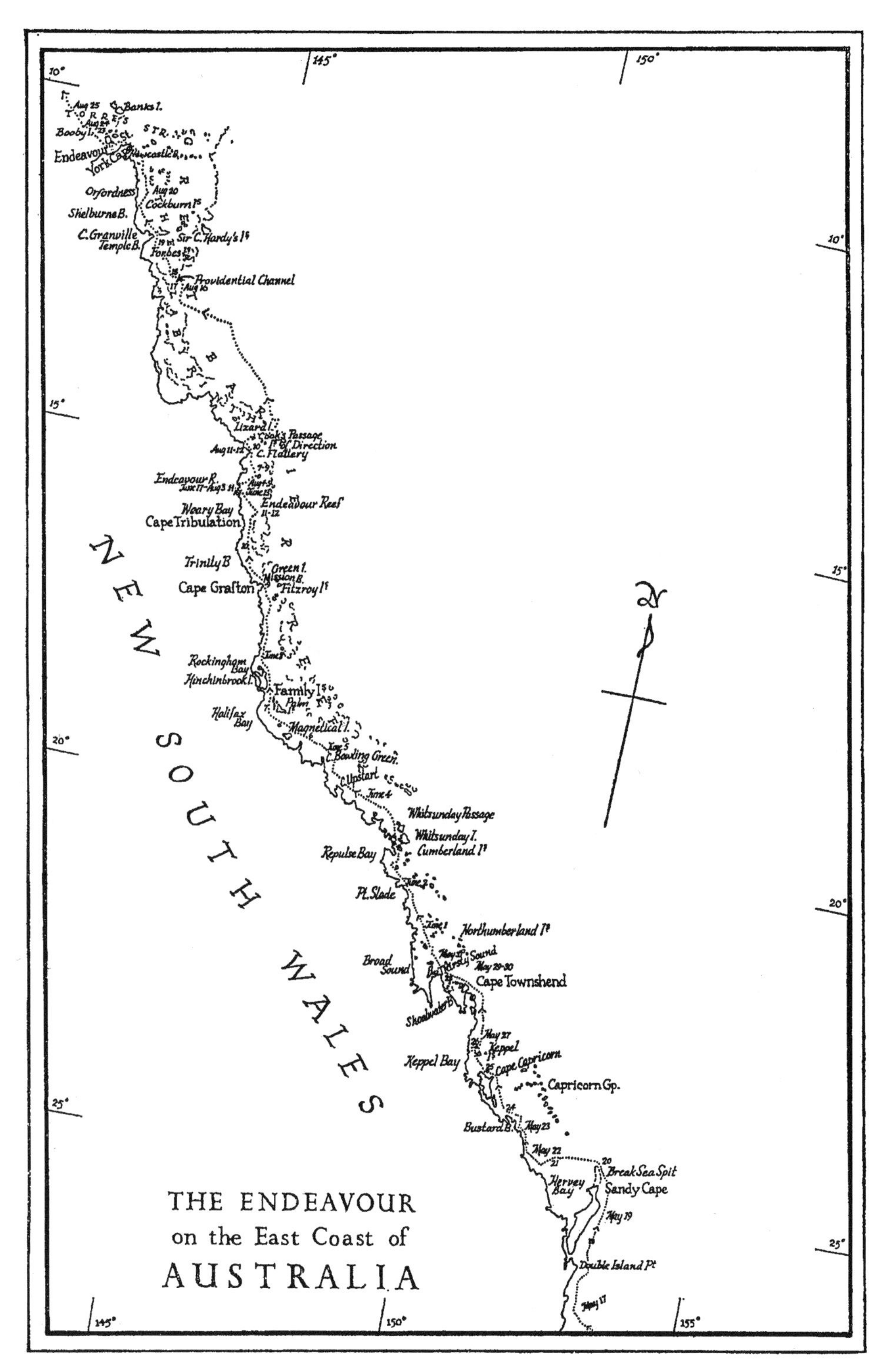

iii. Track of H.M.S. *Endeavour* exploring part of the coast of New South Wales, 17 May–25 August 1770. Reproduced from *The Endeavour Journal of Sir Joseph Banks, 1768–1771*, J.C. Beaglehole (Ed.), Sydney (1962): vol. II, p. 48.

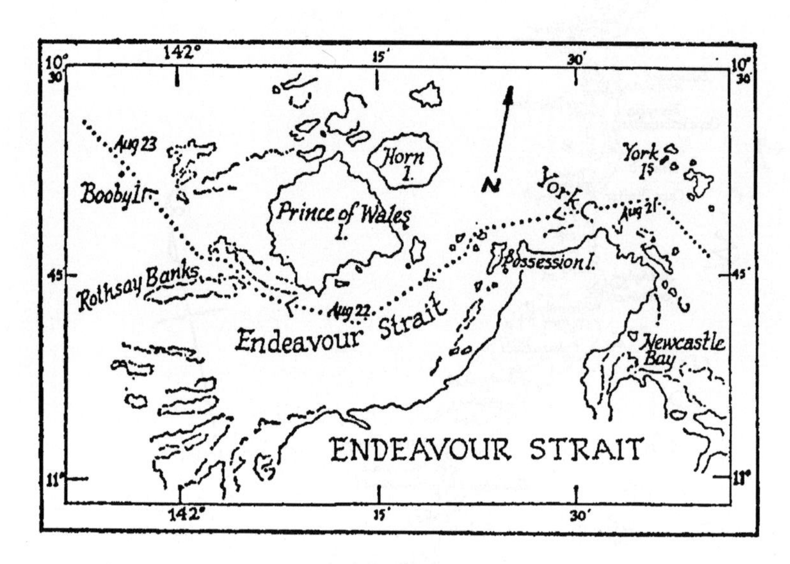

iv. Track of H.M.S. *Endeavour* and the passage through Endeavour Strait, 21 August–23 August 1770. Reproduced from *The Endeavour Journal of Joseph Banks, 1768–1771*, J.C. Beaglehole (Ed.), Sydney (1962): vol. II, p. 68.

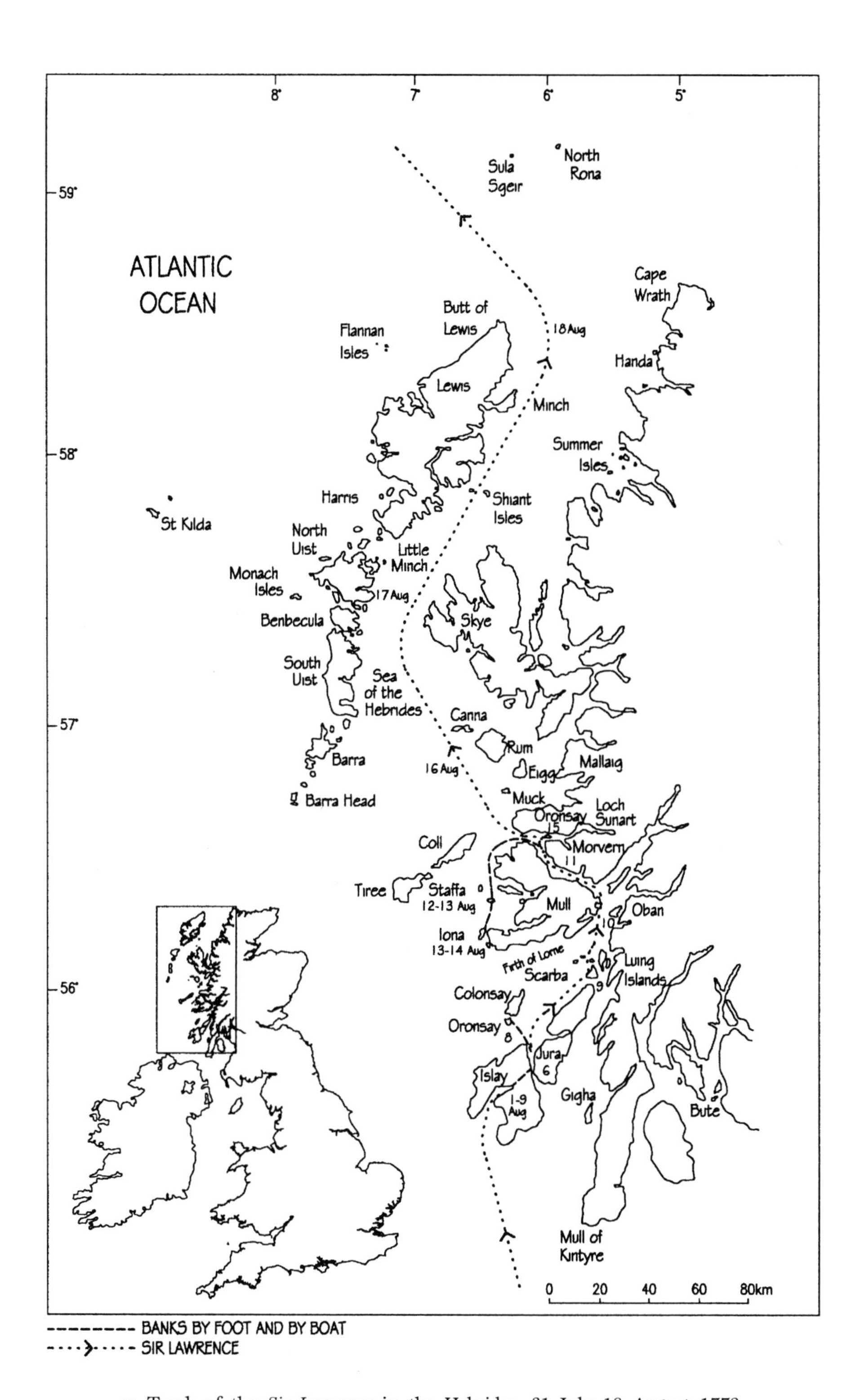

v. Track of the *Sir Lawrence* in the Hebrides, 31 July–18 August 1772.

Select Reading

A Man will turn over half a library to make one book Dr. Johnson

This short list mostly includes useful reading related directly to Banks, but some other biographies and general accounts have also been suggested. It is not an exhaustive list. More volumes and papers are cited in the notes to the letters, and the Biographical Index of Recipients.

BANKS, R.E.R. *et al* (Eds.): *Sir Joseph Banks: A Global Perspective*, Royal Botanic Gardens, Kew (1994).

BEAGLEHOLE, J.C. (Ed.): *The Endeavour Journal of Joseph Banks, 1768–1771*, 2 vols., Sydney (1962).

BLADEN, F.M.: *Historical Records of New South Wales*, 7 vols., Sydney (1893–1901).

BOAS-HALL, M.: *All Scientists Now: The Royal Society in the nineteenth century*, Cambridge (1984).

BREWER, J.: *The Pleasures of the Imagination: English Culture in the Eighteenth Century*, London (1997).

CAMERON, H.C.: *Sir Joseph Banks, KB, PRS: The Autocrat of the Philosophers*, London (1952).

CARTER, H.B.: *His Majesty's Spanish Flock: Sir Joseph Banks and the Merinos of George III of England*, Sydney and London (1964).

CARTER, H.B. (Ed.): *The Sheep and Wool Correspondence of Sir Joseph Banks, 1781–1820*, Sydney (1979).

CARTER, H.B.: *Sir Joseph Banks, 1743–1820: a guide to biographical and bibliographical sources*, Winchester (1987).

CARTER, H.B.: *Sir Joseph Banks (1743–1820)*, London (1988).

DAWSON, W.R.: *The Banks Letters: A Calendar of the manuscript correspondence of Sir Joseph Banks*, London (1958)

DE BEER, Sir Gavin: *The Sciences were never at War*, London (1960).

DUYKER, E.A.J. & TINGBRAND, P.E. (Eds.): *Daniel Solander: Collected Correspondence, 1753–1782*, Oslo, Copenhagen, Stockholm (1995).

EHRMAN, J.: *The younger Pitt: The Years of Acclaim*, London (1984).

EHRMAN, J.: *The younger Pitt: The Reluctant Transition*, London (1993).

EHRMAN, J.: *The younger Pitt: The Consuming Struggle*, London (1996).

FOTHERGILL, B.: *Sir William Hamilton, Envoy Extraordinary*, London (1969).

GASGOIGNE, J.: *Joseph Banks and the English Enlightenment: Useful Knowledge and Polite Culture*, Cambridge (1994).

GASGOIGNE, J.: *Science in the Service of Empire: Joseph Banks, The British State and the uses of Science in the Age of Revolution*, Cambridge (1998).
GLEASON, M.L.: *The Royal Society of London: Years of Reform, 1827–1847*, London (1991).
HERMANNSSON, H.: *Icelandica Vol. XV: 'Sir Joseph Banks and Iceland'*, New York and London (1928).
JARRETT, D.: *Britain 1688–1815*, London (1965).
LYONS, Sir Henry: *The Royal Society, 1660–1940: A History of its Administration under its Charters*, Cambridge (1944).
LYSAGHT, A.M.: *Joseph Banks in Newfoundland and Labrador, 1766 (Ed.): His Diary, Manuscripts and Collections*, London (1971).
MABBERLEY, D.J.: *Jupiter Botanicus: Robert Brown of the British Museum*, Braunschweig (1985).
MACKANESS, G.: *The Life of Vice-Admiral William Bligh R.N., F.R.S.*, 2 vols., Sydney (1931).
MACKANESS, G.: *Sir Joseph Banks, His Relations with Australia*, Sydney (1936).
MACKAY, D.: *In the wake of Cook: Exploration, Science and Empire, 1780–1801*, London (1985).
MAIDEN, J.H.: *Sir Joseph Banks: 'The Father of Australia'*, Sydney and London (1909).
MILLER, P. & Reill P.H.: *Visions of Empire: voyages, botany and representations of nature*, Cambridge (1996).
O' BRIAN, P.: *Joseph Banks*, London (1987).
SCOTT, E.: *The Life of Matthew Flinders, R.N.*, Sydney (1914).
SMITH, B.W.: *European Vision and the South Pacific*, London (1985).
SMITH, E.: *The Life of Sir Joseph Banks, President of the Royal Society*, London (1911).
SMITH, Lady (Ed.): *Memoir and Correspondence of the late Sir James Edward Smith M.D.*, London (1832).
STAFLEU, F.A.: *Linnaeus and the Linnaeans: The spreading of their ideas in systematic botany, 1735–1789*, Utrecht (1971).
VAN DOREN, C.: *Benjamin Franklin*, New York (1938).

Theses

AGNARSDÓTTIR, A.: *Great Britain and Iceland, 1800–1820*, London School of Economics and Political Science, Department of International History (1989).
FARNSWORTH, J.: *A History of Revesby Abbey, 1764–1820*, Faculty of the Graduate School of Yale University (1955).
HUNT, W.M.: *The Role of Sir Joseph Banks, K.B., P.R.S., in the Promotion and Development of Lincolnshire Canals and Navigations*, Open University (1986).
MACKAY, D.L.: *Exploration and Economic Development of Empire, 1782–1798, with special reference to Joseph Banks*, University of London (1970)
RUTHERFORD, H.V.: *Sir Joseph Banks and the Exploration of Africa, 1788–1820*, Graduate Division of the University of California (1951).

Biographical Index of Recipients

The Biographical Index of Recipients provides brief details of the life and career of each of Banks's correspondents in this volume. Where helpful, remarks on the official positions held, and particulars of the individual's relationship with Banks are also included. However, the accounts given are not intended to be comprehensive.

AITON, William Townsend (1766–1849): gardener and botanist; eldest son of William Aiton (who worked as a gardener at Kew, 1759–1793, and assumed control of Richmond Gardens, 1784); superintendent at Kew following his father, 1793–1841; in charge of the gardens at Kensington and St. James's Palace from 1804; Royal Gardener to George IV at the Royal Lodge gardens, Windsor Great Park. William Townsend Aiton did much to perpetuate his father's excellent management of the Royal Forcing and Pleasure Gardens at Kew. He worked closely with Banks, and together they raised the gardens to a pre-eminent position in the horticultural and botanical worlds. Banks, Jonas Dryander and Robert Brown assisted Aiton with a second, enlarged edition of his father's *Hortus Kewensis*, London (1810–1813). Banks and Aiton were among the founder members of the Horticultural Society in 1804. Aiton retired from his duties in 1845.

Letter 123.

ALLEN, John (b.1775): miner; traveller. Allen was a Derbyshire miner, who came from Ashover. Banks arranged for him to be the miner on board H.M.S. *Investigator* through William Milnes, steward of the Banks family estates at Overton. Allen received a salary of £105 for this. He returned from the mission in 1804, and afterwards went to work in Ireland.

Letter 84.

ALSTRÖMER, Johan F.R.S. (1742–1786): Swedish merchant; industrialist; collector, mainly of natural history; President of the Royal Swedish Academy of Sciences; director of the Swedish East India Company. Alströmer visited England, 1777–1778. It was then that he met Banks in person. He was elected a Fellow of the Royal Society before departing in 1778.

Alströmer maintained a friendly correspondence with Banks, to whom he wrote with condolences on the death of Dr. Daniel Solander in 1782: Alströmer to Banks, 24/5/1783, B.L. Add. MSS. 8095 117–118. Some time later Banks wrote back with a moving tribute to Solander, which was published in Swedish in *Upfostrings Sälskapets Tidningar*, Pt. 14, Stockholm 21 February 1785, pp. 105–110. His original letter has not been located, but the published translations remain. Alströmer's last letters to Banks speak of financial ruin, although his interest in natural history seemed to persist: Alströmer to Banks, 28/5/1786, B.L. Add. MSS. 8096 207–212.

Letter 23.

AZAMBUJA, Rolim de Moura, Conde de (1709–1782): soldier; diplomatist; captain in the Portuguese royal infantry; appointed to the governorship of Mato Grosso, 1749; promoted to Captain-General of Bahia, 1764; created Conde de Azambuja, with the rank of Brigadier and Commander in the Order of Christ; appointed Portuguese Viceroy of Brazil, 1767. Azambuja had a distinguished career in the Portuguese diplomatic service, but for various reasons regarded the arrival of H.M.S. *Endeavour* at Rio de Janeiro in 1768 with suspicion. On 14 November he refused Banks permission to land and explore. He also placed restrictions on James Cook, which effectively meant Cook was only allowed ashore with sufficient men to obtain supplies. The Viceroy did not believe repeated explanations from both Banks and Cook that their journey was to observe the Transit of Venus. Indeed, he seemed unable to comprehend what such an event might be. He certainly could not understand why anyone would travel so far merely to botanize. Instead, the British boat seemed to be engaged in spying or smuggling, and Azambuja had no intention of allowing either. H.M.S. *Endeavour* was therefore closely attended by guard boats until she weighed anchor on 7 December. However, Banks and members of his party landed secretly to collect specimens.

Letter 2.

BANKS, Sarah Sophia (1744–1818): virtuoso; Joseph Banks's sister. Sarah never married, and remained a life-long companion of her brother and sister-in-law, Dorothea. She amassed impressive collections of ephemera, books, coins and medals. Her coins and medals are now at the British Museum.

Letter 9.

BARROW, Sir John F.R.S. (1764–1848), 1st Baronet: traveller; author; Admiralty official; comptroller of the household, Macartney Embassy to China, 1792–1794; private secretary to Sir George Macartney at the Cape of Good Hope, 1797–1803; 2nd Secretary of the Admiralty, 1804–1845 (with one brief interruption); African Association member; founder member of the Royal Geographical Society, 1830; knighted, 1835. Barrow was an able man, who took an early interest in science and travel. He was comptroller of the household on the Macartney Embassy to China, and Sir George subsequently took Barrow to the Cape of Good Hope as his private secretary. Barrow published accounts of these events in *Travels into the interior of Southern Africa…*, London (1801) and *A Voyage to Cochin China, in the years 1792, and 1793…*, London (1806). His work in South Africa provided useful experience when he later joined societies concerned with exploration of the continent. One was the African Association, of which Banks was perhaps the most influential figure. Following Banks's death, the Association declined somewhat, and in 1831 it merged with the Royal Geographical Society. Barrow had done much to found the Royal Geographical Society a year earlier.

Barrow was a long-serving Secretary at the Admiralty, a capacity in which Banks also knew him well. Both men promoted Arctic exploration, Barrow having spent a season in Greenland and Spilzbergen waters on the whaler *Peggy* when a young man. In 1817 Banks approached the First Lord of the Admiralty, Robert Saunders Dundas, with his views on the matter. Like Barrow, Banks felt that the recent thaw in polar ice might open an Arctic route to the Pacific, and a greater understanding of the region in general. That year two naval expeditions were mounted. One was to the north of Greenland and Spitzbergen, under Captain David Buchan and Lieutenant John Franklin. The other was to Baffin's Bay, under Captain John Ross and Lieutenant William Edward Parry. Barrow chose this as a suitable time to publish *A Chronological History of Voyages into Arctic Regions...*, London (1818).

Letter 115.

BAUER, Ferdinand Lucas (1760–1826): natural history artist; traveller. Bauer travelled in Greece and Asia Minor, 1785–1786, with John Sibthorp F.R.S. (1758–1796) before being appointed as natural history artist for the *Investigator* expedition. Sibthorp was the Sherardian Professor of Botany at Oxford, and he wanted Bauer to make drawings for the illustrations in his *Flora Graeca...*, London (1806–1840). Bauer's impressive training, considerable skill and substantial experience made him an excellent choice for the *Investigator* voyage under Mathew Flinders, 1801–1803. He worked closely with Robert Brown to collect and record Australia's rich flora on this mission. Good opportunities to land and explore were provided at King George Sound, New South Wales and the Gulf of Carpentaria, before H.M.S. *Investigator* became unseaworthy, and the mission broke up. Brown and Bauer stayed in New South Wales. Bauer then visited Norfolk Island from 1804 to 1805.

He returned to England with Brown late in 1805, and Banks excitedly wrote: 'Our travellers, Brown and Bauer, arrived safe, and have brought with them unspeakable treasures so great that I scarce know how to estimate their value': Banks to Unknown Correspondent, 20/9/1806, M.L. Banks MSS. Miscellaneous Correspondence 1766–1818 A 80/4.* Bauer set to work on his drawings and engravings. He published three small fascicles: *Illustrationes Florae Novae Hollandiae: sive icones generum quae in Prodromo Florae Novae Hollandiae et Insulae Van Diemen descripsit Robertus Brown/Ferdinandi Bauer*, London (1806–1813). However, few copies were sold. In 1814, Bauer returned to Austria, where he continued to portray plants and prosper. Most of the completed illustrations he made from the *Investigator* expedition are at The Natural History Museum, London. Bauer's original drawings of Australian and Norfolk Island plants are in the Naturhistorisches Museum, Vienna.

Letter 84.

BILLARDIÈRE, Jacques Julien Houttou de la (1755–1834): botanist; explorer; travelled in the Near East, 1786; naturalist on the French voyage to the South Pacific under Joseph-Antoine Bruni d'Entrecasteaux, 1791–1794; director at the Jardin des Plantes, Paris; from 1828 at Rouen. La Billardière met Banks in London on a visit to England in 1785. At the time, a French mission to the South Pacific under Jean-François de Galaup de la Pérouse was planned, and Banks sent gifts through La Billardière. Banks also assisted in the preparations for a French mission to search for La Pérouse when he did not return. Banks provided instructions on plant collecting in the Pacific for La Billardière, who went as a naturalist.

However, La Billardière was interned at Java, along with his colleagues and collections, when the expedition collapsed. The officers, who remained loyal to the French monarchy, objected to the republican views of the scientists, and turned them over to the Dutch authorities. La Billardière's collections of plants, minerals and insects were removed without his permission, and offered to Louis XVIII. Louis then gave them to Queen Charlotte of England. La Billardière eventually returned to France, where he argued the collections had not legitimately been in the gift of the exiled French King. It fell to Banks to ensure that the collections were all returned to their owner at the Jardin des Plantes, Paris. La Billardière was indebted to Banks once again, and published *Relation du voyage à la recherche de La Pérouse...*, Paris (1799), which was followed by the series *Novae Hollandiae plantarum specimen...*, Paris (1804–1807).

Letters 62, 71.

BLAGDEN, Sir Charles F.R.S. (1748–1820), 1st Baronet: physician; physicist; army medical officer until 1814; assistant to Henry Cavendish, 1782–1789; knighted, 1792; Secretary of the Royal Society, 1784–1797. Blagden was elected Secretary of the Royal Society on 5 May 1784 following the Royal Society 'dissensions' of 1783–1784. Paul Henry Maty F.R.S. (1745–1787), an opponent to Banks, had previously resigned the post. Blagden supported Banks throughout this episode, and Banks proposed him for the vacant position. Blagden liked to travel abroad, and towards the end of his life passed much of each year in France. The letters between Banks and Blagden are rapid reports of science and society in Europe, and reveal a long but sometimes fraught relationship. Some 450 remain, which makes this the largest series with one person in the Banks correspondence. It covers the years 1773 to 1820.

Letters 15, 17, 20, 21, 37.

BLIGH, William F.R.S. (1754–1817): navigator; cartographer; master, H.M.S. *Resolution*, 1776–1780; on active service as a junior lieutenant, 1781–1783; merchant captain in the West Indies, 1783–1787; commander on first bread-fruit voyage, H.M.S. *Bounty*, 1787–1789; commander on second bread-fruit voyage, H.M.S. *Providence*, 1791–1793; naval captain, various battleship commands, 1791–1805; Governor of New South Wales, 1805–1810; promoted Rear-Admiral, R.N., 1811; promoted Vice-Admiral, R.N., 1814. Bligh accompanied Cook around the world on his third Pacific voyage, 1776–1780. He went as a master on H.M.S. *Resolution*. Afterwards, Bligh made important hydrographic surveys prior to taking part in the battle off Doggerbank, 5 August 1781. His reputation as a navigator was well established by the time he took command of H.M.S. *Bounty* for the expedition to transplant bread-fruit trees from the Society Islands in the Pacific to the British West Indies, 1787–1789. Banks organized the mission, which ended with the infamous mutiny led by Fletcher Christian. Blame for the mutiny must be shared. Bligh was certainly an abrasive man, but he lacked loyal support from some of his crew on what was a hard voyage. Furthermore, Christian might have been unstable, and the fate of the mutineers showed they were not faultless either. Bligh was cast adrift in a 23-foot-long open boat along with 18 others, and made the journey 3,700 nautical miles to Timor in the vessel. This was perhaps his most remarkable achievement at sea. A second attempt was made to transplant the bread-fruit, once again with Bligh in command, 1791–1793. It was successful.

In 1805, Banks offered Bligh the governorship of New South Wales, which was accepted. This proved a disastrous appointment for Bligh. Unlike some of his predecessors, he was determined to follow instructions strictly. In particular, he sought to curb the illegal trade in spirits, which was crippling the colony's economy. This brought him into conflict with some influential figures, such as John Macarthur. Bligh also antagonized the New South Wales Corps, which had a monopoly on the importation and sale of strong liquor. On 26 January 1808, Bligh was arrested by Major George Johnston of the New South Wales Corps, who assumed the title of Lieutenant-Governor. Johnston was eventually recalled to England for this, and tried at Chelsea Hospital, where he was cashiered. Macarthur sailed to England as an elected delegate for the rebels, but found he had less influence in London than in New South Wales. He failed to vindicate Johnston's actions, and was not allowed to go back to New South Wales for eight years. Bligh was not popular on his return either, but continued to be promoted. His days at sea were over now, and he settled at Farningham Manor, Kent.

Letter 98.

BLUMENBACH, Johann Friedrich F.R.S. (1752–1840): comparative anatomist; anthropologist; appointed curator of natural history collections and extraordinary professor at University Göttingen, 1776; appointed professor of medicine at University Göttingen, 1778. Blumenbach studied and taught at Göttingen. He showed early promise as a student when he started impressive anthropological and ethnological collections. His influence at the university and beyond grew with the publication of important works, which included *De generis humani varietate nativa liber...*, Göttingen (1776), and *Handbuch der Naturgeschichte...*, Göttingen (1779). These established a lasting scientific reputation for him.

Blumenbach wrote to Banks with details of, and requests for crania, especially from the Pacific. Along with these, Banks supplied published papers and news of scientific progress in England. Their later correspondence shows a growing interest in Africa, and, once again, Blumenbach's concern with anatomical and racial studies. It was Blumenbach who recommended Friedrich Conrad Hornemann and Johan Ludwig Burckhardt (1784–1817) as explorers for the African Association.

Letter 54.

BOULTON, Matthew F.R.S. (1728–1809): engineer; manufacturer; entrepreneur. In 1761 Boulton began building his factory at Soho, north of Birmingham, where he manufactured a wide variety of metal products. In 1775, he went into partnership with James Watt (1736–1819), and they secured an extended 25 year patent to develop the steam engine commercially. It was a lengthy process, and Boulton was over 60 years old before significant profits were made. He also used steam power in coining machines, supplying coinage to foreign governments, to the Honourable East India Company, and to the Royal Mint. It was on such business matters that Banks and Boulton generally corresponded, and there are letters between them for most years from 1772 to 1806. Boulton established the Lunar Society of Birmingham with Erasmus Darwin, and William Small (1734–1775). This society included such men as Joseph Priestley, and Josiah Wedgwood.

Letter 19.

BOUVERIE, William Pleydell (1779–1869), 3rd Earl Radnor: politician; entered parliament as M.P. for Downton, 1801, and the following year was returned for Salisbury borough, which he represented as Viscount Folkestone; succeeded to the title of Radnor, 1828, and entered the House of Lords. He was an active Whig politician, who strongly advocated liberal principles. Bouverie attacked corruption, oppression of most kinds, and attempts to influence the press. He censured many leading public figures. For instance, he sought an inquiry into Arthur Wellesley's conduct in India. He opposed the proposal to pay William Pitt the younger's debts, and attacked Lord Castlereagh in 1819 for his coercive policies. Bouverie fiercely resisted the imposition of the corn laws in 1815. In the upper house, as Lord Radnor, he concentrated on university reform, with unsuccessful bills to abolish subscription to the Thirty-nine Articles (in 1835), and to revise the statutes of Cambridge and Oxford universities (in 1837). In later life, Radnor devoted himself to agricultural pursuits and work as a country gentleman.

Letter 106.

BRIQUET (née Brennier), Marguerite Ursule Fortunée (1782–1815): poet; biographer. She was born at Niort, where her father was a lawyer. Her first poetical works appeared in *L'Almanach des muses…*, L'École centrale des Deux-Sèvres (1798). However, she is remembered for *Dictionnaire historique, littéraire et bibliographique des femmes français et des étrangères naturalisées en France…*, Paris (1804), which was considered rather superficial.

Letter 88.

BROWN, Robert F.R.S. (1773–1858): plant taxonomist and physiologist; traveller; naturalist on the voyage of H.M.S. *Investigator*, 1801–1803; librarian and curator to Banks, 1810–1820; librarian, Linnean Society, 1806–1822; Keeper of the Botanical Collection, British Museum, 1827–1858. Brown was educated in medicine at Edinburgh University, and was subsequently a surgeon in a Scottish infantry regiment. His real interest lay in botany though, and in 1791 he submitted his first paper to the Natural History Society of Edinburgh. It concerned plants he had collected in Scotland, and gained him a reputation among naturalists. Late in the 1790s, Brown met Banks, and thus gained access to the library and herbarium at Soho Square. Now fully committed to the study of botany, Brown resigned from the army, and was appointed naturalist on H.M.S *Investigator* under Matthew Flinders. Brown made remarkable collections on this voyage, returning to London in 1805. By 1810 he had completed *Prodromus Florae Novae Hollandiae…*, London. In the same year he became librarian and curator to Banks, who bequeathed Brown the use of the collections at Soho Square for life. These were eventually moved to the British Museum, and Brown was made curator of the botanical collections there.

Brown's discovery in 1827 of the movement of suspended particles had wide implications for science. He observed movement within the very fine pollen grains of the plant *Clarkia pulchella* when they were suspended in water. This was announced in the *Edinburgh New Philosophical Journal*, 'A brief account of microscopical observations made in the months of June, July and August 1827 on the particles contained in the pollen of plants, and on the general existence of active molecules in organic and inorganic bodies' (1828). The notion of 'Brownian movement' or 'motion' had been introduced. Research on fertilization in *Orchidaceae* and *Asclepiadaceae* established a 'nucleus' which is fundamental to plant growth, and this term is still used. Brown also identified the difference between gymnosperms and angiosperms.

Letters 84, 91, 97.

BURGES, Sir James Bland (afterwards Lamb), (1752–1824), 1st Baronet: politician; writer; entered parliament as M.P. for the borough of Helston in Cornwall, 1787; Under-Secretary of State, Foreign Office, 1789–1795. Burges was educated at Westminster School and University College, Oxford. In 1777, he toured parts of Europe, and in the same year was called to the bar at Lincoln's Inn, where he was appointed a commissioner in bankruptcy. Having made William Pitt the younger's acquaintance, and then gained his confidence, Burges was able to save the Mutiny Bill, which was introduced by Pitt's administration against determined opposition in parliament. Burges cleverly observed that the Mutiny Bill was not necessarily a money bill, and many bills of this kind had been introduced through the House of Lords anyway.

In 1787, Burges was elected M.P. for the borough of Helston in Cornwall, and to Pitt's displeasure sided with Warren Hastings (1732–1818), who had been impeached. Burges and Pitt did not remain on bad terms though. Burges subsequently supported William Wilberforce (1759–1833) in the struggle to abolish slavery, and he also prepared a bill to improve the condition of prisoners for debt. In 1789 Burges was appointed Under-Secretary of State at the Foreign Office, but lost his seat in the general election of 1790 when there was a double return for it. However, his position at the Foreign Office was not affected. He continued to assist Pitt in public affairs, and, like Pitt and Evan Nepean, Secretary of the Admiralty, arrived at his office for governmental work throughout the London riots of 1795. Burges retired from the Foreign Office that year, and spent more of his time in literary pursuits.

Letter 59.

CHARRETIÉ, Jean (or Josef), (fl.1796–1797): French Commissary for Prisoners of War, London, c.1796–1797. Charretié helped with the arrangements to return Jaques Julien Houttou de la Billardière's collections in 1796. Much of the botanical material had been given to Queen Charlotte in April in the mistaken belief that everything belonged to King Louis XVIII. However, La Billardière argued that this was his property, a view Banks agreed with. By August, Banks had arranged for the collections to be sent to France, and Charretié was invited to Soho Square to discuss the best way to pack and transport them.

Early in 1797, Banks approached Charretié again, this time with the idea of exchanging scientific journals. This was a difficult business. Revolution and war had interrupted communications between Britain and France, and scientists in France suffered greatly during the Reign of Terror, 1793 and 1794. For instance, Antoine-Laurent Lavoisier F.R.S. (1743–1794) was executed in France on 8 May 1794 for a 'conspiracy' that amounted to little more than corresponding with his colleagues in Britain. Furthermore, strict British laws prevented contact with the French after war was declared in 1793. Nevertheless, Banks suggested to Charretié that exchanges between the Institut National and the Royal Society might be renewed: 'I have taken all possible pains to find a way of obtaining from Paris the Scientific Journals published there…I trouble you therefore with this letter, thinking it not improbable that by your intercourse, leave may be granted for these Journals to be sent to me…', Banks to Charretié, 4/2/1797, S.L. Banks MSS. A. 4: 55. Banks also sought an interview with the Prime Minister, William Pitt the younger, to discuss the matter.

Letters 68, 70.

COUTTS, Thomas (1735–1822): banker; member of the African Association, active 1790–1819. Thomas founded the banking house of Coutts & Co. in the Strand along with his brother James. On the death of his brother in 1778, Thomas became the sole partner in the company. He was banker to George III, and a large number of the aristocracy, including Banks.

Letter 94.

DAVY, Sir Humphry F.R.S. (1778–1829): chemist; P.R.S., 1820–1827; knighted, 1812. Davy began his researches in electro-chemistry following Alessandro Volta's announcement of the voltaic cell in 1800. Using electrical methods, Davy conducted a series of experiments which resulted in the discovery of sodium, potassium, magnesium, and chlorine, which was used by the textile industry for bleaching. In 1801, Davy moved to the Royal Institution, where he did much to promote science with brilliant lectures and experiments. As always, his work had practical uses, such as that to do with agricultural chemistry. Davy took on Michael Faraday as his assistant in 1813, and Faraday accompanied him on a tour of Europe. Davy invented the famous safety lamp for miners when he returned in 1815. He is also remembered for his preparation of nitrous oxide (dinitrogen monoxide) by heating ammonium nitrate in 1799. He found it caused intoxication. Davy was elected President of the Royal Society in November 1820 following a brief term in that post for Banks's successor, William Hyde Wollaston.

Letter 125.

DEVAYNES, William (d. 1805): director of the Honourable East India Company. William Devaynes was active as a deputy chairman and chairman of the Honourable East India Company. During his career he was deputy chairman on five occasions: 1777–1778; 1779–1780; November 1783–1785; December 1788–1789, and 1790–1791. He was also chairman four times: 1780–1781; 1785–1786; 1789–1790, and 1793–1794.

Letter 35.

DOLOMIEU, Dieudonné (called Déodat) de Gratet de (1750–1801): geologist; soldier; traveller; Knight of Malta; professor, École des Mines, Paris. Dolomieu had a troubled career as a Knight of Malta, but was a distinguished geologist. In 1768 he killed a fellow member of the order in a duel, and was sentenced to life imprisonment. However, Pope Clement XIII (1693–1769) obtained his release. Dolomieu then travelled widely in the Alps, Italy, Sicily, Spain and Portugal, amassing an impressive collection of minerals. His pro-Revolutionary views led to further problems with the Knights of Malta, who thought he tried to undermine them. The excesses of the French Revolution, 1789–1794, repelled Dolomieu though. His relatives were executed or imprisoned, and his family wealth was lost. Dolomieu therefore decided to teach in Paris, and by 1796 he was lecturing in physical geography at the École des Mines.

When Napoleon Bonaparte attacked Egypt in 1798, Dolomieu accompanied the army as part of a scientific expedition to study Egyptian history and culture. He was used by Napoleon in negotiations before the fall of Malta, and soon afterwards left Egypt due to 'illness'. It seems more likely that he objected to being involved in the negotiations. A storm forced his ship into Taranto on the way home, and he was made a prisoner by the Neopolitans. Dolomieu was carried to Messina, where he was accused of assisting the French in taking Malta. He received little mercy from his captors, among whom were Knights of the Order of Malta. They placed him in solitary confinement for 21 months. During this time intellectuals across Europe strongly objected to such treatment, with Banks privately leading those in Britain. Dolomieu was released in March 1801, but died in November. The mineral dolomite, widespread in the Alpine regions, is named after him.

Letter 85.

DOUGLAS, James F.R.S. (1702–1768), 14th Earl of Morton: invested with the order of the Thistle, 1738; appointed Lord of the Bedchamber, 1739; 1st President of the new Society for Improving Arts and Sciences, 1739; appointed Lord Clerk Register of Scotland, 1760; P.R.S., 1764–1768. Douglas was educated at King's College, Cambridge, where he graduated M.A. in 1722. He travelled on the continent for some years afterwards, and studied physics. On his return to Scotland, Douglas took an active part in remodelling the Medical Society of Edinburgh. This became the Society for Improving Arts and Sciences, with Douglas as President.

Douglas was interested in astronomy as well, and submitted papers on this subject to the *Philosophical Transactions*. He was elected President of the Royal Society in 1764, and later assisted in preparations for the *Endeavour* mission to observe the Transit of Venus in 1769. Douglas was also one of the first Trustees of the British Museum, and a Commissioner of Longitude.

Letter 4.

DOUGLAS, Sylvester F.R.S. (1743–1823), Baron Glenbervie: lawyer; politician; colonial governor; author; Chief Secretary to the Lord Lieutenant of Ireland, 1794–1795; sworn a member of the Irish Privy Council; entered Irish parliament as the member for St. Canice, or Irishtown, Kilkenny, 1794; sworn a member of the English Privy Council, 1794; entered English parliament as M.P. for the borough of Fowey, 1795; Commissioner of the Board of Control, 1795–1806; Lord of the Treasury, 1797–1800; Joint Paymaster-General, 1801–1803; Vice-President of the

Board of Trade, 1801–1804; Surveyor-General of the Woods and Forests, 1803–1806 and 1807–1810; member of the African Association, active 1802–1810; 1st Chief Commissioner of Land Revenue, Woods and Forests, 1810–1814. Glenbervie was educated at the University of Aberdeen, where he was a distinguished scholar of science and the classics. In 1800 he was appointed Governor of the Cape of Good Hope, but decided not to take up the position, preferring political life in England. On 29 December 1800, he was created Baron Glenbervie in the peerage of Ireland. However, this title became extinct on his death.

Letter 83.

DUNDAS, Robert Saunders F.R.S. (1771–1851), 2nd Viscount Melville: statesman; administrator; entered parliament as the member for Hastings, 1794; during his early years in parliament he acted as private secretary to his father, Henry Dundas; appointed a Keeper of the Signet for Scotland, 1800; sworn a member of the Privy Council, 1807; President of the Board of Control, 1807 and 1809; Irish Secretary, 1809; Lord Privy Seal of Scotland, 1811; a governor of the Bank of Scotland; First Lord of the Admiralty, 1812–1827 and 1828–1830; elected Chancellor of the University of St. Andrews, 1814; made a Knight of the Thistle in 1821. Dundas was judicious, sensible and perceptive, and proved an efficient adminstrator of the navy. He was interested in exploration too. Thus, when Banks suggested in 1817 that expeditions into the Arctic ought to be launched, the First Lord of the Admiralty agreed. Melville Sound and Melville Island commemorate Dundas's time at the Admiralty, and the voyages he took part in promoting.

Letter 132.

FALCONER, Thomas (1736–1792): classical scholar; barrister, Lincoln's Inn. Falconer had a family connection to Thomas Pennant by marriage, and was also a friend of Johann Reinhold Forster F.R.S. (1727–1798). Falconer left Brasenose College, Oxford, without a degree, and was called to the bar at Lincoln's Inn on 20 June 1760. He did not practise at the bar due to poor health. He lived at Chester, where he pursued antiquarian and literary interests instead. Falconer corresponded with Banks in relation to the *Endeavour* voyage, 1768–1771, before exchanging letters about Iceland.

Letters 8, 11.

FLINDERS (*née* Chappell), Ann (1770–1852): Matthew Flinders married Ann on 17 April 1801. She was the daughter of a sailor, who died in command of a ship engaged in the Baltic trade. A rector at Brothertoft, near Boston, became her stepfather. It seems that Matthew knew and felt for Ann as early as 1794. He sailed for Australia in 1795, and three years later explored Bass's Strait, where he named a hill in Kent's Group, Mount Chappell. In 1799 he called a small group of islands Chappell Isles as well.

Flinders decided to keep his marriage to Ann a secret from Banks, and he even wanted to take his wife to Sydney on H.M.S. *Investigator*. However, Banks learned of the marriage through a newspaper announcement on 21 May 1801. He advised against taking Ann on the expedition, especially as he sensed disapproval at the Admiralty. Ann therefore remained in England while her husband left for the South Seas. He did not return for six and a half years. The Flinders had one daughter, Anne (1812–1892).

Letters 100, 107, 109, 111, 113.

FORSYTH, William (1737–1804): gardener; founding member of the Horticultural Society, 1804. Forsyth moved from Scotland to England in 1763, and worked under Philip Miller at the Chelsea Physic Garden. In 1770 he succeeded Miller, and in 1774 he started construction of the first rock garden in Britain, using old stone from the Tower of London, flint and chalk, and lava brought from Iceland by Banks two years earlier. Forsyth was appointed to His Majesty's Garden at Kensington in 1784.

Letter 86.

FRANKLIN, Dr. Benjamin F.R.S. (1706–1790): statesman; writer and legislator; natural philosopher; Clerk to the Pennsylvania Assembly, 1736–1751; Member for Philadephia, 1751–1764; appointed Deputy Postmaster of Philadelphia, 1737–1753; appointed Joint Deputy Postmaster-General for the Colonies, 1753–1774; signatory of the American Declaration of Independence, August 1776; American Commissioner in Paris, 1776–1785. Benjamin Franklin gained an early reputation as a journalist when he bought the *Pennsylvania Gazette* in 1729. He started printing his famous *Poor Richard's Almanack* in Philadelphia in 1732. Franklin was also a skilled negotiator. Much later he obtained British recognition of American Independence in 1783 when he was American Commissioner in Paris. Moreover, he managed to obtain passports for British ships of discovery to sail without being attacked by French vessels. Afterwards he was President of the Pennsylvania Executive Council on three occasions, working hard to abolish slavery. Franklin was an important figure in the Federal Constitutional Convention of 1787.

Banks maintained a warm correspondence with Franklin on matters of science, both men having been friendly in London before politics separated them. Franklin was a great scientist, who introduced the terms 'negative' and 'positive' to describe electrical charges. He also proved that lightning is electrical in nature, and went on to invent the lightning conductor. He was awarded the Copley Medal of the Royal Society in 1753 for this work. Franklin founded the American Philosophical Society in 1743, and an Academy for the Education of Youth, opened in 1751, which later became the University of Pennsylvania.

Letters 13, 16, 18.

GILBERT (formerly Giddy), Davies F.R.S. (1767–1839): landowner; natural philosopher; politician; author; entered parliament as M.P. for the borough of Helston, 1804; High Sheriff of Cornwall, 1792–1793; P.R.S., 1827–1830. Gilbert (as he was known after marrying into the Gilbert family in 1808) had literary tastes at an early age, and enjoyed the company of men of letters. He had wide interests in natural history too, and was a Fellow of the Linnean Society. Gilbert also promoted the Geological Society of Cornwall, becoming its president. Thomas Beddoes (1760–1808) dedicated *Observations on the Nature of Demonstrative Evidence...*, London (1793) to Gilbert. Humphry Davy was encouraged by him as well.

Gilbert worked extremely hard in parliament, and was a member of many committees. As a wealthy landowner, with large estates in Eastbourne and Cornwall, he took a keen interest in money and the economy. Gilbert was particularly concerned at the effect of high gold prices on currency values. In 1811 he produced *A Plain Statement of the Bullion Question...*, London. He later served on a committee to consider ways of preventing the forgery of bank notes. This was chaired by Banks at 32 Soho Square in 1819, as was another committee on which Gilbert served to investigate weights and measures. Soon after Banks died in 1820, Humphry Davy was elected President of the Royal Society, and Gilbert was appointed Treasurer. However, by 1827, Davy was too ill to continue as President, and so Gilbert took his place.

Letter 129.

GOOD, Peter (d.1803): gardener; traveller; assisted Christopher Smith in the transfer of plants from the Royal Botanic Gardens at Kew to Calcutta, 1795, and returned with plants for Kew; employed by Lieutenant-General William Wemyss (1760-1822) at Wemyss Castle, Kilmarnock, Scotland until his appointment to H.M.S. *Investigator* in 1801. Good was the gardener on board H.M.S. *Investigator*, and received a salary of £105. He died of dysentery on 12 June 1803, and so did not survive the voyage. His journal is a good record of his work on the voyage though: P.I. Edwards (Ed.), *Bulletin of the British Museum (Natural History), Historical Series*, 'The Journal of Peter Good, Gardener on Matthew Flinders Voyage to Terra Australis 1801–1803', vol. 9, London (1981). The manuscripts of this journal are at The Natural History Museum, London.

Letter 84.

GOULBURN, Henry (1784–1856): statesman; entered parliament by petition as M.P. for Horsham, 1808; Under-Secretary for the Home Department, 1810–1812; Under-Secretary for War and the Colonies, 1812–1821; commissioner negotiating peace with America, 1814; sworn a member of the Privy Council, 1821; Chief Secretary, Ireland, 1821–1827; Chancellor of the Exchequer, 1828–1830 and 1841–1846; Home Secretary, 1834–1835. Goulburn was an effective minister. He introduced such useful bills as the Irish Tithe Composition, 1823. However, he also resisted the Roman Catholic Disability Removal Bill, and maintained his opposition to the relief of Roman Catholics in later years. As Chancellor, Goulburn achieved good financial management of the economy. Only a few letters between Banks and Goulburn survive. These concern exploration and plant collecting in the main, but also reveal a warm relationship between the two men.

Letter 120.

GRENVILLE, William Wyndham (1759–1834), 1st Baron Grenville: statesman; entered parliament as M.P. for the borough of Buckingham, 1782; Chief Secretary to the Lord Lieutenant of Ireland, 1782–1783; sworn a member of the Irish Privy Council, 1782; appointed Paymaster-General in 1783; sworn a member of the English Privy Council, 1783; Joint-Paymaster-General, 1784–1789; appointed Commissioner of the Board of Control in 1784; Vice-President of the Committee of Trade, 1786–1789; Speaker of the House of Commons, 1789; Secretary of State for the Home Department, 1789–1790; President of the Board of Control, 1790–1793; Secretary of State for Foreign Affairs, 1791–1801; Auditor of the Exchequer, 1794; Prime Minister, 1806–1807. Grenville feared revolutionary principles, and advocated strong resistance to France in war. After the Irish rebellion of 1798, Grenville supported plans for a legislative union, to be followed by Catholic emancipation. The union came in 1800, but when George III opposed Catholic emancipation in 1801, Grenville resigned along with William Pitt the younger, and several other cabinet members. Pitt died in 1806, and Grenville formed a brief coalition of 'All the Talents.' It ended with a confrontation between Grenville and the King over, once again, the Catholic question. However, this was not before the abolition of the slave trade in June 1806.

Letter 63.

GREVILLE, Colonel Robert Fulke F.R.S. (1751–1824): diarist; soldier; officer of the 10th Dragoons; Equerry to George III, 1781–1797. Greville was appointed equerry to George III without having made an application. His good nature made him popular at court, and Greville accompanied the King in his daily routines. Greville kept a diary in which he described the unusual events of George III's illnesses: *The Diaries of Robert Fulke Greville*, (Ed.) F.M. Bladon, London (1830). Arrangements were made for Banks to accompany the King on walks in Richmond and Kew Gardens as the King recovered from his first serious attack of porphyria: 'The Second Diary' (1788–1789), pp. 75–261.

Greville also assisted in the plan to import merino sheep to England, and when he informed the King of Banks's willingness to act in the matter, "The King instantly replied 'Sir Joseph Banks is just the Man. Tell Him from Me that I thank Him, & that his assistance will be most welcome.' ": 'Importation of Merino Sheep', pp. 71–73. Greville resigned as equerry in 1797 to marry Louisa Cathcart, Countess of Mansfield, (1758–1843).

Letters 102, 103.

GREVILLE, Henry Francis (1760–1816): Lieutenant-Colonel in the Army. Greville married twice, firstly on 18 August 1791 to Catherine Graham (d.1803) of Norton Conyers. His second wife was a widow, Sophia Lambert (*née* Whyte), and they were married on 25 February 1805. There were children from this marriage. Sophia survived her husband, dying in 1839. Only two known letters between Banks and Greville survive.

Letter 108.

HAMILTON, Lady Emma (*née* Emily Lyon), (c.1765–1815): wife of Sir William Hamilton, and Horatio Nelson's mistress. Emma's youthful beauty won her admiration, but in later years she was involved in affairs which caused controversy. Sir William Hamilton was first introduced to Emma by his nephew, Charles Francis Greville, who had been living with her for four years. Hamilton was greatly attracted to the pretty 'tea-maker of Edgware Row'. In 1786 he took Emma and her mother to Naples. Five years later Hamilton and Emma were married. Lady Hamilton first met Horatio Nelson in 1793, and they became lovers. She bore him a daughter, Horatia (1801–1881). Lady Hamilton became bankrupt after the death of her husband and Nelson, and in 1813 was arrested for debt. The next year she fled to Calais, where she died. The majority of Banks's letters to and from Naples were with Sir William.

Letter 80.

HAMILTON, Sir William F.R.S. (1730–1803): diplomatist; collector and antiquary; archaeologist; pioneer in the scientific study of volcanoes; Ambassador at the Court of Naples, 1764–1800; sworn a member of the Privy Council, 1791; member of the Society of Dilettanti. During his years as British Envoy in Naples, Hamilton collected pictures and ancient works of art, mainly Etruscan, Greek and Roman antiquities. He gained an international reputation as a scholar and arbiter of taste. Hamilton was also a pioneering natural historian, who studied Vesuvius and Etna. He ascended the former 22 times, and contributed much to early knowledge of volcanoes. In 1791 he married Emma Hart, who later became Horatio Nelson's mistress. Hamilton was made a Knight of the Bath in 1772.

Banks's correspondence with Hamilton was varied. It concerned natural history, and Hamilton's work on volcanoes. Banks wrote: 'That I envy you your situation within two miles of an Erupting Volcano, you will easily guess. I read your Letters with that Kind of Fidgetty anziety which continually upbraids me for not being in a similar Situation. I envy you. I pity myself. I blame myself, & then begin to tumble over my Dried Plants in hopes to put such wishes out of my head, which now I am tied by the leg to an arm chair I must with diligence suppress.' Banks to Hamilton, 4/12/1778, Egerton MSS. 2641, ff. 130–1. As President of the Royal Society, Banks encouraged Hamilton to submit papers to the *Philosophical Transactions* on the subject instead. He also kept Hamilton informed about social and scientific news in England. In return, Hamilton used his influence at the Court of Naples for Banks. Both men shared an interest in the customs and remains of Italy too.

Letters 14, 27, 79.

HAWLEY, Sir Henry (1745–1826) 1st Baronet of Leybourne Grange, Kent; knighted, 1795. Hawley was Banks's first cousin, and their correspondence relates to family life in the main. The management of property, social matters and some contemporary affairs are discussed. Hawley was an executor of Banks's will.

Letters 58, 61.

HOME, Sir Everard F.R.S. (1756–1832), 1st Baronet: surgeon; 1st President of the Royal College of Surgeons in 1821; member of the African Association, active 1802–1831, and afterwards a Fellow of the Royal Geographical Society. Home trained under, and then assisted John Hunter, the surgeon and anatomist who married his only sister. In 1790, Home lectured for Hunter at St. George's Hospital, and by 1792 had succeeded him as lecturer on anatomy. When Hunter died in 1793, Home was elected surgeon to St. George's Hospital, a post he held until 1827. In due course, Home became a keeper, and afterwards a trustee of the Hunterian collection. By 1821 he had risen through the College of Surgeons to the position of President, and in the same year he was made surgeon at Chelsea Hospital.

Home had become a successful lecturer, and a surgeon of considerable professional standing with a large surgical practice. Home made a prodigious number of contributions to the *Philosophical Transactions*, but his earlier papers were the most important ones. However, there remained a question over the extent to which Home relied on Hunter's papers, which he kept for 30 years after Hunter's death, and then destroyed. Banks maintained a friendly and open relationship with Home, who acted as his personal physician, discussing, among other things, intimate details of personal health.

Letters 73, 114, 119, 135.

HOOKER, Sir William Jackson F.R.S. (1785–1865): botanist; botanical artist; author; traveller; Regius Professor of Botany, Glasgow, 1820–1841; Director of the Royal Botanic Gardens, Kew, 1841–1865; made a Knight of Hanover, 1836; a founder of the Wernerian Society, Edinburgh; knighted, 1836. Hooker devoted himself to natural history at an early age. In 1806 he botanized in Scotland with his future father-in-law, Dawson Turner. Three years later, and on Banks's advice, Hooker travelled to Iceland to collect in all branches of natural history. He narrowly escaped with his life when the ship caught fire as he returned. His collections were destroyed. In 1814, Hooker toured in France, Switzerland and Northern Italy, meeting many botanists who would later prove useful contacts. Like Banks, he had by now gathered together an impressive library and herbarium. Moreover, his reputation was growing along with his carefully managed correspondence. Banks suggested Hooker accept the Regius Professorship of Botany at Glasgow in 1820, where Hooker lectured with great success. Many former pupils swelled a global network of letter writers on a scale to match Banks's example. Hooker also maintained links with the Admiralty, colonial and India offices.

When William Townsend Aiton retired from the Royal Gardens at Kew in 1841, Hooker was appointed Director. The gardens grew significantly under his management, and became a national institution. Public access was allowed, and Hooker promoted a wide system of plant exchange similar perhaps to the one Banks had fostered. In addition to these achievements, Hooker produced some one hundred volumes on systematic and economic botany. He readily assisted other botanists too. When Hooker died, his library and herbarium were purchased by the nation, and they form the basis of the collections at Kew.

Letter 116.

HUNTER, John (1738–1821): Vice-Admiral, R.N.; second captain of H.M.S. *Sirius* in the 'First Fleet' under Commodore Arthur Phillip, 1787–1788; captain of H.M.S. *Sirius* on voyage from Port Jackson to Cape of Good Hope by Cape Horn, 1788–1789; Governor of New South Wales, 1795–1801. Hunter was steadily promoted as he sailed on various ships of the Royal Navy. In 1786 he was advanced to post rank, and appointed Second Captain of H.M.S. *Sirius*. H.M.S. *Sirius* was part of the 'First Fleet', which sailed under Commodore Arthur Phillip in 1787 to found a colony at New South Wales. In 1790, Hunter lost his ship on a coral reef near Norfolk Island, where he remained with a large group of convicts for a year before being relieved. However, he was acquitted of responsibility for the wreck on his return to England in 1792. Hunter served on H.M.S. *Queen Charlotte* as a volunteer, seeing action in 1794. In 1795 he was appointed Governor of New South Wales.

Governing distant colonies proved a difficult task, not least because of the illegal trade in spirits maintained by profiteers and monopolists. Some men grew powerful through it, and ensured that Hunter's name was blackened in England. Since Hunter received limited official support when he complained about the situation, his position was weak. Nevertheless, he encouraged exploration, and did what he could to establish the colony on a firm basis.

Letters 69, 75.

HUTCHINSON, James (1752–1793): physician; army surgeon; Surgeon-General of Pennsylvania, 1778–1784; physician of the Port of Philadelphia; staff member at the Pennsylvania Hospital, 1777–1778 and 1779–1793; trustee of the University of Pennsylvania, 1779–1781; made a member of the American Philosophical Society, 1779; a Fellow and incorporator of the College of Physicians; Secretary of the

American Philosophical Society, 1782–1793; held chair of materia medica at the University of Pennsylvania, 1789–1791; professor of chemistry, University of Pennsylvania, 1791–1793. Hutchinson studied in Philadelphia and London. He carried important dispatches from Benjamin Franklin to Congress on his return to America from Europe via Paris. Once in Philadelphia again he joined the army as a surgeon, and later became Surgeon-General of Pennsylvania. After establishing a successful practice in Philadelphia, Hutchinson became a trustee and professor at the University of Pennsylvania. He was a good physician and teacher. He was also interested in local politics, and a member of the Whig party.

Letter 29.

JENKINSON, Charles (1727–1808), 1st Baron Hawkesbury and 1st Earl of Liverpool: statesman; administratior; entered parliament as the M.P. for Cockermouth, 1761; appointed Under-Secretary of State, 1761; Secretary to the Treasury, 1763–1765; made a Lord of the Treasury, 1767; made a Lord of the Admiralty, 1768; appointed Vice-Treasurer of Ireland, 1772; sworn a member of the Privy Council, 1772; Master of the Mint, 1775; appointed Secretary at War, 1778; President of the Board of Trade, 1786; Chancellor of the Duchy of Lancaster, 1786. Banks and Liverpool had common interests in agriculture, trade and coinage. These are reflected in their correspondence, which frequently touched on official matters in a detailed way. After 1783 Liverpool spoke little in parliament, except on commercial affairs. By 1796 he had almost retired from public life, and from 1800 his knee joints deteriorated so much he could not walk.

Letters 25, 66, 78.

JUSSIEU, Antoine Laurent de F.R.S. (1748–1836): botanist; graduated Medical Faculty, Paris, 1770; appointed sub-demonstrator, Jardin du Roi, 1773; Hospital of Paris (in charge), during the French Revolution, 1789–1794; professor of botany, Museum d'Histoire Naturelle, 1793–1826. Jussieu promoted a 'natural' classification system which distinguished relationships between plants by reference to a large number of characters. It was an advance on the Linnean system, which relied on only a few. His great publication was *Genera Plantarum...*, Paris (1789). This provided a thorough summary of the current knowledge of plant taxonomy, with excellent generic descriptions. Jussieu had drawn on the correspondence and collections of men like Philibert Commerson, Banks and Dr. James Edward Smith to produce it. He distinguished 15 classes, and 100 families. The lasting value of his contribution is clear, for 76 of the 100 families remain in botanical nomenclature today. In a few years his work had influenced most leading European botanists. Robert Brown was one of Jussieu's leading proponents in Britain. The letters between Banks and Jussieu mostly concern botany, but references to scientific societies and public events are also sometimes made.

Letter 31.

KING, John (1760–1815): state official; Permanent Under-Secretary of State, Home Office, 1792–1800. The correspondence between Banks and King concerned official business, especially to do with home affairs. The disturbances in Lincolnshire late in 1796 caused a flurry of letters between them. Their correspondence continued into 1797, and the Lincolnshire supplemental militia was the main subject then. In 1798 both men exchanged interesting letters to do with the colony at New South Wales, and arrangements for the voyage of H.M.S. *Porpoise*. A good example was: Banks to King, [20]/6/1798, M.L. Banks MSS. Series 19.03. Banks proposed exploration of the interior of Australia in this letter, and recommended Mungo Park to undertake it. By the turn of the century Banks had arranged for a mission to Australia, with Matthew Flinders in command, and Robert Brown leading the natural historians and collectors.

Letter 65.

KNIGHT, Thomas Andrew F.R.S. (1759–1838): plant physiologist; horticulturist; President of the Horticultural Society, 1811–1838. His brother was Richard Payne Knight. By 1795, Thomas Andrew Knight was submitting papers to the Royal Society on grafting and diseases among fruit trees. In 1803, Banks introduced him to Humphry Davy, and the two men became close friends. Knight was an inaugural member of the Horticultural Society in 1804, as was Banks. In 1811 he became its President.

Banks's correspondence with Knight was rich and interesting, and led to a number of papers Banks published in the *Transactions of the Horticultural Society*, while more than 20 pieces by Knight appeared in the *Philosophical Transactions*, 1795–1818. Banks and Knight discussed various topics in horticulture and botany, including the influence of gravitation on the direction of growth in plants, the movement of sap, and new varieties of fruit and vegetables. Indeed, Knight raised new apples, cherries, nectarines, plums, cabbages, peas and potatoes, many of which bear his name.

Letters 126, 127, 131.

KÖRTE, Dr. Friedrich Heinrich Wilhelm (1776–1846): teacher; bibliophile. Körte was born in Achtersleben, where his father was an Archdeacon. He was first educated in the town, and then from 1792 at the cathedral college in Halberstadt. Afterwards, Körte studied law in Halle, but his personal interests drew him instead to literature and art. In 1799 he returned to Halberstadt, where he became headmaster of a school recently founded by his uncle, the poet J.W.L. Gleim. However, Körte's methods caused controversy, and disagreed with those his uncle envisaged. A long legal battle was therefore fought in which the school was closed, and Gleim's legacy taken from Körte. A new school was established without Körte, who did not teach again. He received substantial compensation from the Gleim fund though, and this allowed him to lead an independent life in Halberstadt studying literature and art. He inherited his uncle's library and papers, and also those of his father-in law, Heinrich von Kleist. These provided ample material from which Körte published a number of volumes, mainly biographies, letters and collected works.

Letter 128.

LAURAGUAIS, Louis Léon Félicité, Comte de (1733–1824), later Duc de Brancas: French author; bibliophile, savant and patron; member of the Académie des Sciences, 1758. Lauraguais was born at Versailles, and was known both for his wit, and for his liberalism. He was a friend to Banks as a young man, and they travelled together on the *Sir Lawrence* as far as Deal in 1772. Lauraguais went ashore there, and Banks proceeded on his voyage to the Hebrides and Iceland.

In December 1771, Banks wrote a long letter to Lauraguais describing the *Endeavour* voyage. He called it an 'abstract'. Lauraguais was greatly impressed by the account, and had it printed along with one of his own letters. He wrote to Banks explaining what he had done: Lauraguais to Banks, 17/[3]/1772, N.H.M. B.L. D.T.C. I 31. However, Banks immediately stopped the publication 'in the Press'.

Letter 6.

LESLIE, Sir John (1766–1832): mathematician; natural philosopher; elected to the chair of mathematics, University of Edinburgh, 1805; elected to the chair of natural philosophy, University of Edinburgh, 1819; knighted, 1832. Leslie was educated for the church at the University of Edinburgh, but preferred science to theology. In 1792 he moved to Etruria, Staffordshire, to stay with the Wedgwood family as a tutor. Leslie had a wide range of scientific interests, including electricity, on which he published in 1791. However, he is best remembered for *Experimental Inquiry into the Nature and Propagation of Heat*, which appeared in 1804, and this was dedicated to Thomas Wedgwood (1771–1805). Leslie was awarded the Rumford Medal a year later for his work. In 1805, ministers in Edinburgh opposed Leslie's election to the chair of mathematics at the university because of his approval of David Hume's ideas on causality. Nevertheless, he was elected. Leslie published a number of treatises on mathematics while in this post, and continued his researches on heat. In 1810 he used the absorbent powers of sulphuric acid to freeze water under the receiver of an air pump. It was the first recorded artificial congelation.

Letter 99.

L'HÉRITIER DE BRUTELLE, Charles Louis F.R.S. (1746–1800): botanist; author; appointed Superintendent of the Waters and and Forests, Paris Region, 1772; appointed Counsellor at the Cour des Aides, Paris, 1775; after the French Revolution, at the department of Justice. L'Héritier became interested in botany when he was Superintendent of the Waters and Forests, Paris Region. Although he studied independently, L'Héritier sought contact with the botanists at the Jardin du Roi, and by 1775 was rising in French society. In 1785 he commenced publication in Paris of the *Stirpus novae* series, which described many plants that had recently been introduced to Paris gardens. Banks provided considerable assistance to L'Héritier with this work, both by written letter, and through the specimens and plants he sent from 1783 onwards.

In 1786 L'Heritier fled France with Joseph Dombey's (1742–1794) herbarium. It had been gathered in Peru and Chile, and by agreement with the Spanish government could not be published. However, L'Héritier obtained the collection from Buffon, director of the Jardin du Roi, with the intention of including material from it in the *Stirpus novae*. The French government could not allow such a breach of promise, but failed to prevent L'Héritier crossing the Channel with the collection. He made his way to London, and Soho Square. Banks was annoyed at, and suspicious of L'Héritier, who arrived in September, but Banks did not feel the Frenchman could be turned away. Banks explained this to Jonas Dryander in a series of letters at: N.H.M. B.L. D.T.C. and Dryander Corr. So, L'Héritier had protracted use of Banks's library and herbarium, which, along with the Royal Gardens at Kew, contributed to his *Sertum Anglicum*..., (1788). This included only a small number of species from Dombey's herbarium. In December 1787, L'Héritier returned to Paris, but subsequently lost most of his fortune in the French Revolution, 1789–1794. He continued his botanical research, but could not now afford to publish anything. He was murdered on 16 August 1800, but the crime was never solved.

Letter 30.

LIND, Dr. James F.R.S. (1736–1812): physician; traveller. In 1766, Lind visited China in an East Indiaman. He graduated M.D. at Edinburgh in 1768, and observed the Transit of Venus at Hawkhill the next year. Lind submitted a paper to the *Philosophical Transactions* on the event, which was printed. He settled at Windsor in about 1771, and subsequently became physician to the Royal household. However, his love of the East, and the pleasure he obtained from experiments and tricks, might not have enhanced his reputation as a physician. Lind accompanied Banks on his voyage to Iceland in 1772, where Lind measured the height of the Great Geysir at Haukadal using a quadrant. In 1792, Banks tried to arrange for Lind to join the Macartney Embassy to China, 1792–1794, but Lind declined. Lind was also a friend and 'mentor' to Percy Bysshe Shelley (1792–1822).

Letter 36.

LINNÉ (Linnaeus), Carl von, the younger (1741–1783): naturalist; professor of medicine and botany at University of Uppsala, 1777–1783. Carl, an only son, was educated at Uppsala, where he later became a university professor. However, he did not inherit his father's passion for botanical study, and published little. Perhaps his best known work was *Supplementum plantarum*..., Brunsvigae (1781), which was unexceptional. Some have described his attractive appearance and reputation as a 'ladies' man', but he never married. His father's herbarium was left to the wife, and not son. There was even the stipulation that Carl should have no access to it. Carl's mother allowed the collection to deteriorate before her son eventually managed to sort through everything, discarding much of the damaged material. Carl visited Banks at Soho Square, and was present when Dr. Daniel Solander died there in May 1782. Just over a year later, he was dead too, and his mother offered the herbarium to Banks. Banks declined to buy it, but suggested Dr. James Edward Smith should.

Letter 12.

LLOYD, John F.R.S. (1749–1815): lawyer; landowner of Denbigh, Wales; called to the bar, 1781; Bencher of his Inn, 1811, and then Reader, 1815; entered parliament by petition as M.P. for Flintshire, 1797. Lloyd was descended from two ancient families of Wigfair and Hafodunnos. He was known as 'The Philosopher' because of his wide range of varied interests, and substantial library. The library comprised some 10,000 books, manuscripts and maps. He also possessed a large collection of scientific apparatus. As a landowner and intellectual, Lloyd had much in common with Banks. Their correspondence reflected this, and contained letters to do with natural history, local politics and estate or farm business. Banks invited Lloyd to visit Soho Square and Revesby Abbey, although Lloyd did not always come.

Letter 32.

LJUNGH, Sven Ingemar (1757–1828): civil servant; naturalist; collector; bibliophile; author; agriculturist; schooled at Jönköping; attended the gymnasium in Växjö, 1774; at Uppsala University, 1775–1777, taking a degree in theology, 1776; received academic stipend in 1778, and was given a position in the Kammarkollegium, which was the Swedish civil service; clerk or Kammarskrivare in the Kammarrevisionen, a department with judicial and accounting functions, 1778; appointed Deputy Crown Bailiff in North and South Vedbo, 1779, becoming Crown Bailiff there, 1780–1793; Fellow of the Patriotic Society, 1806; member of the Academy of Science, 1808; member of the Science and Antiquarian Society of Gothenburg, 1808; founding member of the Jönköping County Economic Society, 1814; corresponding Fellow of the Agricultural Academy, 1812. Ljungh was a very active man, and his interests were similar to Banks's in many respects. Ljungh was well educated. He was proficient in Latin, modern languages and botany. He studied natural sciences, and visited Carolus Linnaeus's home, where he received a warm welcome, and was taught by Linnaeus the younger. In 1777, Ljungh started medical studies, which were interrupted when he fell ill with malaria. Following his recovery, Ljungh became a civil servant. He was steadily promoted until in 1779 he was appointed Deputy Crown Bailiff in North and South Vedbo, becoming Crown Bailiff there in 1780. Towards the end of the decade, Ljungh's health failed due to overwork, and so he went on an extended tour with his wife to recuperate. His travel diary is now at Malmö City Archive. Ljungh retired from public service with the title 'Landskamrer' in 1793.

He then dedicated himself exclusively to natural history, and to the large gardens and farms which he cultivated in Skareda, in Lommaryd parish, and at Skärsjö in Bälaryd. Ljungh carried out a successful regime of experimental agriculture on both of these estates. He exchanged, purchased and collected insects, birds, shells, plants and minerals. In addition, he assembled a fine library containing some 5000 largely scientific volumes. All of these resources were available for others to use. Ljungh's own research produced essays, most of which were traditional species descriptions, and these were often published. He described exotic animals from Java, Ceylon and the Cape of Good Hope, small mammals from Småland, and new insects from his collection, which grew to nearly 6000 species. Furthermore, Ljungh wrote on agricultural discoveries he made, crop yields and meteorology. He was rewarded with the Vasa Order for his agricultural work.

Letter 65.

MACARTNEY, Sir George (1737–1806), 1st Earl Macartney: diplomatist; colonial governor; author; Envoy-Extraordinary to St. Petersburg, 1764–1767; entered the English parliament as M.P. for Cockermouth, 1768; entered Irish parliament as the member for Antrim, 1768; Chief Secretary for Ireland, 1769–1772; Knight Order of the Bath, 1772; appointed Governor of Toome Castle, 1774; Captain-General and Governor for Grenada, the Grenadines and Tobago, 1775–1779; Governor and President of Fort St George, Madras, 1780–1786; sworn to the Irish Privy Council, 1788; received Irish peerage, 1792; Ambassador Extraordinary and Plenipotentiary for the first British Embassy to China, 1792–1794; Governor at the Cape of Good Hope, 1796–1798. Macartney was an agreeable man, and an impressive ambassador. He was awarded the Polish Order of the White Eagle for the commercial treaty he negotiated as Envoy-Extraordinary to St. Petersburg. Macartney's tact and firmness served him well in Ireland too. In 1779 he saw action against the French when Grenada was attacked. Macartney defended his post gallantly as Captain-General and Governor, but was captured and then released. He served with particular distinction in Madras next, where he dealt with difficult military subordinates, and secured notable victories over the Dutch.

Macartney was appointed to lead the Embassy to China late in 1791, and Sir George Leonard Staunton was made Secretary. Banks was consulted about the personnel for this mission, and offered advice on Chinese natural history and culture. The Embassy sailed on 26 September 1792, and eventually reached the Imperial Court at Chengde almost a year later. However, the experience proved all too brief and uninformative. The Embassy departed after a stay at Macau, arriving home in September 1794. Banks assisted with the arrangement and description of the plant specimens collected from the Far East and China. He also helped Staunton with the official account of the voyage. For his part, Macartney published political accounts and journals based on his diplomatic experiences.

Letter 45.

MAITLAND, John (1754–1831): wool merchant of Basinghall Street, London, and of Woodford Hall, Essex; M.P. for Chippenham, Wiltshire; Chairman of the Committee of the Woollen Trade. Maitland was an old friend to Banks. For many years they worked together on the plan to establish the Spanish merino sheep in England. Banks sent five merino fleeces to Maitland in 1787 to be manufactured, and experiments like these showed that fine wools produced in England were comparable to those imported from Spain. From this modest beginning, based on a few animals Banks had obtained, emerged a close relationship, a Royal flock of Spanish sheep, and the dispersal of the breed to farms throughout the country. A supportive merchant like Maitland was essential to this process. He was equally important when the idea to encourage such sheep in New South Wales was considered in London. John Macarthur arrived in the capital in 1803 prepared with wool samples and a proposal of his own. As events took their course, Macarthur drew on Maitland's knowledge, and his influence with Banks. The result of their combined labours is apparent today in the successful flocks of Australasia.

Letter 95.

MALONE, Edmund (1741–1812): critic; editor; author; scholar. Malone was educated at Trinity College, Dublin. He came to London as a student of the Inner Temple in 1763, and became interested in politics and literature. He was introduced to Dr. Samuel Johnson in 1765. After travelling in France, 1766–1767, he arrived back in Ireland to pursue a legal career. This work did not satisfy Malone, who had settled in London permanently by 1777. In 1782 he was invited to join the Literary Club. Malone was later one of the committee members responsible for the erection of Johnson's monument in St. Paul's Cathedral. He is chiefly remembered for his edition of William Shakespeare's (1564–1616) plays, published in 1790. He also researched the order in which Shakespeare wrote the plays. Malone sought to establish Shakespeare's texts as accurately as possible, and he set high scholarly standards for editors of Elizabethan literature in general.

Letter 43.

MANNING, Thomas (1772–1840): traveller; scholar of Chinese language and culture. Manning became interested in Chinese language and culture while living in Cambridge. He was a scholar of Caius College, and a private tutor while there. He went to Paris to study in 1800, but was forced to leave in 1803 when war was resumed between Britain and France. Napoleon Bonaparte signed his passport home. Manning then planned to perfect his Chinese at Canton. In 1806, Banks therefore obtained free passage on an Honourable East India Company vessel for Manning, who was to work as a doctor at Canton. He remained there until 1810, when he travelled to India to explore. By 1811 he had reached the north-east of the country, near Bhután, and late in the year he arrived at the Tibet border. In December, Manning entered the 'forbidden' city of Lhasa. After returning to Canton for a period, Manning joined the Amherst Embassy to China as a junior secretary and interpreter. This was in 1816. In 1817, after departing the Embassy, and following a shipwreck, Manning landed at St. Helena, where he met Napoleon once again. Afterwards, Manning settled in England, but made a trip to Italy in 1827. He then retired. His papers and printed books were given to the Royal Asiatic Society following his death.

Letter 104.

MARCH, Thomas (d.1790): merchant of Lisbon. March was senior partner in Thomas March and Company at Lisbon. He assisted Banks in the plan to obtain merino sheep from Spain, and bring them to England for the royal and national flocks.

Letter 33.

MARUM, Martin (Martinus) van (1750–1837): natural philosopher; plant physiologist; physician; appointed Director of the natural history collections of the Netherlands Society of Sciences, 1777; appointed Director of Teyler's Museum and Library (*Teyler's Stichting*), 1784; appointed Secretary of the Society, 1794. Van Marum graduated from Groningen University, where he had become interested in plant physiology. His dissertation on the circulation of plant juices was published in 1773 as *Dissertatio...de motu Fluidorum in plantis, experimentis et observationibus indagato...*, Groningae. It gained him a doctorate, and in the same year he received a medical degree for a study of comparative animal physiology. Working collaboratively, Van Marum developed an electrical machine in 1776, and remained interested in electricity for the rest of his scientific career. He practised medicine in Haarlem from 1776 to 1780, and was elected to the Netherlands Society of Sciences (*Hollandsche Maatschappij der Wetenschappen*). In 1777, he was made director of the Society's cabinet of natural curiosities.

In 1784, Van Marum was made director of Teyler's Cabinet of Physical and Natural Curiosities and Library, an appointment which kept him very busy. He had a large electrical machine constructed for experiments, and described his work in the periodical *Verhandelingen uitgeeven door Teyler's Tweede Genootschap*, (1785, 1787 and 1795). Van Marum concluded that Benjamin Franklin's theory of a single electric fluid was correct. He also later introduced the term 'Voltaic pile'. Van Marum visited Antoine Laurent Lavoisier in 1785, and convinced himself Lavoisier's combustion theory was valid. Afterwards, he promoted the Frenchman's system of chemistry in the Netherlands. Van Marum analysed carbon dioxide, ammonia, and smelled ozone, which he did not recognize as a form of oxygen. Van Marum was also interested in fossils, which he bought and collected. He was able to arrange the collection of minerals at Teyler's Museum according to the methods of Jean Léopold Nicholas Fréderick Cuvier F.R.S. (1769–1832). In 1803, Van Marum bought a country house with a large garden where he cultivated mainly South African plants.

Letter 81.

MENZIES, Archibald (1754–1842): botanist and collector; naval surgeon, R.N.; exploring naturalist; on H.M.S. *Nonsuch* as assistant surgeon, 1782; on H.M.S. *Assistance* as surgeon, 1784–1786; on *Prince of Wales* (a fur trader) as surgeon, 1786–1789; on H.M.S. *Discovery* II as surgeon and naturalist, 1790–1795; on H.M.S. *Princess Augusta* as surgeon, 1796–1799; on H.M.S. *Sans Pareil* as surgeon, 1799–1802. Menzies trained as a gardener at the Royal Botanic Garden, Edinburgh. Dr. John Hope, Professor of Botany, enabled him to study as a surgeon at Edinburgh University. Menzies made a tour of Scotland and the Hebrides in 1778, collecting plants as he went. Following a number of voyages to the New World and Far East as a ship's surgeon, he was chosen as naturalist and then surgeon on the voyage of H.M.S. *Discovery* II under Captain George Vancouver. This was in 1790, and it was Banks who recommended Menzies. However, Menzies and Vancouver could not work together, and Menzies was placed under arrest for the final months of the expedition. Many living plants under his care were therefore lost, but the dried plants and seeds arrived safely home in 1795. Menzies returned to military service in November 1796, and sailed on H.M.S. *Princess Augusta* and H.M.S. *Sans Pareil* for about three years each. He married in 1802 and settled in London, where he practised surgery. By now his health was poor as he suffered from asthma. Menzies retired in 1826, but continued to correspond with other botanists, particularly on ferns and mosses.

Letter 42.

MERCK, Johann Heinrich (1741–1791): amateur palaeontologist and naturalist; collector; author; essayist; critic; senior civil service official; entered civil service, 1767; appointed Paymaster in the War Ministry, 1768, later achieving the position of Councillor of War at the Court of Landgraf Ludwig IX of Hessen in the city of Darmstadt. Merck was born in Darmstadt, where his father was a pharmacist. He was the youngest of 11 children, and attended the gymnasium there. As a young man, Merck travelled to Geneva, where he met his future wife. He was initially interested in poetry, and wanted to be a writer. He was friendly with eminent men of letters, such as Johann Wolfgang Goethe (1749–1832), but could never rival their talent. He therefore turned to botany and mineralogy, which he studied with dedication.

In due course, Merck started to collect elephant and rhinoceros bones. Most of the sites he visited for material had been exploited, but he assembled a respectable collection of fossil vertebrates by purchasing items as well. Drawings of these were made, many of which were excellent. Some were engraved, but never published because Merck believed the animals still lived. He wanted to publish an encyclopaedia containing only extinct species. His work is important, especially in the way it preceded that of Jean Léopold Nicholas Fréderick Cuvier. Merck took students when his finances deteriorated, and died an unhappy man. He never achieved fame despite much hard work, and committed suicide.

Letter 22.

MILIUS, Pierre-Bernard (1773–1829): Captain, French navy; navigator. Milius took part in the French mission led by Nicolas Thomas Baudin, 1800–1803. Baudin sailed into the Pacific with two ships, *Le Géographe* and *Le Naturaliste.* Milius was 1st Lieutenant on *Le Naturaliste* under Captain Hamelin. *Le Naturaliste* arrived in Port Jackson on 24 April 1802 in considerable difficulty. The officers and crew were suffering from scurvy, but exemplary help was provided by the colony there. Milius was too sick to leave with his ship on 18 May, and stayed behind for further treatment. Baudin arrived in *Le Géographe* on 20 June. Milius departed for Mauritius on her. When Baudin died at Mauritius, Milius took command for the journey back to France.

Milius later commanded *La Didon,* and was captured by the British after a vigorous fight with H.M.S. *Phoenix.* In 1804 and 1805 he sought liberty to take the waters at Bath to improve his health. Banks intervened on his behalf, and by May 1806 Milius was free. It seems clear that Banks wanted to negotiate for English prisoners held abroad in return for such a gesture, particularly Matthew Flinders. However, Flinders was not released, and Banks was castigated at the Admiralty for his efforts.

Letter 105.

MONTAGU, John F.R.S. (1718–1792), 4th Earl of Sandwich: diplomatist; statesman; appointed a Lord Commissioner of the Admiralty, 1744; appointed Captain in the Duke of Bedford's foot regiment, 1745; aide-de-camp and 2nd colonel in the army, 1745; plenipotentiary at the conferences at Breda, 1746, and at Aix-la-Chapelle, 1748; Joint Vice-Treasurer and Receiver of the Revenues of Ireland, 1755–1763; Principal Secretary of State, 1763–1765; Postmaster-General, 1768–1770; First Lord of the Admiralty, 1747–1751, 1763–1765 and 1771–1782; nominated Secretary of State, 1770. Sandwich was an influential friend and patron to Banks as a young man. He was also a controversial figure. Sandwich took part in the prosecution of John Wilkes F.R.S. (1727–1797), and was criticized as First Lord of the Admiralty for the poor state of the navy, particularly during his final tenure. In April 1779 his mistress, Martha Ray, was murdered by James Hackman (1752–1779), whose proposal of marriage she had rejected. Sandwich retired from public life in 1782.

Letter 7.

NELSON, Horatio (1758–1805), 1st Viscount Nelson, Duke of Bronté: Admiral, R.N. Following early naval missions to the Caribbean, Arctic and India, Nelson was made captain of a frigate in the West Indies at the age of 20. He served here during the American War, and in 1784 returned to enforce the Navigation Acts. He married Frances Nisbet (1761–1831) in 1787. Nelson is chiefly remembered for his aggressive tactics, and a willingness to disobey orders in pursuit of victory. Both proved effective against the French, whom he defeated in more than one historic engagement at sea.

Nelson was injured in the wars with France of 1792–1802. He lost his right eye during the attack on Calvi in the Mediterranean, and then his right arm near Santa Cruz in the Canary Islands. By this time he had been made a Rear-Admiral. Nelson defeated a French fleet at Aboukir Bay, near Alexandria, in 1798. This was called the Battle of the Nile, and it isolated the French army in Egypt. Nelson recovered from his wounds at Naples, where he started an affair with Lady Emma Hamilton. In 1801 he was back in action as a Vice-Admiral when he attacked the Danish fleet at Copenhagen. In this battle Nelson disobeyed his commanding officer's order to cease fire, and continued fighting with the Danish until they surrendered. He was appointed Commander-in-Chief of the Fleet following the battle. On 21 October 1805 Nelson was mortally wounded at Trafalgar.

Letter 93.

NEPEAN, Sir Evan F.R.S. (1751–1822), 1st Baronet: Admiralty official; administrator; colonial governor; clerk and purser in the navy, 1776–1782; Secretary to Port Admiral at Plymouth, 1782–1784; Under-Secretary of State, Home Office, 1784–1794; Under-Secretary for War, 1794; knighted, 1802; Secretary of the Admiralty, 1795–1804; Chief Secretary for Ireland, 1804; Lord Commissioner of the Admiralty, 1804–1806; Governor of Bombay, 1812–1819. Nepean was an excellent administrator in government, and therefore a useful contact for Banks. They worked together in planning the First Fleet, which sailed for Botany Bay, 1787–1788. Nepean came to trust Banks, who was allowed to organize voyages such as those mounted under William Bligh to obtain bread-fruit from the Pacific: H.M.S. *Bounty* 1787–1789, and H.M.S. *Providence* 1791–1793. Later on Nepean gave Banks considerable freedom in the preparations for the *Investigator* mission, 1800–1803. The diligent work of both men extended British maritime influence around the world, and their corrrespondence continued when Nepean was appointed Governor of Bombay, India, in 1812: yet another area of mutual interest.

Letter 56.

PENNANT, Thomas F.R.S. (1726–1798): traveller; naturalist; antiquarian. Pennant's reputation among naturalists in the eighteenth century was based on such publications as *The British Zoology*, London (1766), and his *History of Quadrupeds*, London (1781). However, Banks's contribution to Pennant's work has not been widely appreciated. For instance, from 1768 Banks allowed Pennant to study his Newfoundland diary, many bird skins, his zoological notes, and paintings of his Newfoundland specimens by Sydney Parkinson. Pennant made free use of these in his *Arctic Zoology*, London (1784), especially in the sections on Newfoundland.

Furthermore, Pennant published accounts of his own travels, such as *A Tour of Scotland and a Voyage to the Hebrides...*, Chester and London (1774–1776). In this, he included some admirable passages from Banks's unpublished *Journal of a Voyage to the Hebrides begun July 12th 1772 and a small part of the continuing*

Voyage to Iceland and the Faeroes as far as September 6th (Sunday), Blacker-Wood Library, McGill University, Montreal.

Pennant was a very able zoologist, but he could be demanding too. In 1783 Banks and Pennant quarrelled. Banks discovered that John Frederick Miller had engraved and exhibited plates made from drawings of his Iceland expedition, 1772, without permission. When Pennant wanted to use some of them for his *Arctic Zoology* Banks refused. He would not support Pennant's request for drawings from James Cook's last voyage either. These were at the Admiralty. By 1790 both men were again on friendly terms despite these differences.

Letter 5.

PERRIN, William Philip (or Phelp) F.R.S. (1742–1820): naturalist; landowner of Farleigh, Kent, and Jamaica, West Indies. Perrin was one of Banks's friends at Eton, and a fellow undergraduate at Christ Church, Oxford. Perrin's London home was in Bloomsbury Square as his father was rector of St. George's Church, Bloomsbury Way. While at Oxford, Perrin read Law, and in 1761 was admitted as a student to the Inner Temple. In 1774 he moved to Smith's Hall, West Farleigh, Kent, and was High Sheriff for the county in 1776.

Perrin and Banks had much in common. Like Banks, Perrin was interested in botany and gardening. The few letters between the two men which remain concern both topics. Indeed, Perrin accompanied Banks on his earliest expedition collecting plants in the Weald of Kent, 1765. Banks also wrote to Perrin with news of his voyages. These are lively pieces, which reveal something of Banks's style as a youthful correspondent and traveller. In 1772 Perrin was elected a Fellow of the Royal Society, and Banks one of those who nominated him. However, Perrin was really a minor, amateur naturalist, and he did not publish any scientific work of note.

Letters 1, 3, 74.

PHILLIPS, Thomas F.R.S. (1770–1845): painter; elected an Associate of the Royal Academy, 1804; made a Royal Academician, 1808; Professor of Painting in the Royal Academy, 1825–1832. Phillips was born in Dudley, Warwickshire. At the age of 20 he travelled to London, and Benjamin West found him work on the painted-glass windows of St. George's Chapel at Windsor. Phillips became a student of the Royal Academy in 1791, and the next year sent his first picture for exhibition, a 'View of Windsor Castle.' Others followed, and his talent for portraiture emerged, but his subjects were not yet famous men and women. In 1804 he was elected an Associate of the Royal Academy, and in 1808 he became a Royal Academician. Phillips's diploma piece was creative, 'Venus and Adonis.'

By 1806 Phillips was painting the most eminent members of society, such as the Prince of Wales. Two years later it was Banks's turn, and his portrait of 1808 was exhibited at the Royal Academy in 1809, and engraved by Niccolo Schiavonetti, in 1812. More than one copy of this portrait was made, and in 1814 Phillips painted

Banks again for the Corporation of Boston. Banks was shown at the age of 71 in a military uniform, holding a map of a fen drainage scheme to signify his enormous influence in Lincolnshire. Another younger traveller, but in the Romantic tradition, was painted twice by Phillips for the exhibition of 1814: Lord Byron. In one portrait he appeared in an Albanian costume, and the other was a fine likeness for John Murray (1778–1843). Phillips was elected Professor of Painting in the Royal Academy in 1825, and visited Italy and Rome. He resigned the post in 1832. In 1833 he published *Lectures on the History and Principles of Painting*..., London, which may be read alongside many occasional essays on the fine arts.

Letter 110.

PHIPPS, Constantine John F.R.S. (1744–1792), 2nd Baron Mulgrave: Arctic explorer; naval officer; politician; bibliophile; entered parliament as M.P. for Lincoln, 1768; Lord of the Admiralty, 1777–1782; appointed Joint Paymaster-General of the Forces, 1784; sworn a member of the Privy Council, 1784; Commissioner for India, 1784–1791; Lord of Trade and Plantations, 1784–1791. Phipps was one of Banks's friends at Eton, and they sailed together as supernumeraries on H.M.S. *Niger*, on fishery patrol, in April 1766. It was an opportunity for Phipps to rest from full naval duties, and for Banks to explore overseas for the first time as a field botanist. They returned in January 1767, stopping at Portugal on the way. The next year Phipps stood as Member of Parliament for Lincoln, and Banks ensured he was elected by using his influence with local freemen.

In 1773 Phipps commanded H.M. Ships *Racehorse* and *Carcass* on a voyage of Polar exploration, with Horatio Nelson serving as midshipman on the *Racehorse.* However, ice blocked the sea just north of Spitzbergen. Banks and Dr. Daniel Solander classified and described the natural history collections that were made for Phipps's *A voyage towards the North Pole, undertaken by His Majesty's Command, 1773*, London (1774). On page 12 Phipps acknowledged Banks's assistance with 'very full instruction in the branch of natural history'.

Phipps displayed courage when in naval action against the French, particularly off Ushant in 1778, and again in 1781 when he captured the 32-gun frigate *Minerve* off Brest. In 1791 he became too ill to continue with the public duties he took up after his time at sea. Phipps was a member of the Society of Dilettanti, and established a private library which was remarkable for its books on maritime subjects. He also helped establish the Society for the Improvement of Naval Architecture in 1791 with Banks as vice-president. Beforehand, Phipps wrote to Banks: 'If there is to be the comfort of honourable Peace, I hope to spend much of my time in your Society; we are, I believe, the oldest friends to each other, and I can with great truth assure you that the length has only added to the value of such a friendship in my estimation.' 19/10/1790, N.H.M. B.L. D.T.C. VII 169. This was a tribute to a long frienship based on similar interests and drive, and one that lasted into the next generation of the Mulgraves.

Letter 10.

PITT, William, the younger (1759–1806): statesman; orator; entered parliament as M.P. for Appleby, 1781; Chancellor of the Exchequer, 1782; Prime Minister, 1783–1801 and 1804–1806. Pitt was made Chancellor of the Exchequer in 1782, and then, at the age of 24, accepted the premiership. He was Britain's youngest Prime Minister. During his first ministry Pitt introduced important reforms, his policies being influenced by the political economy of Adam Smith F.R.S. (1723–1790). After the Irish rebellion of 1798, he proposed a legislative union, to be followed by Catholic emancipation. The union came in 1800, but Pitt resigned in 1801 rather than confront George III's hostility to emancipation. He returned to office in 1804, when Napoleon Bonaparte threatened Europe with tyranny. Pitt formed a coalition with Russia, Austria and Sweden to contain the threat. In 1805 the French fleet was defeated at Trafalgar by Horatio Nelson, and the possibility of an invasion of the British Isles was ended. Pitt died in 1806, much exhausted by his labours, gout and port. He was a distant relation of Banks.

Letters 38, 39, 40, 53, 57, 67.

PRICE, William (fl.1780–1816): army Major; Equerry to to George III, c.1786–1791; Vice-Chamberlain and Secretary to Queen Charlotte, 1792–1801; landowner of Mancell, Herefordshire. Price was well known to Banks as a landowner with interests in sheep farming. Spanish merino sheep were crossed with the Ryeland breed by tenants on Price's farms in Herefordshire. George III favoured Price with gifts of sheep from the Royal Flock, and in return Price gave some of his own stock to the King. Price's name is therefore found in letters and documents relating to Banks's management of the Royal Flock.

In 1796 Price was a useful intermediary at court when Banks sought the return of Jacques Julien Houttou de la Billardière's botanical collection. Banks explained in letters to Price that material given to the Queen belonged to La Billardière, and should be sent to him in France. The Queen agreed to this, and everything was dispatched to Paris.

Letter 64.

PRIESTLEY, Dr. Joseph F.R.S. (1733–1804): chemist; theologian; member of the Lunar Society of Birmingham. Priestley's early scientific work was influenced by Benjamin Franklin, whom he met in 1766. It concerned electricity and optics. He was most productive from 1772 to 1780, when he discovered several new gases. In 1772 he discovered 'nitrous air' (nitric oxide, or nitrogen monoxide, NO). He left the gas in contact with iron filings and sulphur, and found it decreased in volume. The new gas thereby produced supported combustion. It was nitrous oxide (dinitrogen monoxide, N_2O) — Humphry Davy's 'laughing gas'. In the same year he isolated gaseous ammonia.

In 1772 Priestley found that a sprig of mint revived air, which a candle had exhausted, so that combustion could once more take place. His most famous discovery was that of oxygen. This was made in 1774. Priestley heated red calx of mercury (mercuric oxide, HgO) or minium (red lead, Pb_3O_4). Inhaling the resultant gas caused a strange sensation in the chest, but it kept mice in a sealed container conscious twice as long as normal air. Priestley concluded that he had prepared dephlogisticated air by removing phlogiston, a fiery principle. His adherence to the phlogiston theory is curious since isolating oxygen demolished that theory. However, Priestley was credited with the discovery of oxygen after publishing his findings promptly.

Priestley was an outspoken dissenter, and a supporter of the French revolutionaries. His radical views made him unpopular, and in 1791 rioters attacked his home in Birmingham. In 1794 he emigrated to Northumberland, Pennsylvania, in America, where he died.

Letter 41.

SAINT-AMANS (alias Amand), Jean Florimond Boudon de (1748–1831): French botanist; soldier; with the French expeditionary force in the French Antilles, 1768–1773; from 1773 a private scientist at Agen; sometime professor of natural history at the École Centrale of Lot-et-Garonne; President of the Conseil-General of the Départment Lot-et-Garonne, 1800–1803. Besides botany, Saint-Amans was interested in archaeology and numismatics. Current records show that just three letters between Banks and Saint-Amans have survived. They were exchanged in 1792, with two from Saint-Amans in Agen, and one reply from Banks in Soho Square. Banks wrote the shortest of these. Saint-Amans announced himself in the other two, giving his biographical details, and requesting a regular correspondence with Banks.

Letter 47.

SCORESBY, Reverend William F.R.S. (1789–1857): Arctic explorer; whaler; author; divine; member of the Wernerian Society; Fellow of the Royal Society of Edinburgh, 1819. After early whaling voyages with his father, Scoresby studied natural philosophy and chemistry at the University of Edinburgh. In 1807 Banks met Scoresby, who had been at the siege of Copenhagen, and from then on encouraged his interest in exploration. With Banks's help, Scoresby invented an apparatus called the 'marine diver' for obtaining deep-sea temperatures. Scoresby established that the Arctic Ocean was cooler at the top than the bottom. In 1817 he expressed a strong interest in polar exploration, but did not obtain a place in the Royal Navy vessels which were chosen for the task in 1818. Nevertheless, his scientific interests flourished with papers on the variation of the magnetic needle, submitted to the Royal Society in 1819. In the same year Scoresby superintended the building of *Baffin*, for the Greenland trade. This ship returned from Greenland in August 1820 with the largest cargo ever brought to Britain from that country. In 1820 Scoresby produced *An Account of the Arctic Regions, with a History and Description of Northern Whale-Fishery*, Edinburgh. He decided on a life in the ministry when his wife died in 1822, and therefore entered Queen's College, Cambridge, in 1823. Scoresby married again, and continued to travel and write for the remainder of his very active life.

Letters 112, 130.

SEYMOUR (*née* Douglas Hamilton), Charlotte, (d.1827), 11th Duchess of Somerset: married Edward Adolphus Seymour F.R.S., 11th Duke of Somerset, on 24 June 1800. Charlotte is described as 'bright and clever in conversation, she was very fond of society. In spite of delicate health, which increased as she grew older and caused her much suffering, she was always glad to welcome her friends, even when obliged to receive them while reclining on her sofa in the drawing-room. There she would often be seen surrounded with clever men and women, who found in her not only an agreeable companion, but also a sympathetic and staunch friend.' Lady Guendolen Ramsden (Ed.), *Correspondence of Two Brothers: Edward Adolphus, Eleventh Duke of Somerset, and his brother, Lord Webb Seymour, 1800 to 1819 and after*, London (1906). Banks exchanged affectionate letters with Charlotte in the last decade of his life. They contain fascinating discussions of personal health and Lord Byron's verse. In 1814 Banks became godfather to the Seymour's third son, Algernon Percy Banks Seymour. He succeeded to the title of 14th Duke of Somerset in 1891.

Letters 118, 121, 122, 124.

SEYMOUR, Charlotte Jane (1803–1889): Charlotte was one of five children from the marriage of Edward Adolphus Seymour, 11th Duke of Somerset, to Charlotte Seymour (*née* Douglas Hamilton), 11th Duchess. Banks sent them all gifts, and encouraged their interests.

Letter 134.

SHEPHERD, Richard F.R.S. (c.1732–1809): theologian; poet; miscellaneous writer; probationary fellow of Corpus Christi College, Oxford, 1760; appointed Archdeacon of Bedford, 1783; Bampton lecturer at Oxford, 1788; installed as rector of Wetherden and Helmingham, 1792. He was the son of Henry Shepherd (d.1764), vicar of Mareham-le-Fen, Lincolnshire. Richard Shepherd published on a variety of religious and literary subjects. He also pestered Banks with requests and complaints. These became insistent in 1793 and 1794 when Shepherd found himself in financial difficulty. Even so, he could still claim that the Banks family was 'where I was in a manner cradled': Shepherd to Banks, 27/3/1794, N.H.M. B.L. D.T.C. IX 42–43. Shepherd presented Banks with some of his works, and in 1802 proposed to write a biographical piece on Banks, for which he sought information: Shepherd to Banks, 18/3/[1802], N.H.M. B.L. D.T.C. XIII 55–57.

Letter 72.

SINCLAIR, Sir John F.R.S. (1754–1835): agriculturalist; author; politician; entered parliament as M.P. for Caithness, 1780; knighted, 1786; inaugurated the British Wool Society, 1791; founder, in 1793, and 1st President of the Board of Agriculture, 1793–1798 and 1806–1813; sworn a member of the Privy Council, 1810; appointed Commissioner of Excise in 1811. Sinclair was a hard-working man, who wrote a great deal on agriculture, and tried to improve land use on his estates and throughout the country. He shared an interest in sheep and wool with Banks, and founded the British Wool Society in July 1791. He also devoted tremendous energy to the collection of statistics of many kinds, which were published. Sinclair designed a 'Statistical Account of Scotland' (1791–1799), which included details of the natural history, population, and productions of many parishes.

In 1793 Sinclair embarked on a plan to establish a Board of Agriculture. William Pitt the younger supported him in parliament as a favour, and the Board's charter was signed on 23 August 1793. Sinclair was the Board's first President, Arthur Young was appointed its secretary, and Banks was an official member. Sinclair wanted to produce an account of England similar to the one he had already attempted in Scotland. However, this was too ambitious for the limited funds and staff available, a circumstance both Young and Banks had pointed out. In June 1796 a number of members voted to stop all payments for expensive county surveys initiated by Sinclair as President. In March 1798 Sinclair, who had lost Pitt's support as well, was voted out of the chair. He was replaced by John Southey, Lord Somerville. From then on, the Board was managed more prudently, and Sinclair was re-elected to the chair in 1806.

Letter 52.

SMITH, Sir James Edward F.R.S. (1759–1828), 1st Baronet: botanist; purchaser of the Linnean collections, 1784; founder, in 1788, and 1st President of the Linnean Society of London, 1788–1828; knighted, 1814. He studied medicine at the University of Edinburgh, and was also taught botany by Dr. John Hope. Smith came to London in 1783, and met Banks for the first time. When Carl von Linné the younger died that year, the manuscripts, herbarium and natural history collections made by him and his father were offered to Banks for 1000 guineas. Banks declined, but recommended Smith purchase them. Smith received the collections in October 1784, and devoted himself to botany thereafter. In 1788, after a successful European tour, he took a house in Great Marlborough Street. The first meeting of the Linnean Society was held there on 8 April that year. Smith was elected President, and Jonas Dryander was appointed librarian. Banks joined the Society as one of only three honorary members and a founding Fellow. In 1791 Smith arranged Queen Charlotte's herbarium, and he taught the princesses botany and zoology at Frogmore, Windsor Great Park.

After 1796 Smith spent more of his time in Norwich, and returned to London mainly to lecture at the Royal Institution. He continued his long-standing work on *English Botany,* London (1790–1814), which was illustrated by James Sowerby.

By 1804 Smith had completed his *Flora Britannica...*, London (1800–1804), and was then chosen to edit the *Flora Graeca...*, London (1806–1840), of his friend John Sibthorp, who had died in 1796. This was published by Smith and John Lindley F.R.S. (1799–1865), with 966 plates engraved by Sowerby. Smith also published in two volumes *Florae Graecae Prodromus...*, London, 1806 and 1813. The first edition of Smith's most successful work appeared in 1807 as *The Introduction to physiological and systematic Botany*, London. However, the last seven years of his life were spent on perhaps his best work: *The English Flora...*, London (1824–1836). Smith wrote many essays and articles in a distinguished career publishing on botany.

Letters 89, 117, 133.

SMITH, William (1756–1835): politician; collector of fine art and books; entered parliament as M.P. for Sudbury, 1784; elected member of the Whig Club, 1796. Smith contested a number of seats in his career. He was active in parliament, speaking at great length on a variety of issues, including the repeal of the Test and Corporation Acts, the abolition of the slave trade and parliamentary reform. In 1790 he rose to proclaim himself a dissenter. Smith was also a connoisseur of literature and art. He acquired a fine library, and an impressive collection of pictures.

Letter 101.

SPENCER, George John F.R.S. (1758–1834), 2nd Earl Spencer: statesman; administrator; bibliophile; entered parliament as M.P. for Northampton, 1780; appointed Junior Lord of the Treasury, 1782; Ambassador Extraordinary at Vienna, returning in 1794; First Lord of the Admiralty, 1794–1801; Home Secretary, 1806–1807; President of the Royal Institution, 1813–1815; Trustee of the British Museum; founder and 1st President of the Roxburghe Club, 1812. Spencer was First Lord of the Admiralty during a period of tremendous naval success for Britain in war, due, in part, to the quality of his administration. In 1807 he retired from public service, and devoted himself to county affairs in Northamptonshire. He also concentrated on improving Althorp library, which became one of the greatest private libraries in Europe.

Letters 77, 82.

STAUNTON, Sir George Leonard F.R.S. (1737–1801), 1st Baronet: diplomatist; author; Secretary to the Governor of Dominica, c.1760s; Member of the Legislative Council, and Attorney-General for Grenada, c.1770s; colonel of the militia, and aide-de-camp to the Governor, Grenada, 1779; Secretary to the Governor of Madras, 1781–1784; knighted, 1785; Secretary to the Embassy to China, 1792–1794. Staunton completed his education in France, where he trained in medicine. In 1762 he went to the West Indies, and held official positions there while working as a physician.

Staunton acquired a large fortune, and purchased an estate in Grenada. In 1770 he returned to England, but spent much of the decade back in the West Indies managing his estate. It was also in this decade that Staunton met George Macartney, who became a firm, life-long friend. Both men were in the West Indies at this time, Macartney having been appointed a governor there in 1775. They fought the French when Grenada was attacked in 1779, but were captured. Staunton was ruined by the defeat, his estates being pillaged. He was taken as a hostage to Paris, and there negotiated Macartney's release from parole. In due course, he too was released. When Macartney went to Madras as governor in 1781, he took Staunton as a secretary. Staunton was an excellent secretary, and coped well with more than one difficult situation.

After his return to England in 1784, Staunton could not find public employment, and it was not until 1791 that a place on the Embassy to China was offered him. He accepted, and Banks provided useful advice. For instance, Banks wrote a long paper, which was sent to Staunton, called 'Hints on the Subject of Gardening suggested to the Gentlemen who attend the Embassy to China', August 1792, L.S. MSS 115. However, the Embassy was not a tremendous success. Health problems curtailed further diplomatic appointments in the Far East for Staunton. In 1797 he published: *An authentic account of an embassy from the King of Great Britain to the Chinese Emperor...*, London (1797). Banks assisted in the preparation of this account, as he did in the arrangement and classification of the natural history collections made on the mission. Staunton's only son, Sir George Thomas Staunton F.R.S. (1781–1859), accompanied his father to China as a page to the ambassador. He was a founder of the Royal Asiatic Society in 1823, and donated 3000 volumes of Chinese works to its library.

Letter 48.

THOMPSON, Benjamin F.R.S. (1753–1814), Count von Rumford: American physicist; soldier; founder of the Royal Institution, 1799, and its 1st Secretary, 1799–1802. Rumford was loyal to the Crown in the American Revolution. He took on official work for the British government, and was a major in the New Hampshire militia. He was eventually forced to flee to London in 1776, where he was appointed Lieutenant-Colonel of a New York Regiment. After the war, Rumford lived permanently in exile in Europe. He moved from England to Bavaria in 1784, where he became War and Police Minister, and Grand Chamberlain to the Elector as well. He combined the ability to invent useful devices with the drive to implement social reform. Under Rumford's administration, a workhouse was created in Munich in which the homeless found warmth, food and work. They were fed nutritious soups made according to Rumford's own recipes. These included the potato, a new vegetable for most. Uniforms were manufactured cheaply for the army in the workhouse, and this reduced the military budget, much of which had previously been spent on clothing. In 1791 Rumford was made a Count of the Holy Roman Empire, taking the title of 'Rumford' from his early home in America (now called Concord).

Rumford's main scientific interest was in heat and light. In 1798 he showed that heat is a form of motion, and not a fluid, by studying the way drills reacted as they bored iron blocks to make cannons. Most types of technology interested him when it came to heat. He looked at the best way to prevent heat loss with clothing; he devised the domestic range and its utensils; he even produced designs for open fires, which included the smoke shelf and damper. Rumford also invented the Rumford shadow photometer, and established the standard candle. In 1796 he arranged for the prestigious Rumford Medal to be awarded by the Royal Society, and the Academy of Arts and Science, Boston. With Banks's help, he founded the Royal Institution in 1799. Their aim was to promote science and technology, and to benefit society through both. In 1801 Rumford settled in Paris with the widow of Antoine Lavoisier, whose caloric theory of heat he had disproved. The relationship did not last, but Rumford lived on in Paris until his death.

Letter 96.

TYSSEN, Samuel Jnr. (1786–1845): Norfolk landowner. Banks was a conscientious trustee of the Tyssen family estates in Norfolk following the death of Samuel Tyssen (1756–1800), who was Dorothea Banks's second cousin. Banks's letters to Tyssen Jnr. contained kind advice and invitations to join the Banks family in London and Lincolnshire.

Letter 90.

VOLTA, Alessandro Giuseppe Antonio Anastasio F.R.S. (1745–1827): Italian physicist; appointed principal lecturer at the Gymnasium, Como, 1774, and in 1775 made professor of experimental physics there. Volta's early interest was in electricity, which had been famously investigated by Benjamin Franklin. His reputation grew with experiments on static electricity until, in 1774, he took a post at the Gymnasium, Como, where he was appointed professor of experimental physics. In 1775 he invented the electrophorus, a practical device for the production of charges. He also developed a simple yet sensitive electrometer using straws. Volta did not confine himself to electricity though, and in 1776 he discovered methane using marsh gas. Furthermore, he discovered the proportion of oxygen in air.

Some of Volta's greatest work was announced in Britain in the *Philosophical Transactions* through letters he sent to Banks, whom he admired. In 1792 he showed that Luigi Galvani's explanation of a 'nerveo-electrical' fluid in animals was wrong. Volta determined that the true source of electricity which stimulated the nerves and muscles of dissected frogs was in the junction of two metals touching the animal tissue. In 1800, Volta invented the 'voltaic pile', enabling high electric currents to be produced. These were then applied to electrolysis in chemistry, and led to future developments in electricity.

Letter 50.

WEDGWOOD, Josiah F.R.S. (1730–1795): potter; entrepreneur; member of the African Association, active 1790; member of the Lunar Society. Wedgwood came from a well-established family of potters in Staffordshire. In 1759 he opened his own pottery business at Burslem. He worked and experimented endlessly to improve the quality of his pottery. In 1769 Wedgwood moved his factory to Etruria to be closer to the Grand Trunk Canal, which linked Staffordshire with ports on the east and west coasts of England. This was the same canal he fought to have made in the years 1764–1766. Wedgwood was inspired by antique models, and invented unglazed black basalt ware, and blue jasper ware with raised designs in white. He also cared about the social welfare of his employees, and built them a village at the Etruria works.

Letter 46.

WESTALL, William A.R.A. (1781–1850): landscape artist; traveller. Westall was appointed landscape artist on the *Investigator* expedition to Australia under Matthew Flinders, 1801–1803. During the voyage around Australia, Westall produced pencil sketches, some with wash, of the coast, landscapes, aborigines and a few natural history subjects. Some of these were damaged when the *Porpoise* was lost on Wreck Reef in August 1803. However, many were eventually delivered to Banks in London. Meanwhile, Westall had travelled to China and India, and so reached London early in 1805. He then visited Madeira and Jamaica, 1805–1806. Once Westall had returned to England from these places, he concentrated on painting and exhibiting his work. In 1811 Westall and Banks chose nine sketches to be used in *A Voyage to Terra Australis...*, London (1814) by Flinders. Westall produced oil paintings of them, which were engraved for the publication. These oil paintings hang in Admiralty House, London.

Letter 82.

WILES, James (1768–1851): gardener; botanist; gardener to Richard Anthony Salisbury; gardener under William Bligh on H.M.S. *Providence*, 1791–1793. Rather than return to England with H.M.S. *Providence*, Wiles took the option of remaining at the garden at Liguanea, Jamaica, 1794–1811. He became Head Gardener there, and edited *Hortus Eastensis...*, Jamaica (1806).

Letter 44.

WINDHAM, William (1750–1810): statesman; Chief Secretary to the Lord Lieutenant of Ireland, 1783; entered parliament as M.P. for Norwich, 1784; Secretary of State for War, 1794–1801; Secretary of State for War and the Colonial Office, 1806–1807. Windham was a fine scholar of Greek and Latin, and a close friend of Dr. Samuel Johnson. He joined the Literary Club, and attended many meetings. Windham advocated strong resistance to France in war, and favoured the use of naval power in the struggle. As Secretary of State for War, he raised the pay of troops, established pensions and saw the Royal Military Asylum founded. However, his opposition to peace, and to reform in general, brought unpopularity. Windham admired William Cobbett, whom he helped to set up the *Political Register*. Banks and Windham disagreed more than once, although their relations remained cordial. In 1773 Windham accompanied Constantine Phipps on an Arctic voyage, but was forced to turn back at Norway because of sea sickness.

Letter 51.

WOLLEY, Adam, (1758–1827): attorney, Matlock, Derbyshire; antiquary; genealogist; topographer; Steward of the Manor of Wirksworth and of the Barmote Court, 1807–1822; administrator of charities in Matlock parish. Wolley came from an ancient Derbyshire family. He trained in the law, and established a considerable practice. He collected many documents and records concerning Matlock and the county of Derbyshire. Wolley seems to have acquired quantities of them in his legal work so that more than fifty volumes of manuscripts were bequeathed by him to the British Museum in 1837. These form the Additional Manuscripts 6666 to 6718, and include the 'Mineralia', legal and ecclesiastical documents, charters, wills, abstracts of title deeds, and miscellaneous papers. Wolley visited Banks at his Overton estate on a number of occasions.

Letter 76.

YONGE, Sir George (1731–1812), 5th Baronet: statesman; entered parliament as M.P. for Honiton, 1754; Lord of the Admiralty, 1766–1770; Vice-Treasurer for Ireland, 1782; Secretary for War, 1782–1783, and 1783–1794; Master of the Mint, 1794–1799; Governor of the Cape of Good Hope, 1799–1801. Despite his considerable experience, Yonge was too old or inflexible by the time he went to the Cape of Good Hope. His governorship there was marked by disagreements with subordinates, and allegations of various other abuses. However, he returned in 1801, and paid little attention to his critics.

Letters 26, 28.

YOUNG, Arthur F.R.S. (1741–1820): agriculturalist; traveller; author; Secretary of the Board of Agriculture, 1793–1820. In 1763 Young had no means of support, and took one of his mother's farms at Bradfield, Berkshire. Despite his inexperience, he started to publish on the subject of agriculture: *The Farmer's Letters to the People of England...*, London (1768). Young experimented in farming, and toured rural England to gain more knowledge. Through the 1770s he produced a string of publications on contemporary affairs and agriculture. In 1784 Young commenced his *Annals of Agriculture, and other useful arts...*, London (1784–1815), a monthly publication, for which he provided nearly a third of the contents. Banks, a friend and substantial landowner, also contributed ten pieces. They unsuccessfully opposed the Wool Bill of 1788, when Banks acted as one of the Lincolnshire Wool Committee, and Young represented the wool-growers of Suffolk. Both Banks and Young obtained detailed facts about the woollen industry, which were then used to counter the arguments of merchants and manufacturers who demanded protectionism.

Young travelled widely in France, surveying the economy and society there. In 1792 he published *Travels in France...during the Years 1787, 1788 and 1789*, Bury St. Edmunds. In 1793 Young was the obvious choice for Secretary of the Board of Agriculture, of which Banks was also a member. Young founded the Farmer's Club a year later. He was distraught after the death of his daughter, 'Bobbin' (1783–1797), but continued to publish until his sight failed in 1811. Young was one of the greatest English writers on agriculture. He argued in his books and articles that large-scale farming, using enclosure, the latest techniques and plentiful capital would greatly increase production. His writings provide a rich source of information about the social and economic life of the eighteenth century, as do the manuscripts of Sir Joseph Banks.

Letter 92.

General Index

This index includes the people, places and subjects which occur in the correspondence. Each reference is to the number of a letter. Notes are designated by 'n'. Modern and correct spellings for terms have generally been used. Older names, spellings or misspellings are given in inverted commas following their modern or correct equivalent. Latin and technical terms have not been included here unless they appear in the notes, few being present in the letters. For the same reason no separate index of plants or animals was felt necessary.